Plant Pathology at a Glance

The Authors

Dr. R.P. Singh is presently serving as Scientist-Plant Protection in the Krishi Vigyan Kendra, Post Graduate College, Ghazipur (UP). He did M. Sc. (Ag.) and Ph D. in Plant Pathology from Narendra Deva University of Agriculture and Technology, Kumarganj, Faizabad, Uttar Pradesh. He is recipient of 'Gold Medal' honour of M.Sc. (Ag.) degree and has qualified ICAR-NET in 1996, 1997 and 2001 organized by Agricultural Scientist Recruitment Board (ASRB), New Delhi. He has so far published 04 books, 09 book chapters, 61 research papers, review articles in National and International journals and 80 popular articles in reputed magazines.

Dr. Mamta Singh is working as Scientist-Plant Breeding and Genetics in Jawaharlal Nehru Krishi Vishwa Vidyalaya Jabalpur, Madhya Pradesh. She has published 02 books, 37 research papers, review articles in National and International journals, 09 book chapters, 01 manual and 40 popular articles in reputed magazines. She received M. Sc. (Ag.) Genetics and Plant Breeding from NDUA&T, Kumarganj, Faizabad. After this, she obtained her Ph. D. degree in Plant Breeding from CCS, HAU, Hisar, Haryana. She has qualified ICAR-NET in 1998 and Senior Research Fellow, organized by Agricultural Scientist Recruitment Board (ASRB), New Delhi.

Dr. Smita Puri is working as a Scientist Plant Pathology in Jawaharlal Nehru Krishi Vishwa Vidhyalaya, Jabalpur, Madhya Pradesh. She has published more than 25 research papers and review articles in reputed national and international journals, 8 book chapters, 2 laboratory manuals and 20 popular articles in different magazines. She did her M.Sc. (Ag) and Ph.D. in Plant Pathology from G.B.Pant Univ. of Agriculture and Technology, Pantnagar in the year 2006 and 2010, respectively. She qualified ICAR-NET in 2008 and also received Graduate research assistantship, University research fellowship and senior research fellowship etc.

Plant Pathology at a Glance

(For ICAR's Exams, AIEEA-PG, JRF, SRF, NET, ARS, IARI, Ph.D., SAU's, UPSC, and Allied Agricultural Examinations)

R.P. Singh
Mamta Singh
Smita Puri

2017
Daya Publishing House®
A Division of
Astral International Pvt. Ltd.
New Delhi – 110 002

Cataloging in Publication Data--DK
Courtesy: D.K. Agencies (P) Ltd. <docinfo@dkagencies.com>

Singh R. P. *(Scientist-plant protection)*, **author.**
Plant pathology at a glance : for ICAR's exams, AIEEA-PG, JRF, SRF, NET, ARS, IARI, Ph. D. SAU's, UPSC, and allied agricultural examinations / R.P. Singh, Mamta Singh, Smita Puri.
pages cm
ISBN 9789390384426 (PB)

1. Plant diseases--India--Examinations--Study guides. I. Singh, Mamta (Scientist-plant breeding and genetics), author. II. Puri, Smita, author. III. Title.

SB605.I4S56 2017 DDC 632.0954 23

Published by : **Daya Publishing House®**
A Division of
Astral International Pvt. Ltd.
– ISO 9001:2008 Certified Company –
4760-61/23, Ansari Road, Darya Ganj
New Delhi-110 002
Ph. 011-43549197, 23278134
E-mail: info@astralint.com
Website: www.astralint.com

Digitally Printed at : **Chawla Offset Printers**

त्रिलोचन महापात्र, पीएच.डी.
एफ एन ए, एफ एन ए एस सी, एफ एन ए ए एस
सचिव एवं महानिदेशक
TRILOCHAN MOHAPATRA, Ph.D.
FNA, FNASc, FNAAS
SECRETARY & DIRECTOR GENERAL

भारत सरकार
कृषि अनुसंधान और शिक्षा विभाग एवं
भारतीय कृषि अनुसंधान परिषद
कृषि एवं किसान कल्याण मंत्रालय, कृषि भवन, नई दिल्ली 110 001
GOVERNMENT OF INDIA
DEPARTMENT OF AGRICULTURAL RESEARCH & EDUCATION
AND
INDIAN COUNCIL OF AGRICULTURAL RESEARCH
MINISTRY OF AGRICULTURE AND FARMERS WELFARE
KRISHI BHAVAN, NEW DELHI 110 001
Tel.: 23382629; 23386711 Fax: 91-11-23384773
E-mail: dg.icar@nic.in

Foreword

Plant Pathology is the science of plant disease management that attempts to improve the chances of survival of plants under unfavorable environmental conditions and infestation with pathogenic microorganisms. Plant diseases are therefore challenging and concerted research efforts are critical in averting food/crop loss. It has been estimated that 10-16% of total food losses are caused due to diseases and that the total annual worldwide crops loss from plant disease is about US$ 220 billion. Realising the scope of this discipline, several students are attracted to this "subject of choice", in both PG and Doctoral programmes. The present publication, "Plant Pathology at a Glance" dwells upon subjective and objective forms of questions-answers on the pattern of most competitive examination. Besides being a good resources document in plant pathology. The information in this book gives ample guidance to enable the candidate to appear in examination like JRF, SRF, NET, ARS and entrance tests of various agricultural universities/Institutions in the country.

I appreciate the efforts of the authors in bringing out this useful publication. I am sure, the students will take advantage of this document for successful progress in their respective careers.

(T. MOHAPATRA)

Acknowledgements

We wish to express our sincere thanks to all the authors of plant pathology books which are mentioned in this book. It would not have been possible to complete this book without help rendered by the authors.

R.P. Singh

Mamta Singh

Smita Puri

Preface

Plant Pathology or Phytopathology consists of three Greek words (*phyton*-plant; *pathos*-ailments or diseases and *logus*-discourse or knowledge). Plant Pathology is the science which deals with causes, symptoms, etiology, perpetuation, losses caused by plant diseases and their management. Due to considerable progress made in many aspects of plant diseases during last few years, Plant Pathology could receive relatively more attention and importance in syllabi of undergraduation and postgraduation levels. Students have to answer objective type questions in university examinations and various competitive examinations as well. There are several books of foreign and Indian origin on only descriptive aspects and is beyond the purchasing capacity of most of our students. Lack of genuine study material, proper guidance and above all the dogged determinations are the most commonly cited reasons. Moreover, there is a lack of separate book on objective aspects in these disciplines.

The present book entitled "**Plant Pathology at a Glance** (***For ICAR's Exams, AIEEA-PG, JRF, SRF, NET, ARS, IARI, Ph. D., SAU's, UPSC, and Allied Agricultural Examinations***)" will fulfil a long felt need of Indian students. This book has been designed to cover the courses offered by various Indian Universities at different levels and also the syllabi of various competitive examinations. This will also be useful for those appearing in competitive examinations such as Agricultural Research Service (ARS) of Indian Council of Agricultural Research, National Eligibility Test (NET), State Eligibility Test (SET), Junior Research Fellowships (JRF), Senior Research Fellowships (SRF) and Civil Services Examinations. It may be useful to seek admission in Ph.D. in Plant Pathology through entrance examinations of various institutions.

Precautions have been taken to discuss each topic in systematic manner and avoid mistakes, still omissions and mistake are bound to be found in this attempt. We are prepared to bear full responsibility for making mis-presentation of facts, if any. We wish to express our deep sense of gratitude to the pioneers and authors whose illustrations and reviewed work have been utilized freely in preparation of this book. We are thankful and highly obliged to Dr. R.D. Singh Retd. Professor and Head, Department of Vegetable Science, Narendra Deva University of Agriculture and Technology, Kumarganj, Faizabad Uttar Pradesh and Dr. U.S. Gautam, Director, ICAR-ATARI, Kanpur, UP for being source of inspiration and support.

We are highly grateful for Mr. R.P. Singh, Chairman, Krishi Vigyan Kendra, P.G. College, Ghazaipur who happens to be the driving spirit of the organization. My gratitude also goes to Dr. J.P. Singh, Head, Department of Horticulture, Gochar Mahavidhyalaya, Rampur, Maniharan Saharanpur, Er. Ashish Kumar Bajpai, Dr. Dinesh Singh and Dr. Akhilesh Kumar for their relevant assistance, moral support and motivation which encouraged us to deliver our best.

All suggestions for the future improvement of book will be highly solicited. We are extremely grateful to Dr. Trilochal Mohapatra, Secretary & Director General, Indian Council of Agricultural Research New Delhi for providing a foreword. The cooperation and support of Dr. B.B. Singh, Chief Editor, Astral International Pvt. Ltd., New Delhi is duly acknowledge. We are highly thankful to the Publishers ASTRAL International (Pvt) Ltd for arranging early publication of this book.

R.P. Singh

Mamta Singh

Smita Puri

Contents

Acronyms

CO: Causal organism
DNA: Deoxyribonucleic acid
RNA: Ribonucleic acid
ATP: Adenosine triphosphate
e.g.: Exepli gratia (for example)
IAA: Indole acetic acid
et al.: and others
ADP: Adenosine diphosphate
etc: et cetra (and others, and so on)
PAL: Phenylalanine ammonia lyase
f. sp.: formae specialis
pv. : Pathovars of a bacterium
c.f.: Confer (compare)
pl: Plural
RH: Relative humidity
sing: Singular
ss: Single stranded
ds: Double stranded
MLO: Mycoplasma like organism
sp., spp: Species (singular: sp.; plural: spp)

PPO: Polyphenol oxidase
syn: Synonym
temp.: Temperature
viz.: Videlicet (means: namely)
avr: Avirulence
w.p.: Wettable powder
TMV: Tobacco mosaic virus
CMV: Cucumber mosaic virus
PVX: Potato virus X
PVY: Potato virus Y
TSWV: Tomato spotted wilt virus
ssDNA: Single stranded deoxyribonucleic acid
dsDNA: Double stranded deoxyribonucleic acid
ssRNA: Single stranded ribonucleic acid
dsRNA: Double stranded ribonucleic acid
+RNA: Positive ribonucleic acid
-RNA: Negative ribonucleic acid
TSV: Tobacco streak virus
PLCV: Papaya leaf curl virus
TLCV: Tomato leaf curl virus
ICRSV: Indian citrus ring spot virus
RLB: Rickettsia like bacteria
RLO: Rickettsia like organism
Hfr: High fertility rate
pth: Pathogenicity
dsp: Disease specific gene
EF-G (GDP): Elongation factor G (guanosine diphosphate)
GMO: Genetically modified organism
tRNA: Transfer ribonucleic acid
RILs: Recombinant inbred lines
CHO: Carbohydrate (Carbon, Hydrogen, Oxygen)
PPM: Part per million
TPTA: Triphenyltin acetate
VR: Vertical resistance

HR: Horizontal resistance
PME: Pectin methyl acetate
IPA: Isopentenyladenosine
HCN: Hydrocyanides
PPO: Polyphenol oxidase
DCINA: Dichloro-iso-nicotinic acids
LRR: Leucine-rich-repeats
NBS: Nucleotide binding sites
LZ: Leucine zippers
PG: Polygalacturonase
INA: Ice-nucleation active
λ: Lambda
H_2S: Hydrogen sulfide
VAM: Vesicular-arbuscular mycorrhiza
α: Alpha
β: Beta
PCNB: Pentachloronitrobenzene
pH: 'Pouvour hydrogen' (Hydrogen power)
USA: United State of America
ABA: Abscisic acid
GA: Gibberellic acid
satRNA: Satellite Ribonucleic acid
BC: Before Christ
AD: Anno Domini
'f': In modern Greek means phi or 'ph' (a plane angle)
RT-PCR: Reverse transcription polymerase chain reaction
CTV: Citrus tristeza virus
NASBA: Nucleic acid sequence based amplification
IgG: Immunoglobulin class G (a type of antibody)
mRNA: Messenger ribonucleic acid

Chapter 1

Fungal Plant Pathogens

Queries with Timeline of Fungal Plant Pathogens

1. The attention of man to plant diseases and the science of plant pathology were drawn first only in the European countries. Greek philosopher Theophrastus (370-286 B.C.) recorded some observations on the plant diseases in his book **'Enquiry into Plants'**about 2400 years ago. His experiences were mostly based on imagination and observation but not on experimentation. He had mentioned that the plants of different groups have different diseases which were spontaneous.

2. Indeed, mildews and blights were well known in India even in the Vedic period. Plant diseases were recorded in the Vedas (1200 BC). One finds mention of types of symptoms, diseases and their control in **'Vrikshayurveda'**. The definite mention of plant diseases can be seen in the Buddhist literature of 500 BC (Kullavagga, X, 1, 6). Symptoms of plant diseases are cited in other ancient Indian literatures *viz.*, **Jataka** of Buddhism, **Raghuvamsha** of Kalidas etc.

3. Surapala (1000 AD) wrote **'Vriksha Ayurveda'**which the first book in India where he mentioned details account on plant diseases and their control. In this book, plant diseases were categorized into two groups, internal (probably physiological diseases) and external (probably infectious diseases). Tree surgery, hygiene protective covering with paste, use of honey, plant extracts, oil cakes of mustard, castor, sesamum etc. are some of the disease management practices recorded in the book.

4. In India, the information on plant diseases is available in ancient literature such as rigveda, Atharvaveda (1500-500BC), arthasashtra of Kautilya (321-

186 BC), Sushruta sanhita (200-500AD), Vishnupuran (500AD), Agnipuran (500-700AD), Vishnu dharmottar (500-700AD), etc. In Rigveda, not only the classification of plant diseases has been given but the germ theory of disease was also advocated.

5. D.D. Cunningham and A. Barclay, during 1850-1875, started identification of fungi in India itself. D. D. Cunningham (1889) identified the casual organism of red rust of tea in Assam caused by *Cephaleuros virescens.*

6. The Europeans started systematic study of fungi in India during 19th century. They collected the fungi and sent to the laboratory in Europe for identification. A proper record on the plant disease and their causal organisms were maintained only after development of compound microscope by the Dutch worker Antony von Leeuwenhoek in 1675. Robert Hook (1635-1703) also developed simple microscope which was used to study of minute structure of fungi.

7. Pier Antonio Micheli, an Italian botanist is the founder of the science of mycology. He published **'Nova Plantarum Genera'** in 1729, and was the first scientist who studied fungi and saw their spores in 1729. He also proved that spores placed on pieces of a fruit grow into new thallus of fungus.

8. Matheiu Tillet (1755), the French Botanist, published a paper on bunt or stinking smut of wheat. He proved that the seeds containing black powder on their surface produce more diseased plants than clean seeds and treated seeds with fungicide reduced disease incidence. He believed that the smut dust contained poisonous substance rather than living microorganisms that caused the disease.

9. French scientist Benedict Prevost working on bunt of wheat in 1807, proved that diseases are caused by microorganisms. He studied the life cycle of bunt fungus and germination of its spores. He also reported the fungicidal and fungistatic properties of chemical treatments.

10. The foundation of modern experimental Plant Pathology was laid by German scientist Anton de Bary (1831-1888). He confirmed the findings of Prevost in 1853. He finally proved in 1861, that the late blight of potato was caused by *Phytophthora infestans*. He studied many other diseases like smuts, rusts, downy mildews and rots. He also discovered the heteroecious nature of rust fungi and detailed life cycle of downy mildew fungi with their parasitism. In 1886, his last work pioneered research on the physiology of parasitism. De Bary worked on *Sclerotinia sclerotiorum* and the diseases it caused to carrots and other plants. He also suggested the role of enzymes and toxins in the interactions between pathogens and the hosts. Anton de Bary also regarded as the Father of Plant Pathology, wrote a book named **'Morphology and Physiology of Fungi, Lichens and Maxomycetes'** in 1866.

11. Christian Hendrick Persoon (1761-1831) first time published observations mycologicae and in 1801, he published **'Synopsis Methodica Fungorum'** for nomenclature of Ustilaginales, Uredinales and Gasteromycetes. He also

published 'Mycologica Europica' in 1822. He gave the name to rust pathogen of wheat as *Puccinia graminis*.

12. Elias Magnus Fries (1794-1878) published three volumes of **'Systema Mycologium'** for nomenclature of hymenomycetes from 1821-1832.
13. R.H. Biffen (1905) reported that resistance to yellow rust in wheat was genetic and governed by a recessive gene.
14. Orton (1909) working with *Fusarium* wilt of cotton, watermelon and cowpea, made distinction among disease resistance, disease escape and disease endurance.
15. American geneticist Blackeslee (1904), first time reported Heterothallism in mucor (*Mucor tennis*) due to dissimilar nuclei in the life cycle of fungi.
16. L.R. Jones (1905) reported the involvement of cytolytic enzymes produced by bacteria in several soft rot diseases of vegetables and did pioneering work on pectic enzymes secreted by soft rot bacteria.
17. William Brown (1915) recognized the significance of pectic enzymes in pathogenesis of some fungi on plants.
18. L.R Jones (1926) at Wisconsin initiated research on the effect of soil temperatures, soil moisture and other environmental factors upon the epidemiology of the disease.
19. Sorauer born in Germany in 1839 initiated studies in physiological diseases.
20. Prevost (1807) demonstrated experimentally and clearly the indirect or secondary causal role of the environment in the etiology of parasitic diseases.
21. Tanaka (1933) reported that the most plant diseases were caused by toxins secreted by pathogens.
22. Swedish scientist Erikson (1894) was the first to discover variability among the pathogen, (fungi). He found five varieties in *Puccinia graminis viz. tritici* on wheat, avenae on oat, secalis on rye, agrostidis on *Agrostis sp*. and poae on *Poa pratensis* and other grasses and termed as *"formae speciales"*.
23. The variability in rust fungi was later studied in detail by E.C. Stakman (1914). He discovered that within a pathogen species, physiological races of the pathogen exist that are morphologically indistinguishable but different in their ability to infect a set of host differential varieties. He described numerous physiological races of *Puccinia graminis tritici* on wheat. He demonstrated by crossing formae speciales (f. sp.) and pathotype of *Puccinia graminis tritici,* that the virulence of the pathogen is also inherited according to Mendel's laws.
24. In 1911, Barrus demonstrated that there is genetic variability within a pathogen species; there are different pathogen races which are restricted to certain varieties of a host species.

25. J.C. Walker (1923) thought that resistance to a pathogen was a result of the presence of a toxic substance in the plant.

26. In 1946, Flor explained host parasite interaction in flax rust caused by *Melampsora lini*. He gave the gene – for – gene hypothesis to explain this reaction.

27. In 1968, J. E. Van der Plank suggested that there are two kind of resistance, one is known as vertical resistance (VR) and the other is known as horizontal resistance (HR) in plants. Terms compound interest and simple interest diseases were given by Van der Plank (1963) in his book **'Plant Disease Epidemics and Control'**.

28. In 1946, Gaumann proposed that in many host pathogen combinations, plants remain resistant through hypersensitivity.

29. K.O. Muller (1961) and Cruickshank (1963) proposed that disease resistance is often brought about by phytoalexins that is antimicrobial plant metabolites that are absent, or present at non-detectable levels in healthy plants, but accumulate to high levels as a result of some pathological stimulus.

30. In1956, J.G. Horsfall published an authoritative book entitled **'Principles of Fungicidal Action'**.

31. Robert Hartig born in Germany in 1839. He worked on forest pathology and aptly called as the 'Father of Forest Pathology'.

32. Fry (1982) published a very useful book **'Principles of Plant Disease Management'**.

33. In (1858), Julius Gotthelf Kuhn published his famous text book for farmers entitled **'The Diseases of Cultivated Crops: Causes and their Control'**.

34. In (1912), Burgeff discovered within a cell of the fungus fusion between dissimilar nuclei can occur. This phenomenon is called as heterokaryosis.

35. Hansen and Smith in 1932, first time demonstrated origin of physiological races through heterokaryosis in *Botrytis cineria*.

36. Plant Pathology in India began with the establishment of the Indian (then Imperial) Agricultural Research Institute at Pusa, Bihar (now at New Delhi) in 1905 and the appointment of Edwin John Butler (later Sir Edwin) as the first Imperial Mycologist. The credit for laying the foundation of Plant Pathology goes to him and he may aptly be called as the **'Father of Indian Plant Pathology'**.

37. Edwin John Butler started the systematic study on Indian fungi and the diseases caused by them. He was sent by British Government to India in 1900 for the study of Indian fungi and diseases cause by them. He is considered as the father of modern Plant Pathology in India. He studied the wilt of cotton, arhar, different diseases of rice, pea, sugarcane, potato and rust of cereals. He wrote a book **'Fungi and Diseases in Plants'** published in 1918 from Mumbai. He

also wrote another book **'Fungi in India'** (with B.R. Bisby), **'Plant Pathology'** (with S.G. Jones) and a monograph **'Pythiaceous and Allied Fungi'**.

38. The first Indian Universities that were established in 1857 at Mumbai, Madras and Mumbai, emphasized on study of fungal taxonomy. Plant Pathology as a University Science became established at Lucknow, Allahabad and Madras Universities (founded in 1921, 1887 and 1857, respectively) only in the 1930s. K. R. Kirtikar was the first Indian scientist who collected and identified the fungi in India.

39. Jahangir Ferdunji Dastur (1886-1971) a colleague of Butler was the first Indian plant pathologist who has studied in details of fungi and plant diseases. He specially studied the diseases of potato and castor caused by genus *Phytophthora* and established the species *P. parasitica* from castor in 1913. In recognition of his command in Plant Pathology, he was promoted to the Imperial Agricultural Science in 1919.

40. G.S. Kulkarni, a student of Butler, published detail information on downy mildew and smut of jowar and bajra in India. Another student S.L. Ajrekar studied wilt disease of cotton, sugarcane smut and ergot of jowar.

41. In 1948, B.B. Mundkur started work on control of cotton wilt through vertical resistance. He started 'Indian Phytopathological Society in 1948' with its journal 'Indian Phytopathology'. He also wrote textbook entitled **'Fungi and Plant Diseases'** published in 1949.

42. Raghubir Prasad (1907-1992) trained under K.C. Mehta, contributed to the identification of physiological races of cereal rusts and life cycle of linseed rust. Subsequently, L.M. Joshi at IARI conclusively studied various aspects of wheat rusts *viz.*, chief foci of infection of rusts, dissemination of rust pathogens in India. Later on, S. Nagarajan and L.M. Joshi developed most useful mathematical models in 1978 to predict appearance of stem and leaf rust of wheat.

43. S.R. Bose was taxonomist, mainly worked on the classification of Polyporaceae and isolated "polyporin" from *Polyporus*.

44. Manoranjan Mitra was considered as one of the most critical plant pathologist worked on *Helminthosporium*. He first reported Karnal bunt of wheat in 1931 from Karnal in Haryana.

45. M.J. Thirumalachar (1914-1999) reported 20 new genera and 300 new species of fungi, monographed genera of Uredinales of the world and Ustilaginales of India.

46. Meehan and Murphy in 1947 showed the attacks and cause in blights on oats of the variety victoria and produced toxin named victorin.

47. In India, works on fundamental Plant Pathology, especially the biochemistry of host-parasite relationship were started at Lucknow and Madras (Chennai) lead by Sachindra Nath Dasgupta (1904-1990) and T.S. Sadasivan (1913-2001), respectively.

48. T.S. Ramakrishnan, a mycologist to Madras Government cultivated ergot diseased rye for toxin production. He published two books entitled **'Diseases of Millets'** (1963) and **'Diseases of Rice'** (1971). Renowned plant pathologists *viz.*, G Rangaswami and R. Ramakrishnan were his students.
49. Y.L. Nene's contributions have been well remembered particularly the viral diseases of pulses and the 'Khaira' disease of rice caused by Zinc deficiency. He wrote a book **"Fungicides in Plant Disease Control"**.
50. In 1926, Kurosawa showed that excessively tall slender seedlings of the rice bakanae or foolish seedlings diseases caused by the fungus *Gibberella* (*Fusarium*), the disease could also be reproduced by treating healthy seedling with sterile culture filterates of the fungus.
51. In 1927, Craigie and Dodge enabled the genetic study of pathogenic fungi.
52. In 1939, Bagchee was the first to review the work on Indian forest mycology and Plant Pathology in India.
53. In 1922, S. Sundaraman reported *Pyricularia oryzae* on rice for the first time from India.
54. Karam Chand Mehta (1894-1950) of Agra had contributed a lot to Plant Pathology of India. He published a monograph entitled **'Further Studies on Cereal Rusts in India'** and did pioneering work on the wheat rust problem in our country and disease cycle of cereal rust in India.
55. Hyphae of *Rhizoctonia solani* often aggregrate to form an infection cushion from which multiple penetration occurs by means of appressoria and penetration peg.
56. Sporangia of Chytridiomycetes, Oomycetes and parasitic slime moulds produce zoospores in wet conditions and germs tube in dry conditions.
57. Sporangia of *Plasmopara viticola* appear to germinate by producing zoospores, which always penetrate the stoma by germ tubes (other *e.g. Pernospora* sp.).
58. Teliospores of uredinales and ustilaginales normally germinate by producing a promycelium on which basidiospores (sporidia) are borne.
59. Zoospores of a species of *Phytophthora* infecting fennel (*Foeniculum vulgare*) almost always enter through stomata while *Phytophthora infestans* almost always enters plants directly but at times through stomata after the formation of an appressorium.
60. *Mycosphaerella musicola* responsible for sigatoka disease of bananas, enter leaves in a similar manner as do the rust pathogen (through stomata).
61. *Cladosporium fulvum* infects tomato, *Cladosporium cucumerinum* infecting cucumber and *Stemphylium solani* infecting tomato, generally enters through stomata.

62. Members of sphaeropsidales entering through stomata are *Phoma trifoli* and *Phoma herbarum* var. *medicaginis* on *Trifolium pratense* and *Medicago sativa*, respectively. These fungi also enter by direct penetration under certain conditions.

63. In India, the 1942 Bengal famine was perhaps largely due to the *Helminthosporium oryzae* (brown spot diseases of rice).

64. In 1946-47, the wheat rust epidemic was responsible for food shortage as well as seeds in Madhya Pradesh.

65. German scientist de Bary (1886) was first to study critically the nature of direct penetration by *Sclerotinia sclerotiorum*.

66. The smut fungi usually penetrate directly into young ovaries; shoot bud, young seedlings or young tissues.

67. The term appressorium was introduced by Frank (1883) for spore like 'organisms' formed on the germ tubes of *Colletotrichum lindemuthianum*, *Polystigma rubrum* and *Fusicladium tremulae*.

68. Appressoria produced by germinating urediniospores of rusts and *Mycosphaerella citri* penetrate through stomata.

69. *Thielaviopsis basicola* and *Fusarium solani* f. sp. *phaseoli* enters through wound.

70. *Heterobasidion annosus*, a destructive pathogen of conifers, naturally colonizes wounds caused by high winds, snow or other natural agencies where it enters through the stumps left after felling.

71. *Cephalosporium gramineum* causes Cephalosporium stripe of wheat seedlings and other grasses. It enters in the host plants through wounds.

72. In the ergot diseases of rye caused by *Claviceps purpurea*, conidia are disseminated by species of Diptera and Coleoptera.

73. *Fusarium* species which causes rot in potato tubers may enter through wounds caused by *Spongospora subterranea* and *Phytophthora infestans*.

74. The apple canker is caused by *Nectria galligena*, may enter through lesions on a stem caused by *Venturia inaequalis*.

75. Potato tubers highly resistant to *Phytophthora infestans* become susceptible when the parasite enters through wounds caused by *Spongospra subterranea*.

76. The pathogen *Ceratocystis ulmi* is carried and transmitted to elm trees by feeding beetles (*Scolytus scolytus*, *S. multistriatus*).

77. *Fusarium oxysporum* f. sp. *vasinfectum* gains entry into the roots of susceptible cotton plants through the injuries caused by the root –knot nematode, especially *Meloidogyne* species.

78. *Plasmodiophora brassicae* (finger and toe diseases of crucifers), *Phymatotrichum omnivorum* (texas root rot of cotton), *Fusarium oxysporum* f. sp. *lini* (flax wilt),

Fusarium oxysporum f. sp. *conglutinans* (cabbage yellow) enters through the roots hair.

79. Vascular pathogens such as *Fusarium, Verticillium* produce pectolytic enzymes. *Fusarium oxysporum* f. sp. *lycopersici* produces a pectin depolymerase, pectin methyl esterase and cellulase and these appear to be involved in pathogenesis.
80. Juice extracted from stems of tomato plants infected with *Verticillium albo-atrum* contains high amount of PME (pectin methyl esterase).
81. *Puccinia graminis tritici* produces pectic enzymes which are thought to be involved in the growth of the hyphae between host cells and in the penetration of the epidermal cell wall.
82. *Rhizoctonia solani, Fusarium moniliforme* produce both cellulolytic and pectic enzymes.
83. *Sclerotium rolfsii* produces cellulolytic enzymes and has the capacity to cause soft rot of many plants and in leaf spot diseases.
84. The production of proteolytic enzymes by the cotton wilt fungus (*Fusarium oxysporum* f. sp. *vasinfectum* and *Helminthosporium oryzae*, is responsible for the brown leaf spot of paddy.
85. Strong exocellular amino acid oxidase enzymes are present in the *Pyricularia grisea, Fusarium oxysporum* f. sp. *vasinfectum* and *H. oryzae*. Both amino acid oxidases and proteases help the pathogen to invade the tissues easily and effectively.
86. Phospholipase 'B' enzyme is produced by *Thielaviopsis basicola, Sclerotium rolfsii* and *Botrytis cinerea*. These enzymes and proteases affect cell membrane.
87. The take all disease of oats caused by *Ophiobolus graminis*, it produces a specific enzyme (glucosidase) which hydrorlyzes avenacin.
88. *Colletotrichum lindemuthianum* is capable of producing a variety of polysaccharide degrading enzymes and the virulence isolate has ability to produce α-galactosidase.
89. Piricularin toxin produced by *Pyricularia grisea* is a typical pathotoxin.
90. Fusaric acid was first isolated as the metabolic product of *Fusarium heterosporum*, by Yabuta *et al.* in 1934. Fusaric acid is called a wilt toxin as it is now known to have a pathogenic role in the wilt of tomato and cotton.
91. Fusaric acid (vivotoxin) involved in the panama disease of banana caused by *Fusarium oxysporum* f. sp. *cubense*.
92. Lycomarasmin is a wilt toxin (vivtoxin) involved in wilt disease of tomato caused by *Fusarium oxysporum* f. sp. *lycopersici*.
93. Ten toxins are a cyclic tetrapeptide and involved in energy transfer into chloroplasts and also interfere or inhibit the light dependent phosphorylation of ADP to ATP.

94. T-toxin is produced by race T of *Heminthosporium* (*Cochliobolus*) *heterostrophus* (southern corn leaf blight) and acts specifically on mitochondria of susceptible cells and inhibits the synthesis of ATP.

95. In monocyclic pathogens, the primary inoculums are the only inoculums available for the entire season, since there are no secondary inoculums and no secondary infection.

96. In polycyclic fungal pathogens, the primary inoculums generally consists of the sexual (perfect) spore or in fungi that lack the sexual stage, some other hardy structure of the fungus such as sclerotia, pseudosclerotia or mycelium in infected tissues.

97. The rate of inoculums or disease increase (r) has been calculated for many diseases and varies from 0.1 to 0.5 per day for polycyclic foliar diseases such as southern corn leaf blight, potato late blight, gram rusts and tobacco mosaic, and 0.02 to 2.3 per year for polyetic diseases of trees such as dwarf mistletoe of conifers, Dutch elm disease, chestnut blight and peach mosaic.

98. Cutinases enzyme breaks cutin molecules and release monomers as well as oligomers of the component fatty acid derivatives from the insoluble cutin polymer.

99. Pectin degrading enzymes involved in the production of many diseases, particularly those characterized by soft rot of tissues and also leads to tissue maceration.

100. Pectic enzymes involved in the induction of vascular plugs and occlusions in the vascular wilt diseases.

101. Gibberellins were first isolated from the fungus *Gibberella fujikuroi*, the cause of the foolish seedling disease of rice.

102. The first compound with cytokinin activity to be identified was kinetin, which however was isolated from herring sperm DNA. Several cytokinins such as zeatin and isopentenyladenosine (IPA) isolated from plants and are responsible for pathogenesis.

103. Cytokinin activity increases in club root galls, in crown galls, in smut and rust infected bean leaves.

104. Ethylene also causes increased permeability of cell membranes, which is a common effect of infections.

105. *Puccinia graminis* f. sp. *tritici* (stem rust of wheat) and *Mycosphaerella pinodes* (leaf spot on pea), produce substances called suppressors, that act as pathogenicity factors by suppressing the expression of defense responses in the host plant.

106. The suppressor molecule suppresses the activity of phenylalanine lyase (PAL) and the normal development of defense responses.

107. The *Mycosphaerella* suppressers seem to reduce to proton pumping activity of the host cell membrane ATPase and thereby temporarily lower the ability of the cell to function and to defend it.

108. Hypertrophy may appear as tumors, galls, knots, abnormal increase in organs size, witches broom etc.

109. Hyperauxinity (excess of IAA) has been detected in many other diseases such as rust of *Euphorbia cyparissias* caused by *Uromyces pisi*, powdery mildew on wheat (*Erysiphe graminis*) and white blister of *Brassica napus* (*Albugo candida*).

110. Increase in the level of kinin –like substance in leaves of bean attacked by *Uromyces phaseoli* and leaves of *Vicia faba* attacked by *Uromyces fabae* is reported to the formations of 'green islands' around the infections center. The green island phenomenon where higher RNA content occurs in this area as compared to the non-infected tissue between the infected sites.

111. Wilt of tomato is caused by *Fusarium oxysporum* f. sp. *lycopersici* produce ethylene and in most cases, the production of ethylene is by the damaged tissues. Necrosis induced by toxic chemicals also results in increased ethylene evolution.

112. The obligate fungal parasites such as rust and mildew fungi are known to cause accumulation of carbohydrates and minerals at the site of infection.

113. The toxins of fungi can affect the chloroplasts and rate of photosynthesis. Occlusions of vascular elements in wilt diseases caused by species of *Fusarium* and *Verticillium* cause water stress and inhibition of CO_2 intake.

114. *Pyricularia oryzae* (*P. grisea*) causing blast of rice, directly penetrates the epidermis.

115. Lignifications, suberisation, deposition of carbohydrates etc. to strengthen the internal tissues are also mechanical barriers to infection.

116. Red scales of onion contain protocatechuic acid and catechol which may exude in drops and impart resistance to attack of *Colletotrichum circinans*. These phenolic substances inhibit spore germination of the fungus.

117. The wrinkled seeded varieties of pea are susceptible to seed and root rot because the seed exudes very high quantities of sugars which encourage the growth of *Pythium*.

118. Certain varieties of linseed (flax) resist wilt caused by *Fusarium oxysporum* f. sp. *lini* through the presence of hydrocyanides (HCN) in their root exudates. This substance is extremely toxic to the wilt pathogen and reduces its infectivity around the roots.

119. Elicitor as a term was used by Keen *et al.* (1972) to replace the term inducer earlier used by Cruickshank and Perrin (1968).

120. Systemic signal transduction that leads to systemic acquired resistance is thought to be carried out by salicylic acid, oligogalacturonides released from the plant cell walls, jasmonic acid, systemin, fatty acids, ethylene and others.

121. The most common new cell functions and compound include a rapid burst of oxidative reactions, increased ion movement especially of K^+ and H^+ through the cell membrane.

122. Jasmonic acid is the precursor of the wound hormone traumatin, appears to induce numerous protein changes and act as a signal transducer of the defense reaction in plant- pathogen interactions.

123. Systemic acquired resistance (SAR) genes are β-1, 3- glucanases, chitinases and cysteine rich protiens related to thaumatin and PR-1 proteins, have direct antimicrobial activity or closely related to classes of antimicrobial proteins.

124. Tobacco plants transformed with a chitinase gene from bean became resistant to infection by the soil borne fungus *Rhizoctonia solani.*

125. Transgenic tobacco plants expressing a PR-1 protein gene were resistant to the blue mold fungus *Peronospora tabacina* and plants expressing the systemic acquired resistance gene SAR 8.2 were resistant to the black shank fungus *Phytophthora parasitica.*

126. In 1880, H. M. Ward emphasized the role of the environment in the epidemiology of coffee rust. He considered as father of tropical Plant Pathology. He was also demonstrated the effect of copper fungicides against endophytic fungal pathogens.

127. Yabuta and Hayashi in 1939 showed that active component of filterate responsible for the excessive elongation was growth regulator gibberellins.

128. Skin scab of potato is caused by *Oospora pustulans* and it enters in the tubers through lenticels.

129. *Penicillium expansum* (storage rot of apple) and *Gloeosporium perennans* (storage rot pathogen) enters through stomata or lenticels.

130. *Nectria galligena,* the cause of apple canker, can enter twigs through lenticels. *Botrytis cinerea* enters through hydathodes of *Methiola incana.*

131. *Armillaria mellea, Spongospora subterranea, Phymatotrichum omnivorum* enter through wound or lenticels.

132. Increased transpiration rates lead to increased ion uptake, such as K^+ which accumulates at the infection sites. In take all diseases of wheat, caused by *Gaeumannomyces graminis* var. *tritici,* infected plants are seen to lead to the reduction of K^+, P and Ca^{++}.

133. Rusted wheat leaves and leaves of bean infected with *Uromyces phaseoli* retained chlorophyll in regions termed as 'Green Island' at the edge of the infection.

134. Auxins or kinetins may also induce 'green island' formation, in barley kinetin induced island contain more starch but less nitrogen than mildew induced island.

135. The Embden–Meyerhof–Parnas (EMP pathway) is the principal glycolytic pathway in fungi (accounts for 50 to 100 per cent glucose dissimilation). However, EMP is secondary in importance to Hexose monophosphate pathway (HMP) and enter-doudoroff (ED) pathway found in a few fungi or totally lacking, as in the case of *Caldariomyces fumago*.

136. Powdery mildew of oak is caused by *Microsphaera alphitoides*.

137. In *Arabidopsis* leaves infected by *Albugo candida*, the reduction in the rate of photosynthesis is paralleled by a decrease in the amounts of rubisco (ribulose biphosphate carboxylase) protein present in the host.

138. Starch content in banana leaves is reduced and sugar content is increased in groundnut leaves infected by *Cercosporidium personatum*.

139. In 1953 and 1954, Allen was first pointed out that the disease was the result of disturbances in host metabolism induced by the diffusible substances or toxins produced by invading pathogens.

140. The level of phenylalanine ammonia lyase (PAL) activity increases to a peak within 24 hours of infection by *Ceratocystis fimbriata*.

141. Polyphenols such as chlorogenic and caffeic acid, known to be powerful inhibitors of TAA-oxidase.

142. Onion varieties resistant to *Colletotricham circinans* due to flavones, anthocyanine and simple phenols such as protocatechuic acid and catechol in the dried bulbs scales and leaves of apple resistance to *Venturia inaequalis* due to phloretin.

143. Cell elongation is thought to be brought about by the removal of Ca^{++} and Mg^{++} cation bridges from the pectate network.

144. A ten-fold increase is found in the IAA level of susceptible potato tubers tissues infected by *Phytophthora infestans* and in corn infected by *Ustilago zeae*.

145. The fist cytokinin discovered was kinetin by Skoog and Miller. Cytokinin delay senescence in plant tissues. The action of cytokinin is dependent on auxins. They are derivatives of adenine.

146. Cytokinin content increase in tumours caused by *Ustilago zeae, Synchytrium endobioticum, Plasmodiophora brassicae* and *Taphrina deformans*.

147. Ethylene has an important role in regulation of plant growth and is implicated in a number of disease symptoms such as tissue swelling, stimulation of root formation, leaf abscission, epinasty and fruit ripening.

148. The ethylene treatment increases the level of cellulase and β-1,3-glucanase which are implicated in the degradation of cell wall polymers.

149. Ethylene stimulates those enzymes in plants which are active in the synthesis of phytoalexins, for instance, PAL, peroxidase, polyphenol oxidase and pectin esterase. These enzymes oxidize phenolic compounds and are involved in resistant reactions of the host plants.

150. Wilt of tomato is caused by *Fusarium oxysporum* f. sp. *lycopersici*, ethylene production by the pathogen is sufficient to account for the epinastic symptoms of a disease.

151. The defoliating symptom caused by *Verticillum* wilt of hop plants is attributed to increased level of ethylene in infected hop plants.

152. In *Verticillium* wilt disease ethylene has a dual role, acting directly as a toxin on plant tissues covering leaf drop, epinasty and premature senescence and also inducing disease resistance in infected plants.

153. Transcription is a process by which gentic information encoded in DNA is copied during the process of mRNA synthesis. For example, obligate parasites such as rusts and powdery mildews affect the transcription process in infected cells.

154. *Pyricularia grisea* causes rice blast disease. The outer walls of most of the epidermal cells of rice leaves are lignified and the entry of the pathogen is generally takes place through the thin -walled motor cells having pectin rather than lignin.

155. In the *Rhizoctonia* disease of potato tubers, following infection, cork layer are produced just below the areas of infection. This layer prevents further invasion by the pathogen. Cork layer are also formed between healthy and diseased areas on plum leaves attacked by *Coccomyces prunophorae*.

156. In wilt of sweet potato, caused by *Fusarium oxysporum* f. sp. *batatas*, tyloses formed abundantly, and results in greater resistance as it prevents the further spread of the pathogen.

157. Malformed tissues have lower levels of reducing, non-reducing and total sugars, starch and carbohydrates than the healthy ones.

158. Shot holes of peach caused by *Clasterosporium carpophilum*.The infection results in formation of abscission layer which prevents further spread of the pathogen and this pathogen also causes necrotic ring spot on cherry and also formed abscission layer.

159. In the silver leaf disease of plums caused by *Stereum purpureum*, apple twig infected with *Physalospora cydoniae* and blast of rice caused by *Helminthosporium oryzae*, gum deposits are formed and constitutes a type of mechanical resistance.

160. Fungitoxic substances detected in plant waxes, and cutin acids of citrus lime are toxic to *Gloeosporium limetticola*.

161. *Trichoderma viride* was earlier reported to produce two antibiotic substances, gliotoxin and viridin, but now it is known that the antagonistic effect of

T. viride is not due to these two antibiotic susbtances which, in fact, it does not produce. The antibiotic action of *T. viride* is due to the production of a volatile antibiotic substance known as trichodermin.

162. In 1957, Buxton has done classical work on the resistance of pea cultivars due to root exudates. Exudates from the roots of one pea variety reduce the germination of the spores of a race of *Fusarium oxysporum* f. sp. *pisi*, non pathogenic to this variety.

163. Oat leaves and roots contain fluorescent glucoside (avenacin) which is inhibitory to the growth of several fungi.

164. Phenolic compounds, polyphenol oxidase (PPO) are important in disease resistance because it can oxidize phenolics to quinones which may be more toxic.

165. PPO produced by the pathogen may oxidize the host polyphenols to more highly fungitoxic substances which may prevent further development of the pathogen. For example- PPO inhibitor reduces the toxicity of catechol to *Cochliobolus miyabeanus* (leaf blight of rice) and prevents melanin development, presumably by preventing enzymatic conversions of catechol into fungitoxic quinones.

166. The phenolic content of rice is responsible for resistance to *Pyricularia grisea* (blast of rice), and polyphenols have implicated in resistance to *Venturia inaequalis* (apple scab) and *Venturia pirina* (pear scab).

167. The high sugar contents of tissues fauours some pathogens, such as rusts and powdery mildew, whereas the reverse is true for other, including *Alternaria solani* (early bight of potato) and *Ceratostomella ulmi* (Dutch elm disease).

168. Boron facilitates the translocation of sugars and accumulates in the leaves of boron deficient plants and the later are apparently resistant to *Cochliobolus sativus* (barley leaf spot), a low sugar pathogen, but susceptible to *Erysiphe graminis*, a high sugar pathogen.

169. In 1940, K.O. Muller and H. Boerger, worked on late blight of potato and the term "phytoalexins" was introduced. Phytoalexins inhibit the development of the fungus in hypersensitive tissues, is formed or activated oniy when the host cells come into contact with the parasite.

170. Pathogenesis related proteins (PR proteins) are generally a group of plant proteins that are toxic to invading fungal pathogens. They are either extremely acidophilic or extremely basic and are highly soluble and reactive.

171. Salicylic acid, acetyl salicylic acid, dichloro-iso-nicotinic acids (DCINA), ethylene xylanase acid are the signal molecules which are involved in the induction of plant defense against infection.

172. Healthy plants may contain several PR-proteins in trace amounts; attack by plant pathogens or treatment with defense inducers or wounding or stress induces transcription of a battery of genes that code for PR-proteins.

173. The defense gene are involved in the production of antimicrobial compounds such as phenolics, phytoalexins, PR- proteins and active oxygen species or involved in the reinforcement of the cell wall by accumulating hydroxy-proline-rich glycoproteins, callose, lignin and wall- bound phenolics.

174. Several cloned resistance genes such as leucine-rich-repeats (LRR), leucine zippers and nucleotide binding sites (NBS), detected in barley, lettuce, wheat, soybean, pepper, chickpea, common bean and *Brassica napus*.

175. Monogenic resistance (single gene) is governed by one gene with 3:2 mendelian ratio, it can be monohybrid dominant in resistance of oats to *Helminthosporium victoriae*, resistance of sorghum to *Periconia circinata*, resistance of cabbage to *Fusarium oxysporum* f. sp. *lycopersici*, resistance of onion to *Pyrenochaeta terrestris* and cucumber to *Cercospora*.

176. Monogenic resistance can be monohybrid recessive *e.g.* powdery mildew of barley, sorghum smut, yellow rust of wheat.

177. Oligogenic resistance (governed by several genes) is determined by two to several genes. Thus, it can be dihybrid, trihybrid etc. *e.g.* anthracnose of bean is governed by 3 genes.

178. Polygenic resistance (governed by many genes) involves many genes which are more difficult to analyse and known to involve in a large number of diseases such as cotton wilt.

179. Horizontal resistance is called lateral or generalized, non-specific, or field resistance (minor gene), less strong.

180. Vertical resistance is also called perpendicular or specific (major gene), more strong.

181. Heterokaryosis may take place at the time of spore formation in many fungi, such as inclusion of + and - nuclei in the same ascospores of *Neurospora tetrasperma* or *Podospora anserina*.

182. In *Puccinia graminis trtici* (stem rust of wheat), the haploid basidiospores can infect barberry but not wheat, and the haploid mycelium can grow only in barberry, however, the dikaryotic aeciospores and uredospores can infect wheat but not barberry and the dikaryotic mycelium can grow in both barberry and wheat.

183. In 1952, Pontecorvo and Roper discovered sexual reproduction in the imperfect fungus *Aspergillus nidulans* and they called it 'parasexual cycle' (parasexual recombination).

184. Parasexuality is a novel type sexual reproduction which produced new races of several fungi such as *Fusarium oxysporum* f. sp. *pisi*, *Ascochyta imperfecti* (black stem of alfalfa), *Fusarium oxysporum* f. sp. *cubense* (banana wilt), and *Cochliobolus sativus*.

185. Somatic recombination may occur in smut and rusts as in *Ustilago maydis* and *Puccinia graminis* and *Puccinia recondita*.
186. Cytoplasmic inheritance has been attributed to plasma genes. Cytoplasmic RNA is the genetic material detached from the chromosomes (episomes).
187. Hybridization can be intraspecific, interspecific or even intragenic and the resulting hybrids may have different pathogenic abilities than the parental races.
188. Often the hybrids are intermediate in pathogenicity between the two parental races but some may be more pathogenic than others.
189. Extensive recombination of genes occurs in autoecious rusts such as *Melampsora lini* (flax rust) in which all the spore stages occur in the same plant.
190. In 1992, the first R gene, the maize Hm1 gene was located, isolated and sequenced and its function was described at the molecular level.
191. Avirulence (avr) genes, first identified by H.H. Flor in 1950s.
192. Plant cell-wall degrading enzymes (cutinase), some toxins (victorin, HC-toxin), hormones (IAA, cytokinin), polysaccharides, proteinases, siderophores, melanin etc. are produced by pathogens in plant- pathogen interactions in which they are essential for the pathogen to infect and cause disease on its host. In those cases, therefore, such factors function as pathogenicity factors.
193. In other plant systems, the same componnds are helpful but not essential for disease induction and development. In these cases, these compounds are considered as virulence factors.
194. *Typhula* and *Fusarium*, which cause snow mold of cereals and turf grasses, thrive only in cool seasons or cold regions. Also, *Phytophthora infestans* (late blight of potato) is most serious only during the winter.
195. Many pathogens such as *Monilinia fructicola* (brown rot of stone fruits), *Colletotrichum* (anthracnose) are favoured by relatively high temperature.
196. *Taphrina pruni* and *T. deformans*, *Spongospora subterranea* is limited to cold areas.
197. *Puccinia striformis* also thrives in the cold and in India it oversummers at heights above 7000 feet above sea level *Puccinia graminis tritici* survives at 5000 feet above sea level.
198. *Cercospora beticola* (sugarbeet leaf spot), needs a temperature 15°C for sporulation, germination and infection.
199. The latent period for *Erysiphe graminis* decrease from about 14 days at 5°C to 3 days at 18-25°C.
200. The latent period for *Puccinia graminis tritici* is 22 days at 5°C, 15 days at 10°C and 5 to 6 days at 23°C.
201. In the sunhemp wilt disease caused by *Fusarium oxysporum* f. sp. *crotolariae*, the optimum growth is accomplished at 25-28°C. Blight of wheat requires

(*Alternaria triticina* and *Helminthosporium sativum*) at a comparatively higher temperature that is 20 to 28°C.

202. *Pythium, Phytophthora, Rhizoctonia, Sclerotinia* and *Sclerotium* usually cause their most severe symptoms on plants when the soil is wet but not flooded.

203. The incubation period and the germination of urediniospores of *Puccinia graminis tritci* is favoured by a temperature of 20°C.

204. *Sclerotium cepivorum* (white rot of onion), *Fusarium solani* (root rot of bean), *F. roseum* (seedling blight), and *Macrophomina phaseoli* (charcoal rot of sorghum) grow fairly well in dry environment.

205. A large number of diseases are favoured by increased humidity such as late blight of potatoes, koleroga of arecanuts, downy mildew of cereals, blast of rice, early blight of potato, scab of apple and circular leaf spot of coffee.

206. *Aphanomyces euteiches* (root rot of pear), *Pythium ultimum, Pythium debaryanum* (damping off) and *Plasmodiophora brassicae* (club root disease of crucifers) are greatly increased by an increase in the moisture content of the soil.

207. *Sphacelotheca reiliana* (head smut of sorghum) is most favoured by moderate soil moisture.

208. *Neovossia indica* (Karnal bunt of wheat) is most favoured by heavy rain and soil temperature around 15-20°C, 2-3 week before flowering time.

209. *Colocasia* blight due to *Phytophthora colocasiae* occurs only in the rainy season, especially when there is a continuous spell of moist and cloudy weather.

210. *Plasmodiophora brassicae* (club root of crucifers) is most prevalent and severe at pH about 5.7 (acidic soil). It totally fails to develop at 7.8 pH.

211. *Phymatotrichum omnivorum* (cotton root rot), *Ophiobolus graminis* (take all of wheat), *Verticillium alboatrum* are favoured by the alkaline conditions of soil (pH 7.2-8.0).

212. *Fusarium oxysporum* f. sp. *lycopersici* (wilt of tomato), and *F. oxysporum* f. sp. *vasinfectum* (wilt of cotton), *Synchytrium endobioticum* (potato wart disease), and *Spongospora subterranea* (powdery scab of potato) are associated with acidic soils.

213. Large amount of nitrogen increases the susceptibility of wheat to rust (*Puccinia*) and powdery mildew (*Erysiphe*).

214. Reduced availability of nitrogen increases the susceptibility of tomato to Fusarium wilt, early blight of potato and tomato (*Alternaria solani*) and damping off of seedlings (*Pythium spp*).

215. *Fusarium* spp, *Plasmodiophora brassicae, Sclerotium rolfsii, Pyrenochaeta lycopersici* are more serious when ammonium nitrogen fertilizer is applied while *Ophiobolus graminis* and *Phymatotrichum omnivorum* are more serious when nitrate nitrogen is applied in the soil.

216. Phosphorus reduces the severity of take all disease of barley (*Ophiobolus graminis*) but increases the severity of glume blotch of wheat caused by *Septoria*.

217. Potassium reduces the severity of blast of rice caused by *Pyricularia grisea* and also increases resistance to frost injury.

218. Calcium reduces the severity of several fungi such as *Rhizoctonia, Sclerotium, Botrytis, Fusarium oxysporum,* while it increases the black shank disease of cotton (*Phytophthora parasitica* var. *nicotianae*).

219. Copper application to the soil reduces take all (*Gaeumannomyces graminis* var. *tritici*) and ergot diseases (*Claviceps purpurea*).

220. Manganese reduced potato scab, late blight of potato, stem rot of pumpkin seedlings and molybdenum also reduced the late blight of potato and blight of beans and peas.

221. Silicon reduces the severity of rice blast (*Magnaporthe grisea*) and rice brown spot (*Bipolris oryzae*) diseases.

222. The Irish potato famine of 1845-46 was caused by *Phytophthora infestans* (late blight) epidemic of potato and the Bengal famine (India) of 1943 was caused by the *Cochlibolus* (*Helminthosporium*) *miyabeanus* (brown spot epidemic of rice).

223. Epidemics cansed by the pathogens that requires more than a year to complete a reproductive cycle are slow to develop *e.g.* cedar apple rust (2 years), white pine blister rust (3-6 years) and dwarf mistletoe (5-6 years).

224. One of the first computer simulation programs called EPIDEM, was written in 1969, and was designed to stimulate early blight epidemics of tomato and potato caused by the fungus, *Alternaria solani.*

225. Computer simulators were written for *Cercospora* blight of celery (CERCOS), for *Mycosphaerella* blight of chrysanthamum (MYCOS), for southern corn leaf blight caused by *Cochliobolus maydis* (EPICORN).

226. Pea root rot is caused by the fungus *Aphanomyces euteiches.*

227. Computer simulation model have developed for epidemics of late blight (BLITECAST) and *Alternaria solani* on tomato (FAST) and (TOM-CAST).

228. Dormant mycelium of *Phytophthora infestans* is carried throngh the potato tubers and smut of sugarcane (*Ustilago scitaminea*) and red rot of sugarcane (*Colletotrichum falcatum*) through the diseased cane setts.

229. Bunt of wheat (*Tilletia foetida* and *T. caries*) is carried throngh the seeds externally; potato wart (*Synchytrium endobioticum*) and black scurf of potato (*Rhizoctonia solani*) disperse by adherent spores on the tuber surface.

230. *Colletotrichum coffeanum* is disseminated by water. *Hemileia vastatrix* (coffee rust) is carried through rain splashes.

231. *Colletotrichum falcatum*, Sclerotial bodies of *Sclerotium rolfsii* and seeds of flowering parasites such as Orobanche, Cuscuta (dodder) and Striga also disseminated by irrigation water.

232. Oak wilt is caused by *Ceratocystis fagacearum* and is carried by nitulid and scolytid beetles.

233. Chestnut blight caused by *Cryphonectria* (*Endothia*) *parasitica* is carried by birds, crawling or flying insect or splashing rain.

234. *Ophiostoma ulmi* (Dutch elm disease) are carried by bark beetle (*Scolytus multistriatus*) and native elm bark beetle (*Hylurgopinus rufipes*).

235. The book **"Plant Disease Epidemiology"** was written by S. Nagarajan (1983).

236. The first example of an important plant disease epidemic in India was the stem rust of wheat in the district of Jabalpur (M.P.), in 1839.

237. In 1993, 1995, Rosa *et al.*, have developed computer simulation model, PLASMO, to control downy mildew of grapes caused by *Plasmopara viticola*.

238. Calvero *et al.*, in 1996 have developed disease forecasting/warning simulation model for blast of rice (*Pyricularia grisea*) and Friendrich (1995) developed disease forecasting simulation model for powdery mildew of wheat (*Erysiphe graminis* f. sp. *tritici*).

239. Blue mold of tobacco caused by *Pernospora tabacina* and downy mildew of cucurbits caused by *Pseudoperonospora cubensis*.

240. Tikka disease of groundnut caused by *Cercospora* (now *Cercosporidium personatum*) *personata* and *C. arachidicola* depends upon the prevalence of high humidity and low temperature.

241. Net-septate (transversely and longitudinally septate) conidia of fungi belong to Dictyosporae.

242. *Uromyces* and *Caryophyllinus* have one-celled teliospore. *Puccinia* and *Phragmidium* have two-celled and many celled teliospore, respectively.

243. Wood staining fungi usually called sapstain or blue stain fungi.

244. Uredia and telia are formed on the cereal host while pycnia and aecia are found on the barberry.

245. *Phytophthora palmivora* (Trade mame - Davine) are used as biocontrol of strangler vine (*Morrenia adorata*) weed of citrus.

246. *Colletotrichum gloeosporioides* f. sp. *aeschynomene* fungus (Trade name - Collego) is used for the control of northern jointvetch weed (*Aeschynomene virginica*) of rice and soybean.

247. *Cercospora rodmanii* fungus (Trade Name - ABG 5003) can be used in biocontrol of water hyacinth (*Eichhorinia crassipes*).

248. *Alternaria macrospora* fungus is used for the control of spurred anoda (*Anoda cristata*) weed of cotton.

249. *Colletotrichum gloeosporioides* f. sp. *cuscutae* (Trade Name- LUBO A-2) is used for the control of *Cuscuta* sp. (Dodder or Amarbel) weed of clover and many oil seed crops.

250. Iprodion sold as a Rovral or chipco-26019 is a broad spectrum, foliage contact fungicide. It inhibits the spore germination and mycelia growth but shows mostly preventive and only early curative activity. It is effective against *Botrytis, Monilinia* and *Sclerotinia* and also against *Alternaria* and *Rhizoctonia*.

251. Thiophanate under the trade name Topsin is effective against several root and foliage fungi affecting turf grasses and vegetable crops.

252. Mycelioid which resemble somatic hyphae in being flaccid and indefinite are characteristics of *Erysiphe, Sphaerotheca* and *Levillula*.

253. *Fusarium sambucinum* and *Dactylium fusarioides* act as mycoparasites of sclerotia.

254. *Absidia corymbifera* as well as some species of *Mucor* and *Rhizopus* are known to cause diseases in humans.

255. The *Ascochyta* blight of chickpea pathogen is externally and internally seed borne, seed treatment is an important control measure. Seeds can be treated with organomercurials (Agallol, Agrosan etc), copper sulphate, thiram, benomyl, or calexin-M. Combination of bavistin and thiram (1:1 or 1:3) at the rate of 3g/kg seed gave 95-100 per cent reduction in disease incidence.

256. Sexual reproduction in the Oomycetes is almost always heterogametangic. In the simplest forms the entire thallus act as a gametangium.

257. *Saprolegnia parasitica* causes diseases of fish and fish eggs. Rhizome rots of ginger caused by *Pythium, Myriotylum*.

258. Sporangial proliferation is an interesting phenomenon in the Saprolegniaceae.

259. The pea powdery mildew conidia germinate quickly in the absence of water around them because of their own relatively large water content.

260. The teliospores of Karnal bunt remain viable in soil for 4-5 years and act as primary source of inoculum.

261. Phospholipase 'A' is not known to be produced by plant pathogens. Phospholipase 'B' is produced by *Thielaviopsis basicola, Sclerotium rolfsii* and *Botrytis cinerea*.

262. *Tolyposporium penicillariae* (smut of bajra) are more severe in wet areas in regions where rain occurs for long durations-during the flowering stage of the crop.

263. Mode of development of conidia of *Verticillum, Trichoderma, Gliocladium* and *Paecilomyces* species are phialidic blastic type.

264. The mycelium of *Pythium* gives rise to sporangia, which germinate directly by producing one to several germtubes or by producing a short hypha at the end of which forms a balloon like sceondary sporangium (Vesicle) contain 100 or more zoospores.

265. Ascomycetous fungi are characterized by the production of a sac-like structure containing ascospores.

266. Yeasts are important for fermentation in baking, brewing, distilleries and related industries. Some yeast such as *Candida albicans* are human pathogens.

267. *Pythium irregulare, P. spinosum* and *P. ultimum* are more damaging at lower temperatures while *P. myriotylum, P. aphanidermatum* and *P. arrhenomanes* are damaging at higher temperatures. *Pythium* spp is more destructive when the soil temperatures ranged between 24-30° C.

268. Sporangia of *Phytophthora infestans* germinate almost entirely by releasing 3 to 8 zoospores at temperatures up to 12 to 15°C, where as above, 15°C sporangia may germinate directly by producing a germ tube.

269. Sporangia of most downy mildews germinate generally by producing zoospores or at higher temperatures by producing germ tubes. In the genus *Bremia, Peronospora* and *Peronosclerospora*, the sporangia germinate only by means of a germ tubes.

270. *Plasmopara viticola* (downy mildew of grape) overwinters as oospores in dead leaf lesions and shoots. Oospores germinate and produced sporangium and disease cycle completes from 5 to 18 days.

271. *Pythium aphanidermatum* and *P. butleri* is soil inhabitant and survive as saprophytes on dead organic matter.

272. Rhizome rot of ginger is caused by *Pythium myriotylum* and *P. aphanidermatum*.

273. Late blight of potato was first introduced into the Nilgiri hills between 1870 and 1880.

274. *Plasmodiophora brassicae* (club root of crucifers) was first described by Woronin in 1877.

275. In 1895, black wart disease of potato (*Synchytrium endobioticum*) was first described from Hungary and in India was first reported by Ganguly and Paul in 1953, form Rangbul in Darjeeling district of West Bengal.

276. *Synchytrium endobioticum* is an obligate parasite. The major symptoms appear in early seasons, the outgrowths are green or greenish-white in colour if exposed above ground and light-cream coloured when present underground on the tubers.

277. Sporangia of *Physoderma zeae maydis* perennate from one season to the next in diseased plant debris or in the soil, viable in soil for more than 4 years and secondary spread is through dissemination of the spore dust by wind.

278. Rhizome and root rot of turmeric (*Pythium* spp) was reported by Ramakrishnan and Soumini in 1954 from Andhra Pradesh and Tamil Nadu. The pathogen is mainly soil borne through oospores and optimum temperature is 28-30° C for growth of the fungus.

279. *Phytophthora infestans* (late blight of potato) sporangia germinate at temperature upto 12 or 15°C and release 3 to 6 zoospores and germinate by a germ tube.

280. The fungus, *Didymella bryoniae* (cucurbit gummy stem blight) produces conidia and ascospores and overwinters in diseased plant refuse as chlamydospores and in or on the seed.

281. *Phytophthora palmivora* (bud rot of palms) species is established by Butler, it perpetuates through mycelium in diseased tissues in leaf axils and crown, and through oospores, where formed, lying in dead plant parts. The rhinoceros beetle and rains are agents of inoculum dispersal and disease spread.

282. *Septoria* overwinters as mycelium and as conidia within pycnidia on and in infected seed and on diseased plant refuse left in the field.

283. Southern corn leaf blight is caused by *Cochliobolus heterostrophus* and brown spot disease of rice, caused by *Cochliobolus miyabeanus*.

284. Net blotch of barley is caused by *Phytophthora teres*. In barley, stripe disease is caused by *Pyrenophora graminea* and tan spot of wheat is caused by *Pyrenophora tritici- repentis*.

285. The main symptom of gummosis is oozing of gum from affected parts. Three species of *Phytophthora viz. P. parasitica*, *P. palmivora* and *P. citrophthora* attack citrus. In south India, the main cause of gummosis and root rot is *P. palmivora*, although *P. parasitica* is also a common incitant of fruit rot in Karnataka.

286. All Nectria species produce two celled ascospores in brightly colored perithecia on the surface of a cushion shaped stroma, while *Nectria galigena* produces single celled microconidia and more commonly two to four celled cylindrical macroconidia of the cylendrocarpon type.

287. Downy mildew and green ear disease (*S. graminicola*) is primarily soil borne. The spores which are abundantly present in the diseased leaves fall on the ground and perennate from 8 months to 10 years.

288. The Fungus, *Diplocarpon rosae*, produces marssonina type conidia in acervuli forming between the outer wall and cuticle of the epidermis and ascospores in tiny apothecia formed in old lesions.

289. Due to hypertrophy and hyperplasia in the tissues the floral parts show swellings and distortions.

290. *Sclerophthora rayssiae* var. *zeae* (brown stripe downy mildew of maize) is an obligate parasite and its oospores present in the soil serve as a source of primary infection and they remain viable in soil for more than three years.

291. High soil moisture and shade with suitable inoculums potential favour severe incidence of the stem gall of coriander (*Protomyces macrosporus*). A pH of 7.4 is most suitable for infection. Fruits show maximum infection at 7.8-8.4 pH.

292. *Taphrina deformans* (peach leaf curl) spores persist during winter on twigs, buds and scales and serve as primary inoculum.

293. *Monilinia fructicola* (brown rot of stone fruits) overwinters as mycelium in mummified fruits on the tree and in cankers of affected twigs or as pseudosclerotia in mummies in the ground.

294. *Verticillium* infection may result in defoliation, gradual wilting and death of successive branches. *Verticillium alboatrum* grows best at 20 to 25°C, whereas *Verticillium dehali* prefers slightly higher temperature (25 to 28°C).

295. *Ophiostoma ulmi* (dutch elm disease) mycelium is creamy white, in the vessels the mycelium produces short branches on which clusters of Sporothrix type conidia are formed. In dying or dead trees, the mycelium produces mostly graphium type's spores on coremia developing on bark.

296. The fungus, *Fusarium solani* generally produces only asexual spores, although under certain conditions it produces its perithecial stage as *Nectria haematococca*. The fungus can overwinter as mycelium or spores in infected or dead tissues or seed.

297. The fungus *Ophiobolus graminis* (take all disease of wheat) overwinters in infected wheat and grass plant root and stems and in the host debris. The fungus is more active at temperatures between 12 and18°C. Take-all is most severe in infertile, compacted, alkaline and poorly drained soils. Its severity increases for several seasons (3-6 years) in the fields cultivated continuously with wheat.

298. *Botrytis* causes the grey molds of fruits and vegetables both in the field and in storage. *Fusarium* causes post harvest pink or yellow molds on vegetables and ornamentals and especially on root crops tubers and bulbs. *Geotrichum* causes the sour rots of citrus fruits, tomato and carrots.

299. *Penicillium* species causes the blue mold rots (*Pythium italicum*) and the green mold rots (*Pythium digitatum*) and at first appears as soft rot, watery, slightly discoloured spots on any part of fruit, the fungus also produces several mycotoxins such as patulin.

300. Post harvest diseases are caused by several Ascomycetes and imperfect fungi such as *Alternaria, Cladosporium, Colletotrichum, Diplodia, Fusarium* and *Cochliobolus*. They attack on grains and legumes in the field and require too high moisture content in the seed (25 per cent) in order to grow.

301. Three groups of toxins, zearalenones, trichothecenes and fumonisins are produced by several species of *Fusarium*, citreoviridin, citrinin and luteoskyrin are produced by species of *Penicillium* grow in stored rice, barley, corn and dried fish. They cause cardiac beri-beri, nervous and circulatory disorders and degeneration of the kidneys and liver.

302. Patulin is produced by *Penicillium* and *Aspergillus* and causes edema and bleeding in lungs and brain, damage to kidneys and paralysis of motor nerves and it also induces cancer.

303. Rusts fungi that produce only teliospores and basidiospores are called micro-cyclic or short cycled. Other rust fungi produce in adition to teliospores and basidiospores, spermatia, aeciospores and uredospores are called macro cyclic or long cycled rusts. *Asparagus* rust is a macrocyclic rusts (autoecious rust).

304. The teliospores serve only as the sexual overwintering stage, uredospores are repeating spores and spermatia are male gametes. Spermatia are unable to infect plants, their function is fertilization of receptive hyphae of the compatible mating type and subsequent production of dikaryotic mycelium.

305. *Puccinia graminis tritici* is a macrocyclic, heteroecious rust fungus producing spermogonia and aecia on barberry and uredia and telia on wheat and cereals and geasses. It is a polymorphic species, spermogonia (formerly known as pycnia) and aecia develop on the barberry and mahonia which are not found in the plains of India.

306. The fungus, *Gymnosporangium juniperi virginianae* (cedar apple rust) overwinters as dikaryotic mycelium in the galls on cedar trees.

307. The fungus *Cronartium ribicola* (white pine blister rust) produces its spermogonia and aecia on white pine and its uredia and telia on wild and cultivated currant and gooseberry bushes (*Ribes* sp.). The pathogen overwinters mostly as mycelium in infected white pines and to some extent on *Ribes* sp.

308. Fusiform rust is caused by *Cronartium quercuum* f. sp. *fusiforme* which produces spermogonia and aecia on pine stems and branches and uredia and telia on oak leaves and over winters as mycelium in the fusiform galls.

309. Yellow or stripe rust of wheat (*Puccinia striformis*) is more destructive than black or stem rust. Spore wall (uredospores) posses 6-16 germ pores and their grass hosts is *Bromus japonicum* found in the plains of India also.

310. *Puccinia recondita* (leaf rust) tolerates warm weather better than the yellow rusts. In cooler areas, late October to December is quite favourable for infection by rust fungi. Due to high temperature, uredospores and teliospores of the fungus get killed in plains. Teliospores have no role to play in the actual disease cycle.

311. *Puccinia striformis* is highly susceptible to high temperature, *Puccinia graminis tritici* can not survive the extreme summer temperature in Indo- Gangatic plains of the North and in plains it appears in first week of February to March, April and in peninsular India it appears in December- January. *Puccinia striformis* is totally absent from south India except in hills.

312. The perfect stage of the multinucleate *Rhizoctonia solani* is *Thanatephorus cucumeris* whereas that of binucleate *Rhizoctonia* is *Ceratobasidium*. A few multinucleate *Rhizoctonia* spp (*R. zeae* and *R. oryzae*) have *Waitea* as their perfect basidiomycetous stage.

313. Some ascomycetes such as *Daldinia, Ustilina* and *Xylaria* causes a relatively slow white rot and other species of *Alternaria, Bisporomyces, Diplodia* and *Paecilomyces* causes soft rots of wood. The soft rot fungi utilize both polysaccharides and lignin.

314. Blue stain fungi are species of *Ceratocystis, Hypoxylon, Xylaria, Graphium, Diplodia* and *Cladosporium. Penicillium* stain is wood green or yellow and Aspergillus stain is wood black or green. *Fusarium* and *Rhizopus* produces red and gray colour, respectively.

315. Banded leaf and sheath blight of maize caused by *Rhizoctonia solani* and vertical banded blight of maize caused by *Marasmiellus paspali.*

316. Rust of pearlmillet and cumin blight is caused by *Puccinia substriata* var. *indica* and *Alternaria burnsii,* respectively.

317. *Sclerotinia sclerotiorum* (collar rot of brinjal) and *Macrophomina phaseolina* are plurivorous. Replant disease is caused by biotic factors.

318. *Sclerotia* of *Rhizoctonia solani* are reported to survive in soil for upto 6 yeans.

319. *Verticilium tenerum* a potential biocontrol agent of *Phytophthora capsici* which causes foot rot of black pepper.

320. Wed blight of groundnut caused by *Thanatephorus cucumeris* and brown blight of tea is caused by *Glomerella cingulata.*

321. Snow mould of wheat is caused by *Microdochium nivale. Typhula incarnata* causes only slight winter injury on crops.

322. *Geotrichum citri-aurantii* causes post-harvest disease of citrus fruit and it invades throngh wounds. Growth of *Penicillium* and *Geotrichum* is suppressed, when fruit temperature is below 10°C.

323. Shot hole of peach, grape anthracnose and Purple blotch of onion is caused by *Wilsonomyces carpophilus, Gloeosporium ampelophagum* and *Alternaria porri,* respectively.

324. Linked genes Lr 34/Yr 18 is for slow rusting to leaf and stripe rusts.

325. Identification of the gene Lr 34 for resistance to leaf rust led identification of a linked gene for yellow rust resistance, designated as Yr18. Yr18 gene alone does not provide a very high level of resistance, but combined with other genes can provide an acceptable level of resistance.

326. The gene Sr 26 is for resistance to black stem rust to wheat from *Agropyron elongatum.*

327. The gene Sr 31 is carried on the rye segment of wheat-rye translocated chromosome 1B-1R in which the short arm of the chromosome is replaced by the rye segment. Sr 31 gene can be treated as single gene and much higher level of resistance than Sr 2 gene against stem rust. The gene Lr 13 did not provide a high level of resistance.

328. Smut of sugarcane caused by *Ustilago scitaminia* Syd. It was first described by Rabenhorst in 1870 as *U. sacchari* and in India, it was recorded by Sydow and Butler in 1906.

329. The smutted, black powder of stinking smut and Karnal bunt of wheat, gives a foul smell caused by the presence of a volatile compound trimethylamine.

330. Nitrogenous fertilizers tend to increase the susceptibility of wheat to rusts. Tilt @ 0.1 per cent is effective in controlling the three wheat rusts.

331. Epidemics of bown rust (*Puccinia recondita*) occurred in north western part of India in 1972-73. The pathogen is heteroecious. The aecial and pycnial stages are produced on *Thalictrum*.

332. Rust of linseed (*Melampsora lini*), rust of beans (*Uromyces phaseoli vignae*) and rust of pea *Uromyces fabae* are autoecious rust.

333. Rust of chickpea (*Uromyces ciceris arietini*) is heteroecious. It's pycnial and aecial stages are so far unknown.

334. Polysaccharides are known to be produced by *Ceratocystis ulmi* (Dutch elm disease). *Ceratocystis fagacearum* (oak wilt), *Verticillium alboatrum* (wilt of tomato) and species of *Fusarium* and *Cephalosporium* induce wilting in plants.

335. Extracellular pectolytic and cellulolytic enzymes are known to be produced by the vascular wilt pathogens *viz. Verticillium alboatrum, V. dahlia, Ceratocystis ulmi, C. fagacearum, Fusarium oxysporum* f. sp. *lycopersici, F. oxysporum* f. sp. *cubense, F. oxysporum* f. sp. *vasinfectum, and F. oxysporum* f. sp. *niveum.*

336. The first studied wilt toxin was lycomarasmin, which was isolated in pure form from culture filterates by Clauson-Kass *et al.* (1944). Fusaric acid was first isolated as the metabolic product of *Fusarium heterosporum* by Yabuta *et al.* (1934). Fusaric acid is called a wilt toxin. It impairs the permeability of plasma membrane and also impairs the energy metabolism.

337. Wilt of pigeon pea is caused by *Fusarium udum* Butler. In 1982, B. Rai and R.S. Upadhyay have reported *Gibberella indica* as the perfect stage of *F. udum.* It is a soil borne saprophyte; survive in the soil from 8-20 years. Upadhatyaya and Rai in 1989 have reported that enzymes pectin-methyl-esterase (PME), polygalacturonase (PG) and cellulase are inovoled in the pathogenesis.

338. *Ophiostoma* (*Ceratocystis*) *ulmi* produces two types of anamorphs, Sporothrix-type, directly on the mycelium and a synnematous anamorph (Graphium- type) that is abundant in pupal chambers and provides much more inoculums that is spread by beetles, while *Ophiostoma ulmi* is the ascigerous stage.

339. Root and stem rot of jute is caused by *Macrophomina phaseolina*. The sclerotial stage of the pathogen is *Rhizoctonia bataticola*.

340. Black scurf of potato is caused by *Rhizoctonia solani* and its basidial stage is *Thanatehporus cucumeris*. The pathogen is soil borne as well as tuber borne.

341. Stem rot of rice is caused by *Sclerotium oryzae* and its conidial stage is *Nakataea sigmoidea* (*Helminthosporium sigmoideum*). It has its perfect ascigerous stage in *Leptosphaeria salvinii* now renamed as *Magnaporthe salvinii*.

342. Stripe disease of barley is caused by *Drechslera graminea* and its perfect stage is *Pyrenophora graminea* Ito and Kuribayashi.

343. Brown spot disease of rice is caused by *Drechslera oryzae* and its perfect stage is *Cochliobolus miyabeanus* (Ito and Kuribayashi) Dickson. This fungus produces Cochliobolin toxin. Most recent of outbreak of brown spot disease was occurred in Bengal in 1942. The optimum temperature required for conidial germination is between 25-30°C with more than 92 per cent relative humidity.

344. Tikka disease of groundnut is caused by *Cercospora arachidicola* and *Cercosporidum personatum* and its perfect stage is *Mycosphaerella arachidicola* and *M. berkeleyii*, respectively. *Cercospora* produces a phytotoxin, Cercosporin. Both pathogens are disseminated by wind. The perfect stage of this pathogen has been found on fallen leaves and stems in late autumn or in early winter.

345. Red rot of sugarcane is caused by *Colletotrichum falcatum* and it's teleomorph is *Glomerella tucumanensis*. First symptoms are seen after the rainy season when plant growth ceases and sucrose formation starts. Cavities filled with grayish or whitish mycelium are found in the pith. Chlamydospores of this pathogen can persist in soil for a very long time. Diseased seed setts are the chief means of survival and spread of the pathogen.

346. If red rot infected shoots remain in the field, the secondary infection is caused by conidia, which are produced in acervuli and transmitted through insects, wind and water. Initial penetration by conidia is facilitated by mechanical injury to the stems and buds and infection by cane borer. High humidity, water-logged condition, lack of proper cultural operations, continuous cultivation of the same variety/susceptible variety in a particular locality leads to the build up of inoculums for epiphytotics.

347. Wart disease of potato (*Synchytrium endobioticum*) was first discovered in Hungary in 1895 and in India, it was reported for the first time by A. Ganguli and D.K. Paul in 1953 from Rangbul in Darjeeling district (West Bengal).

348. Ergot of bajra (*Claviceps fusiformis*) was first reported in India in 1956 from the south satara area of Maharashtra.

349. Eye rot (*Nectria galligena*), blue mold rot (*Penicillium expansum*), grey mold rot (*Botrytis cineria*), black mold rot (*Aspergillus niger*), lenticels rot (*Gloeosporium perennans*) and pink rot (*Trichothecium roseum*) are the major post-harvest diseases of apple and pear.

350. High temperature, soil moisture and excessive rainfall are conducive for high incidence of spongy tissue in mango.

351. Sigatoka or banana leaf spot is a serious fungal disease in banana. It is caused by *Mycosphaerella musicola*. It was first observed in the Sigatoka valley in Fiji

in 1913. Black leaf streak or black sigatoka disease caused by *M. fijiensis* var. *difformis*. It was discovered in Honduras in 1972.

352. Powdery mildew of papaya is caused by *Oidium caricae* in humid tropic and subtropics. Stem or foot rot caused by *Phytophthora aphanidermatum*.

353. Heart rot of pineapple is caused by *Phytophthora parasitica*. The disease becomes severe in wet and dry regions, more predominantly in alkaline soils. Strawberry red stele root rot is caused by *Phytophthora fragariae*.

354. Seedling infection is very common within the grain smut and loose smut of sorghum caused by *Sphacelotheca sorghi* and *S. cruenta*, respectively, bunt of wheat (*Tilletia caries*) and flag bunt of wheat (*Urocystis tritici*) etc.

355. *Uromyces pisi* (rust of pea), *Taphrina cerasi* (witches' broom of cherry), *T. deformans* (peach leaf curl), *Venturia inaequalis* (apple scab) and *Synchytrium endobioticum* (wart of potato) enter through buds.

356. *Fusarium caeruleum*, causing dry rot of potatoes cannot penetrate the uninjuned skin of the potatoes but, it invades through lesions already caused by *Phytophthora infestans*, or *Spongospora subterranea*.

357. *Pediculopsis* mites are thought to be involved in a wheat disease caused by *Nigrospora oryzae*. *Fusariums* on wheat and Sporotrichum in carnation bud rot are transmitted by them.

358. Deficiencies of nutrients lead to inversed susceptibility of soybean to *Rhizoctonia solani* and similarly, there are reports of soil pH affecting the susceptibility of leaves to pathogens, notably powdery midews.

359. The germination of uredospores of *Puccinia graminis tritici* on wheat leaves required darkness and surface wetness whereas penetration of the leaf required light and a slightly high temperature for germinarion.

360. Lignolytic enzymes are of minor importance in plant disease, since lignin is a resistant material and is attacked by few micro-organisms. Certain species of *Fusarium* including *F. lactis* and *F. nivale* can readily breakdown lignin.

361. Enhanced production of ethylene (C_2H_4) was first observed in citrus fruits infected with *Penicillium digitatum*. It plays dual role, not only in the development of disease symptoms in susceptible plants but also converse in resistant reactions where rapid ethylene evolution has been shown to enhance the activity of specific enzymes which are brought to be involved in resistance.

362. Sweet potato tissue infected with the black rot organism *Ceratocystis fimbriata* and cowpea mosaic virus is infected cowpea seedlings showed increased activity or synthesis of peroxidases. This increased activity may be due to increased enzyme synthesis and is suggested by the effect of ethylene on RNA and protein synthesis.

363. Epinasty, abscission of leaves and organs, chlorosis and adventitious root development associated with hyperauxiny or altered growth substance

metabolism are known to be induced by ethylene. It was first observed that petiolar epinasty in tomatoes infected with *Fusarium oxysporum* f. sp. *lycopersici* was due to ethylene.

364. The increase of kinin like substances was reported in *Phaseolus vulgaris* and *Vicia faba* leaves attacked by *Uromyces phaseoli* and *U. fabae*. It was suggested that increased kinin leads to formation of 'green islands' around the infection centers, nutrients accumulating in the green tissue and becoming depleted in the general leaf tissue when they become senescent.

365. In 1911, Tischler was the first to investigate the permeability of diseased plants. Despite this and other reports, Thatcher (1939, 1942 and 1943) stands out as the pioneer of research on permeability in relation to plant disease physiology.

366. Resistance to water movement attributed to tomato leaflets increased with the severity of symptoms of *Fusarium* wilt.

367. The role of peroxidase enzyme in plant resistance is due to its ability to oxidize important metabolites either of the parasite or of the host plant (*e.g.* phenolic substances, enzymes I.A.A. and toxins etc.). The increased peroxidase activity has been studied in connection with the oxidation of phenolic substances in the diseased plants and resistance in the host was attributed to the toxicity of these oxidation products.

368. The physiological role of phenoloxidases is thought to be in their participation in redox reactions within plant cell, in chloroplasts, mitochondria, microbodies, etc. An increase in the phenolase activity in the host tissue, especially at and around infection sites is a response which characterise a large number of plant diseases. The high phenolase activity seems to enhance the formation of the characteristic symptoms of the disease such increased phenolase activity in diseased areas of plant tissues is accompanied by increased concentrations of phenolic substances.

369. An increase in ascorbic acid oxidase has been observed in cereals infected by obligate parasites whereas cabbage leaves infected by *Peronospora parasitica* have a decreased activity.

370. Cytochrome oxidase activity increases with the increased respiration of the host plants. Cytochrome oxidase is an important oxidase of the cytochrome system which is localized in the mitochondria.

371. Dehydrogenases are oxidising enzymes catalizing electron transfer from the donor to an acceptor other than molecular hydrogen. The activity of these enzymes especially of the pentose phosphate pathway and of the Krebs cycle increases with the general increase in the metabolism of diseased plants.

372. Host cell hypertrophy with increase in size of nuclei was reported in several host parasite combinations like *Plasmodiophora brassicae* on cabbage and *Puccinia graminis tritici* on wheat. Increase in size of nuclei and nucleoli were accompanied by a higher content of RNA.

373. During pathogenesis, changes in nucleic acid and composition may be brought about by the infection of plants by different pathogens. This was finst observed in wheat leaves infected with *Puccinia graminis tritici* and barley infected with *Erysiphe graminis*. The total RNA content increases in susceptible leaves from the first appearance of symptoms to a maximum at the beginning of sporulation followed by a continuous decline. This stimulation of RNA synthesis after infection was not only due to the growing pathogen but also due to responses of the host plants.

374. Plant Protection Society of China and Japan Plant Protection Association was founded in 1951 and 1945, respectively. Quebec Society for the Protection of Plants was founded in 1908 (Canada) and the Society of Plant Protection of Korea (South) was established in 1962 and started a periodical '*Journal of Plant Protection*'.

375. Federation of British Plant Pathologists was established in 1966 (United Kingdom). Association of Applied Biologists was started in 1914 with its journal '*The Annals of Applied Biology*'. In 1959, Plant Protection Society of China was started with its journal/periodical 'Plant Protection Bulletin. Establishment of International phyto-pathological organizations :- International Society for Plant Pathology (1968), European Committee for Cooperation in Fruit Tree Virus Research (1955), International Council for the Study of Viruses and of Virus Diseases of the Grapevine (1964), International Organization of Citrus Virologists (1957), International Seed Testing Association (ISTA) Committee on Plant Disease (1953) and International Society of Sugarcane Technologists-Pathology Section (1927). In 1908, American Phytopathological Society was founded and in 1910 started publication of its journal '*Phytopathology*'. In 1947, Indian Phytopathological Society was established and its first journal '*Indian Phytopathology*' was published in 1948. Netherlands Society of Plant Pathology was founded in 1891 and began publication of its periodical '*Netherlands Journal of Plant Pathology*' in 1895. The Philippine Phytopathological Society was founded in 1962 and published its journal '*Philippine Phytopathology*' in 1965. In 1930, Canadian Phytopathology Society was founded and in 1933 started its periodical '*Proceedings of the Canadian Phytopathology Society*'. The Phytopathological Society of Japan was founded in 1916 and started publication of its journal '*Annals of the Phytopathological Society of Japan*' in 1918. Indian Society of Mycology and Plant Pathology were started in 1971 with its journal '*Indian Jouranal of Mycology and Plant Pathology*'. First Internationl symposium on Plant Pathology held in 1966-67.

376. Oomycetes were long regarded as lower fungi due to their filamentous growth habit, nutrition by absorption, and reproduction via spores. But with the increase in understanding about evolutionary relationships between different organisms, it is now clear that this group of organisms is unrelated to the true fungi. The data of phylogenetic analyses using molecular analyses of genes and intergenic regions have confirmed the difference between oomycetes and true fungi.

377. In fact, fungi appear more closely related to animals than to oomycetes, and oomycetes are more closely related to algae (heterokont algae) and to green plants. Heterokont algae include the Phaeophyta or brown green algae and Xanthophyta or the yellow- green algae, Chrysophyta or golden algae and Bacillariophyta or diatoms.

378. Amidst some controversy regarding what to exactly call the oomycetes, most authors referred them to kingdom Chromista, phylum Heterokonta while other place them in kingdom Stramenopila (Straminipila).

379. Characteristics distinguishing oomycetes from fungi: In oomycetes, septa (cell walls) in the hyphae are rare, resulting in a multinucleate condition (known as coenocytic). The nuclei of vegetative cells are typically diploid. The cell wall is composed of β-1, 3, and β-1,6 glucans, and not of chitin (the polymer of N-acetyl glucose amine, found in the walls of true fungi). Many species produce wall-less, biflagellated swimming spores (zoospores) in structures called sporangia.

380. They are found in more than 500 species, including the so-called water molds and downy mildews. They are filamentous protists which must absorb their food from the surrounding water or soil, or may invade the body of another organism to feed.

381. In some oomycetes species, two distinct mating types occur and both are required for sexual reproduction (termed as heterothallic as opposed to homothallic species). In heterothallic oomycetes, the gametangia are produced only in the presence of both mating types due to the fact that a hormone produced by one thallus stimulates the other to produce gametangia.

382. In homothallic individuals, sexual reproduction occurs within a single individual. They do not require distinct mating types, but can reproduce sexually by selfing. All *Pythium* and some *Phytophthora* species are homothallic.

383. Genomes of several oomycete species have been sequenced (*Hyaloperonospora arabidopsidis, Phytophthora sojae, P. ramorum, P. infestans* and *Pythium ultimum*) or are being sequenced (*P. capsici, P. parasitica* and *Saprolegnia parasitica*).

384. According to the several recent developments in oomycetes research at the population level. *Phytophthora ramorum* has emerged as a very important pathogen causing sudden oak death and other diseases with a surprisingly large host range.

385. Recent global migrations of *P. infestans* have changed the life history of that organism in many locations in Europe. Several new, naturally occurring hybrid *Phytophthora* "species" have been identified that include completely new pathogenic and non-pathogenic species such as for example *P. alni*.

386. *Hyaloperonospora arabidopsidis* (previously known as *Peronospora parasitica*) has become a model pathogen because it infects the model host plant *Arabidopsis*.

387. *Phytophthora infestans*, the potato late blight pathogen, is a heterothallic species, with only one mating type (the A1) historically dominating the worldwide population with the exception of populations that existed in Mexico; both mating types (A1 and A2) have existed in central Mexico for a very long time.

388. *Plasmopara viticola*, the cause of grapevine downy mildew, is a heterothallic downy mildew with A1 and A2 mating types.

389. *Phytophthora cinnamomi*, the cause of Phytophthora root rot of many plants (more than 3000 plants species), is heterothallic with A1 and A2 mating types, but sexual recombination is not thought to have a significant role in population diversity.

390. *Phytophthora ramorum*, the cause of sudden oak death, Ramorum blight and shoot dieback. This pathogen has two mating types. To date, sexual reproduction via mating of A1 and A2 mating types has not been documented but is considered possible where both types coexist. The pathogen currently exists as three distinct clones (clonal lineages) that reproduce asexually: EU1, NA1, and NA2. Lineage EU1 (A1) was first found in Europe but has now also been found in the US and Canada in select nursery environments. NA1 (A1) was the lineage found to cause significant mortality on coast live oak and tanoak forests in California and Oregon and has been spread throughout North America. The third clone, NA2 (A2), has been found only in nurseries on the US West Coast and Canada.

391. *Sclerophthora rayssiae* var. *zeae*, the cause of brown stripe downy mildew of maize, was first reported in India, but has now also been reported in Myanmar, Nepal and Pakistan; the disease has been most severe in areas of high rainfall (100-200 cm/yr). In India, annual losses of 20-90 per cent have been reported. Survival of this pathogen is via oospores in infected seeds and soil or plant debris.

392. *Peronosclerospora philippinensis*, the cause of "Philippine downy mildew" of maize and other grasses, is endemic to the Philippines where annual losses of 40-60 per cent have been reported. This oomycete does not produce zoospores, but rather the sporangia germinate directly and have been referred to as conidia.

393. *Pythium insidiosum* causes pythiosis and affects horses, cats and dogs and occasionally humans. Pythiosis is found in moist climates with mild winters.

394. *Aphanomyces euteiches* causes seedling and root-rot diseases on many legumes (alfalfa, clover, dry bean, lentil, faba bean, pea, snap bean, and several weed species). It includes plant and animal pathogens found in both terrestrial and aquatic habitats.

395. *Aphanomyces astaci* is an oomycete pathogen that affects crayfish.

396. *Saprolegnia* is the only genus of oomycete pathogens that does not contain plant pathogens but contains pathogens of different water-borne organisms such as crayfish and fish.

397. Ug99 is a, new virulent race of stem rust was identified from wheat fields in Uganda in 1999 and popularly known as Ug99. Using North American scientific nomenclature, Ug99 is known as race TTKSK.

398. Ug99 (Race TTKSK) is a cause for concern as it exhibits unique virulence patterns, in fact no other race of stem rust has been observed to overcome so many wheat resistance genes, including the very important gene Sr31.

399. Due to mutation, rust pathogens changes rapidly, 13 races of Ug99 have been identified in the Ug99 lineage. All exhibit an identical DNA fingerprint, but differ in virulence patterns.

400. In Addition to this, important resistance genes *e.g.*, Sr24, sr25, sr26, sr27, Sr36 and Sr38 have now been defeated by variants of Ug99.

401. Ug99 or variants are considered a major threat to wheat production with an estimated 80-90 per cent of global wheat cultivars susceptible.

402. Mycorrhizal association leads to three distinct types of mycorrhiza widely distributed in the plant kingdom: arbuscular mycorrhizas, ectomycorrhizas and ericoid mycorrhizas.

403. Arbuscular mycorrhizal fungi (AMF), a symbiotic microorganism survives in both soil and roots. Eighty per cent of the plant roots acts as the host for the AMF and they are known as the component of soil and functional links between soil and plant.

404. Ectomycorrhizas are most common in tree species (including the families Betulaceae, Pinaceae, Fagaceae, Dipterocarpaceae, Leguminaceae and Myrtaceae).

405. Ericoid mycorrhizas are conned to genera within the Ericales, including Ericaceae and Vacciniodeae in the northern hemisphere and Epacridaceae in the southern hemisphere.

406. Although these fungi are obligate symbionts they are not host specific and one species may be found to be associated with various plants in the same locality. Also one host plant can support mixed populations of AMF species.

407. Vesicular-arbuscular mycorrhiza (VAM) enhance plant growth through increased nutrient uptake, stress tolerance and disease resistance.

408. Direct (via interference competition, including chemical inter- actions) and indirect (via exploitation competition) interactions have been suggested as mechanisms by which AM fungi can reduce the abundance of pathogenic fungi in roots.

409. Endophytes are the microorganisms, which colonize symptomless the living plant tissue without causing any immediate, overt, negative effect on the plant.

410. The terms "epiphytes" for fungi that live on the surface of their host and "endophytes" for those living inside the plant tissue were first introduced by Anton de Bary in 1866.

411. Most of the plants in natural ecosystem are symbiotic with mycorrhizal fungi and/or fungal endophytes.

412. Fungal symbionts can alter plant ecology, fitness, and evolution, also shaping plant communities and manifesting strong effects on the community structure and diversity of associated organisms (*e.g.* bacteria, nematodes and insects).

413. Unlike mycorrhizal fungi that colonize plant roots and grow into the rhizosphere, endophytes reside entirely within plant tissues and may grow within roots, stems and/or leaves, emerging to sporulate at plant or host-tissue senescence.

414. Two major groups of endophytic fungi, reflecting differences in evolutionary relatedness, taxonomy, plant hosts, and ecological functions are: the clavicipitaceous endophytes (C-endophytes), which infect some grasses; and the non-clavicipitaceous endophytes (NC-endophytes), which can be recovered from asymptomatic tissues of non-vascular plants, ferns and allies, conifers, and angiosperms.

415. C-endophytes (also known as Class 1 endophytes) represent a small number of phylogenetically related clavicipitaceous species that are fastidious in culture and limited to some cool- and warm-season grasses. Usually, these endophytes occur within plant shoots, where they form systemic intercellular infections.

416. Clay and Schardl (2002) recognized three types of clavicipitaceous endophytes, ranging from symptomatic and pathogenic species (Type I) to mixed interaction and asymptomatic endophytes (Types II and III, respectively).

417. NC-endophytes are highly diverse, representing a polyphyletic assemblage of primarily ascomycetous fungi with diverse and often poorly defined or unknown ecological roles. These endophytes are of three types Class 2, 3 and 4.

418. Entomopathogenic fungi are natural enemies of insects and arachnids and these fungi contribute into the regulation of their host populations. These fungi have potential to act as a biological control agents. Common examples are the species of *Metarhizium anisopliae, Beauveria bassiana, B. brongniartii, Nomuraea rileyi* and *Lecanicillium* spp (formerly *Verticillium* spp) and *Paecilomyces* (now in genus *Isaria) fumosoroseus*.

419. Most entomopathogenic fungal species are either from the fungal divisions Zygomycota in the order Entomophthorales or of Hyphomycetes of Ascomycota which do not reproduce sexually.

420. *Beauveria bassiana* is well-known as an entomopathogenic fungus, with worldwide distribution. It is the anamorph stage of *Cordyceps bassiana*, a teleomorph in the ascomycetous family Clavicipitaceae.

421. *B. bassiana* has potential as a dual purpose microbial control organism, against both insect pests and plant pathogens.

422. *Metarhizium anisopliae* is an imperfect, entomopathogenic fungus found in soils throughout the world and known to control four groups of insect pests (termites, locusts, spittlebugs and beetles). It was first recognized as a biocontrol agent in the 1880s.

423. Species in the genus *Hypocrella* Sacc. *S.L.* (anamorph *Aschersonia* Mont. *S. L.*) is insect pathogens having brightly-coloured stromata, filiform ascospores, and pycnidial to acervular anamorphs.

424. Entomopathogenic fungi produce various extracellular enzymes such as proteases, chitinases and lipases to degrade the major constituents of the host insect cuticle (*i.e.* protein, chitin and lipids) to facilitate hyphal penetration. A number of toxic compounds have also been reported in the filtrate of entomopathogenic fungi such as small secondary metabolites, cyclic peptides and macromolecular proteins.

425. The virulence of fungal entomopathogens involves four steps: adhesion, germination, differentiation and penetration.

Multiple Choice Questions (Choose the correct answer)

1. **Which type of environment is required to inhibit the development of *Penicillium* causing decay of citrus fruit:**

   (A) 30°C and 95 per cent relative humidity for 2-3 days.

   (B) 20°C and 70 per cent relative humidity for 2-3 days.

   (C) 25°C and 50 per cent relative humidity for 2.3 days.

   (D) None of these

2. **Who conclusively proved that bunt of wheat is caused by a fungus:**

   (A) Tillet (B) Brefeld

   (C) Prevost (D) Needum

3. **Koch gave three postulates for determining etiology of a disease. To there a fourth postulate was added by:**

   (A) E.C. Stackman (B) E.F. Smith

   (C) J.G. Horsfall (D) W.C. Paddock

4. **Resolving power of light microscope is:**

   (A) 3.0 μm (B) 0.3 μm

   (C) 6.0 μm (D) 0.6 μm

5. **The pathogen causing stem gall of coriander perpetuates through:**

   (A) Soil (B) Seed

   (C) Soil and Seed (D) Weeds

6. **Rust of aonla is caused by:**
   (A) *Ravenelia emblicae* (B) *Puccinia recondita*
   (C) *Uromyces fabae* (D) *Uromyces caryophylinus*

7. **Wart disease of potato is caused by:**
   (A) *Pseudomonas solanacearum* (B) *Phytophthora infestans*
   (C) *Plasmodiophora* sp. (D) *Synchytrium endobioticum*

8. **Aquatic amphibious and terrestrial fungi associated with the family:**
   (A) Leptomitaceae (B) Pythiaceae
   (C) Saprolegniaceae (D) None of these

9. **In the genus *Saprolegnia* the sporangia are:**
   (A) Proliferative (B) Permanent
   (C) Deciduous (D) None of these

10. **VA mycorrhiza helps is the uptake of nutrients particularly that of:**
    (A) Nitrogen (B) Phosphorus
    (C) Potash (D) Sulphur

11. **The method of sexual reproduction in *Rhizopus* is:**
    (A) Gametangial copulation (B) Somatogamy
    (C) Anisogamy (D) Gametangial contact

12. **True fungi are classified into:**
    (A) Myxomycota (B) Eumycota
    (C) Promycota (D) Neomycota

13. **Non-septate conidia of fungi belong to:**
    (A) Amerosporae (B) Didymosporae
    (C) Phragmosporae (D) Helicosporae

14. **Stellate (star-shaped) conidia of fungi belong to:**
    (A) Scolecosporae (B) Staurosporae
    (C) Dictyosporae (D) None of these

15. **Two to many septate conidia of fungi belongs to:**
    (A) Amerosporae (B) Didymosporae
    (C) Phragmosporae (D) Helicosporae

16. **Thread like conidia of fungi belongs to:**
    (A) Scolecosporae (B) Amerosporae
    (C) Phragmosporae (D) None of these

17. Which of the following fungi is known to have biological races?

(A) *Alternaria alternata* (B) *Pleurotus* sp.

(C) *Albugo candida* (D) *Puccinia graminis*

18. The Indian type culture collection is located at:

(A) Allahabad (B) Mumbai

(C) Kolkata (D) New Delhi

19. Cheese is ripened with:

(A) *Asperigillus* (B) *Bacillus*

(C) *Penicillium* (D) *Rhizopus*

20. The book entitled "Fundamentals of Mycology" is written by:

(A) Alexopolus (B) Bessey

(C) Burnett (D) Gaumann

21. Apothecia are found in:

(A) Didamella (B) Mundkurella

(C) Helvella (D) Truffels

22. Plasmodium of some slime molds contains two contractile proteins which are known as:

(A) Actomycosin (B) Bactomycosin

(C) Slimosin (D) Mycosin

23. The reproductive microscopic, spherical and uninucleate spores of myxomycetes are known as:

(A) Meiospores (B) Myxospores

(C) Microspores (D) Meioticspore

24. Species of Oomycetes which produce only one kind of zoospores, each of which on germination directly develops into apparent plant are called:

(A) Monoplane (B) Monoplanetic

(C) Aplanetic (D) Monotypic

25. Compound teliospore is characteristic of genus:

(A) *Puccinia* (B) *Ravenalia*

(C) *Phragmedium* (D) *Uromyces*

26. In which of the following fungus teleutospores germinate to form septate promycelium and teleutospores are united in layers, crusts or columns:

(A) *Puccinia graminis* (B) *P. recondite*

(C) *Uromyces pisi* (D) *Melampsora lini*

27. In which of the following orders deuteromycotina conidia are formed in typical acervuli:

(A) Monitiales (B) Melanconiales

(C) Sphaeropsidales (D) None of these

28. Which of the following fungi is known as 'kerosene fungus':

(A) *Absidia califormensis* (B) *Amorphotheca resinae*

(C) *Rhizopus arrhizus* (D) *Zygorhynchus moelleri*

29. Oospore of downy mildew of bajra contain:

(A) One wall (B) Two wall

(C) Three wall (D) Four wall

30. The perfect stage of *Amorphotheca resinae* is:

(A) *Cladosporium resinae* (B) *Aspergillus* sp.

(C) *Thermoascus aurantiacus* (D) None of these

31. Red rust of mango is caused by:

(A) *Puccinia* (B) *Albugo*

(C) *Cephaleuros* (D) *Uromyces*

32. Parasexual life cycle was first demonstrated in:

(A) *Aspergillus niger* (B) *A. nidulans*

(C) *A. flavus* (D) *A. terreus*

33. *Allmoyces arbuscules* belongs to:

(A) Blastocladiales (B) Chytridiales

(C) Hyphochytridiales (D) Monoblepharaidales

34. Gametangia of *Synchytricum endobioticum* produce:

(A) Zoospores (B) Planogametes

(C) Gametes (D) Resting sporangia

35. In some fungi formation of male and female structures are controlled by hormones that pass between opposite sexes. Such a kind of sexuality s known as:

(A) Monomorphic (B) Homothallic

(C) Gynandromyxis (D) Ambisexualis

36. Effect of one or more environmental factors which make a plant vulnerable to attack by pathogens is known as:

(A) Symbiosis (B) Predisposition

(C) Susceptibility factor (D) Induction

37. Antimicrobial plant metabolites that accumulate as a result of some pathological stimulus are known as:

(A) Antibody (B) Toxin
(C) Phytoharmones (D) Phytoalexins

38. Man made disease is known as:

(A) Manogenic (B) Phylogenic
(C) Iatrogenic (D) Tetratogenic

39. Solanaceous phytoalexins are:

(A) Terpenoids (B) Isoflavonoids
(C) Phaseolloides (D) None of these

40. Legumes phytoalexins are:

(A) Terpenoids (B) Isoflavonoids
(C) Phaseollenoids (D) None of these

41. Phytoalexin 'capsidiol' produce in:

(A) Cotton (B) Pea
(C) Pepper (D) Potato

42. Phytoalexin 'gossypol' produced in:

(A) Cotton (B) Pea
(C) Pepper (D) Potato

43. Gene for gene concept was given by:

(A) Van der Plank (B) Van der Want
(C) Flor (D) Horsfall

44. Koch gave three postulates for determining etiology of a disease. To these a fourth postulates was added by:

(A) E.C. Stakman (B) E.C. Bawden
(C) W.C. Paddock (D) E.F. Smith

45. The concept of 'horizontal resistance' was put forwarded by:

(A) Berkeley (B) Van der Plank
(C) Walker (D) Johnson

46. If 'A' represents the dominant gene for non pathogenicity and 'R' represents the dominant gene for resistance then which of the following combination will give resistant hypersensitive reaction:

(A) Ar (B) AR
(C) aR (D) Ar

47. Resolving power of light microscope is:

(A) 0.3µm (B) 0.6µm

(C) 6.0µm (D) 3.0µm

48. Downy mildew of bajra is caused by:

(A) *Sclerospora garminicola* (B) *Scleophthora microspora*

(C) *Sclerospora sacchari* (D) *Sclerospora sorghi*

49. Wood staining fungi usually called:

(A) Sapstain (B) Solution

(C) Stain (D) None of these

50. Which type of colour produced by *Fusarium*:

(A) Red (B) Blue

(C) Yellow (D) Green

51. Silver leaf of fruit tree is caused by:

(A) *Chondrostereum* sp. (B) *Peniophora* sp.

(C) *Ganoderma* sp. (D) None of these

52. White heart rot of living trees is caused by:

(A) *Polyporus* (B) *Heterobasidion*

(C) *Inonotus* (D) *Leatiporus*

53. Root and butt rot of living trees is caused by:

(A) *Polyporus* (B) *Heterobasidion*

(C) *Inonotus* (D) *Laetiporus*

54. Which of the following fungus produces alkaloids that are fatal to human beings:

(A) *Microphomina phaseolina* (B) *Sclerospora sorghi*

(C) *Claviceps purpurea* (D) *Claviceps fusiformis*

55. Grain smut of jowar is caused by:

(A) *Sphacelotheca cruenta* (B) *Sphacelotheca reiliana*

(C) *Sphacelotheca sorghi* (D) *Sphacelotheca erianthi*

56. Hypoplasia is:

(A) Excessive multiplication of cells (B) Enlargement of cells

(C) Retardation of normal growth (D) Excessive transpiration

57. *Spongospora subterranea* incites:

(A) Wart disease of potato (B) Powdery scab of potato

(C) White blister of mustard (D) Club root crucifers

58. Blossom blight of pepper is caused by:

(A) *Phytophthora capsici*
(B) *Pernospora destructor*
(C) *Choanephora cucurbitarum*
(D) *Fusarium solani*

59. Sodium and boron amendment of soil is advised for the control of:

(A) Phoma blight and heart rot of beet
(B) White rust of crucifers
(C) Onion smudge
(D) Early blight of tomato

60. Leaf spot of turmeric is caused by:

(A) *Taphrina deformans*
(B) *Taphrina maculans*
(C) *Phytophthora* sp.
(D) None of these

61. Leaf shredding disease of sorghum is caused by :

(A) *Sclerospora graminicola*
(B) *Sclerospora indica*
(C) *Pernosclerospora sorghii*
(D) None of these

62. Bud rot of palm is caused by:

(A) *Phytophthora palmivora*
(B) *Colletotrichum* sp.
(C) *Alternaria alternata*
(D) None of these

63. The first epidemic of apple scab occurred in Kashmir in:

(A) 1963
(B) 1968
(C) 1973
(D) 1983

64. The perfect stage of apple scab belongs to:

(A) Ascomycotina
(B) Basidiomycotina
(C) Zygomycotina
(D) Deuteromycotina

65. Conidial stage of *Venturia inaeqalis* is:

(A) *Spilocaea pomi*
(B) *Venturia pirina*
(C) *Venturia asparata*
(D) None of these

66. Anthracnose of grape is caused by:

(A) *Fusarium oxysporum*
(B) *Sphaceloma fawcetti*
(C) *Collectotrichum gloeosporioides*
(D) *Sphaceloma ampelinum*

67. Conidial stage of *Elsinoe ampelina* is:

(A) *Sphaceloma ampelina*
(B) *Sphaceloma fawetti*
(C) *Sphaceloma australis*
(D) None of these

68. Black spot of rose is caused by:

(A) *Diplocarpon rosae*
(B) *D. earliana*
(C) *D. destructive*
(D) None of these

69. The spore of apple scab pathogen must be wet for the following period to cause infection at 6°C:

(A) 6 hr
(B) 12 hr
(C) 18 hr
(D) 26 hr

70. The white rust fungus perpetuates through:

(A) Oospore in soil
(B) Oospore on seed surface
(C) Oospore in soil and seed surface
(D) Zoospore in soil

71. Which of the following fungi are the most common causes for damping off of seedling:

(A) *Fusarium*
(B) *Phytophthora*
(C) *Pythium*
(D) *Rhizoctonia*

72. For controlling damping off of seedling which of the following treatments is most effective:

(A) Treating seeds with fungicides
(B) Washing seed with soap water
(C) Solarization
(D) Soil drenching with fungicides

73. Which of the following species of *Pythium* cause foot rot of papaya in India?

(A) *P. aphanidermatum*
(B) *P. butleri*
(C) *P. gracile*
(D) *P. monospermum*

74. Optimum condition for the spread of white rust of Brassica is:

(A) Moist and cool (10ºC)
(B) Dry and cool (10ºC)
(C) Moist and warm (10ºC)
(D) Dry and warm (10ºC)

75. Stripe rust if wheat occurs commonly in:

(A) All wheat growing area of India
(B) Wheat growing area of North India
(C) Wheat growing area of Central India
(D) Wheat growing area of South India

76. Optimum temperature range for the germination of spore of *Ustilago scitaminea* is:

(A) 5 -9 ºC
(B) 15 - 25 ºC
(C) 25 - 30 ºC
(D) None of these

77. Spores of loose smut of sorghum germinate to form a:

(A) 4- celled proycelium
(B) 3- celled promycelium
(C) 2- celled promycellium
(D) None of these

78. Optimum temperature ranges for the germination of spore of *Sphacelotheca cruenta* is:

(A) 8 -25 °C
(B) 18 -32 °C
(C) 10 - 15 °C
(D) 30 - 40 °C

79. Which of the following species of *Sphacelotheca* cause loose smut of sorghum?

(A) *Cruenta*
(B) *Reiliana*
(C) *Sorghii*
(D) None of these

80. Which of the following species of *Sphacelotheca* cause head smut of sorghum?

(A) *Cruenta*
(B) *Reiliana*
(C) *Sorghi*
(D) None of these

81. What is the size of urediospores of *Puccinia graminis tritici* in microns?

(A) 19 x 16
(B) 22 x 16
(C) 25 x 17
(D) 32 x 20

82. Who described the method for physiologic races of *Puccinia graminis tritici*:

(A) Stakman
(B) Wilcoxon
(C) Tillet
(D) Mehta

83. A given variety of wheat may be susceptible to a given set of races of the rust fungus in the seeding stage but highly resistant to some races when the plants have matured. Such resistance is known as:

(A) Protoplasmic
(B) Functional
(C) Morphogenic
(D) Systemic

84. Who initiated the study of plant pathology in India:

(A) S.R. Bose
(B) E.J. Butler
(C) B.E. Mundkur
(D) FJ.F. Shaw

85. Where is herbarium *Cryptogamae orientalis* located in India:

(A) Aurangabad
(B) Kolkata
(C) New Delhi
(D) Udaipur

86. From where the *Indian Phytopathology* is published:

(A) Lucknow (B) New Delhi

(C) Varansi (D) Udaipur

87. In which year publication of *Phytopathology* was started:

(A) 1900 (B) 1925

(C) 1945 (D) 1911

88. Annual review of Phytopathology was first published in:

(A) 1963 (B) 1953

(C) 1957 (D) 1977

89. The book entitled "Disease of Cultivated Crops their Causes and their Control" is written by:

(A) Alexopolus (B) J.C. Luthra

(C) Kuhn (D) Burnett

90. From where the *Indian Journal of Plant Pathology* is published:

(A) Lucknow (B) New Delhi

(C) Udaipur (D) Kolkata

91. In which year Indian Society of Mycology and Plant Pathology was started:

(A) 1971 (B) 1963

(C) 1976 (D) 1946

92. In which year the first computer simulation program of plant disease epidemics was published for the fungal induced early blight disease of tomato and potato:

(A) 1963 (B) 1971

(C) 1975 (D) 1969

93. In which year American Phytopathological Society was founded:

(A) 1911 (B) 1908

(C) 1900 (D) None of these

94. Who showed for the first time that DNA carries genetic information:

(A) Avery *et al.* (1944) (B) Hershey and Chase (1952)

(C) Watson and Crick (1953) (D) None of these

95. Who designated each isolate by a third name (Trinomial nomenclature) and called it sub species or "*formae specialis*":

(A) Stakman (1922) (B) Erikson(1891)

(C) Boyer and Cohen (1973) (D) None of these

96. Who demonstrated the first time generic variability within a pathogen species:

(A) Baurrus (1911) (B) Stakman (1914)
(C) Walker (193) (D) Flor (1946)

97. Who proposed that in many host pathogen combinations, plants remain resistant through hypersensitivity?

(A) Gaumann (1946) (B) Mullar (1961)
(C) Van der Plank (1963) (D) None of these

98. Who discovered that within a pathogen species physiological races of the pathogen exist:

(A) Baurrus (1911) (B) E.C. Stakman (1914)
(C) Van der Plank (1963) (D) None of these

99. The book entitled "Fungi Delight of Curiosity" is written by:

(A) Burnet (B) De Barry
(C) Paul Neergaard (D) Brodie

100. What is the size of conidia of *Venturia inaequalis* in microns?

(A) 12-22 x 6-9 (B) 25-30 x 10-12
(C) 28-32 x 11-14 (D) None of these

101. Ascospore of apple scab contains:

(A) One celled (B) Two celled
(C) Three celled (D) Four celled

102. The unequal size of the 2 cell gives the species its name:

(A) *Asparata* (B) *Inaequalis*
(C) *Sclerotiorum* (D) None of these

103. The book entitled "Comparative Morphology and Biology of the Fungi, Mycetozoa and Bacteria" is written by:

(A) Brodie (B) B.J. Butler
(C) De Barry (D) Smith

104. What is the size of conidia of *Peronospora pisi* in microns?

(A) 22-27 x 15-19 (B) 28-35 x 20-25
(C) 30-40 x 25-28 (D) 35-38 x 3-45

105. Optimum temperature range for the germination of spores of *Sphacelotheca sorghii* is:

(A) 10-15°C (B) 20-30°C
(C) 5-10°C (D) None of these

106. White rust of mustard is caused by:

(A) *Colletotrichum falcatum* (B) *Albugo candidia*

(C) *Puccinia graminis* (D) *Erisyphe cichoracearum*

107. Which of following fungus is mycoparasite of sclerotia of *Claviceps microcephala*:

(A) *Fusarium sambucinum* (B) *F. moniliforme*

(C) *Nectria cinnabarina* (D) *Trichothecium roseum*

108. How many triplet codons can be read in four nucleotides?

(A) 24 (B) 48

(C) 64 (D) 56

109. At what stage the crossing over occurs:

(A) Diakinesis (B) Pachytene

(C) Leptotene (D) Metaphase

110. A reducing agent is a substance which can:

(A) Accept electrons (B) Donate electrons

(C) Accept protons (D) Donate protons

111. Phenotypic ratio for a monohybrid cross is:

(A) 1 :1 (B) 1 : 2 :1

(C) 1:3:3:1 (D) 3 :1

112. Useful gene can be transferred to plants through:

(A) Ti-plasmid (B) Genephore

(C) TMV (D) *Pseudomonas solanacearum*

113. Paddy straw mushroom is best cultivated at:

(A) 15-20°C (B) 20-25°C

(C) 30-35°C (D) 25-30°C

114. The repenting spore of the rust fungus is the:

(A) Uredospore (B) Basidiospore

(C) Aeciospore (D) Spermatia

115. Which among the following is a cellulose degrading ascomycete :

(A) *Eremothecium* (B) *Chaetomium*

(C) *Taphrina* (D) *Leptosphaeria*

116. Which among the flowing is a member of Zygomycotina producing light sensitive sporangia:

(A) *Rhizopus* (B) *Cunninghamella*

(C) *Pilobolus* (D) *Mucor*

117. The conidiophore do not form vesicle but their branching may be mono, bi or poly verticellate in:

(A) *Penicillium* (B) *Aspergillus*

(C) *Taphrina* (D) None of these

118. Conidial stage of *Talaromyces* is:

(A) *Aspergillus* (B) *Penicillium*

(C) *Taphrina* (D) None of these

119. Conidial stage of *Eurotium* is:

(A) *Aspergillus* (B) *Penicillium*

(C) *Taphrina* (D) None of these

120. Vertical resistance is against:

(A) Several race (B) Race specific

(C) Few races (D) All of the above

121. Which one among the following is a mycorrhizal fungus:

(A) *Glomus* (B) *Rhizopus*

(C) *Glomerella* (D) *Mucor*

122. The conidia of *Alternaria* sp.

(A) Both longitudinally and transversely septate

(B) Longitudinally septate

(C) Transversely septate

(D) Aseptate

123. Clamp connection are seen in the following:

(A) Ascomycotina (B) Deuteromycotina

(C) Zygomycotina (D) Basidiomycotina

124. In the genus *Phytophthora* the sporangia are:

(A) Deciduous (B) Proliferative

(C) Permanent (D) Not branched

125. Aquatic amphibious and terrestrial fungi associated with the family:

(A) Leptomitaceae (B) Pythiaceae

(C) Saprolegniaceae (D) None of the these

126. In the genus *Saprolegnia* the sporangia are:

(A) Proliferative (B) Permanent

(C) Deciduous (D) None of these

127. Sporangia borne in chains, periplasm conspicuous obligate parasites of plants:

(A) Albuginaceae (B) Pythiaceae

(C) Pernosporaceae (D) None of these

128. Sporangia borne singly, in clusters or in whorls on somatic hypha or on sporangiophores of indeterminate growth, periplasm in thin layer or absent, facultative parasites or saprobes:

(A) Albuginaceae (B) Pythiaceae

(C) Pseudomonadaceae (D) None of these

129. The sexual fruiting body produced by *Puccinia graminis trifolli* on barberry is:

(A) Basidium (B) Perithecium

(C) Aecium (D) Telium

130. Cell wall of *Phytophthora* consists mainly of:

(A) Cellulose (B) Chitin

(C) Chitosan (D) Peptidoglycan

131. Ergotism is due to the fungus:

(A) *Endothia parasitica* (B) *Elsinoe fawcettii*

(C) *Claviceps purpurea* (D) *Valsa eugeniae*

132. Allomyces occurs in:

(A) Moist soil (B) Water

(C) Mud (D) All of the above

133. Yeasts are richest source of:

(A) Vitamin A (B) Vitamin B complex

(C) Vitamin C (D) All of the above

134. Rickettsia is:

(A) Sensitive to penicillin (B) Sensitive to tetracycline

(C) Sensitive to streptomycin (D) All of the above

135. *Plasmodiophora brassicae* produces:

(A) Zoospores (B) Hypnospores

(C) Amoeboid thalli (D) All of the above

136. Urediniospores of *Puccinia graminis* can be carried to long distances by wind up to:

(A) 100 km. (B) 500 km.

(C) 800 km. (D) More than 800 km.

137. Teliospores of *Phragmidium disciflorum* were first revealed in microscope by:

(A) Robert Hook and Micrographia (1665) (B) De Barry (1885)

(C) Erikson (1896) (D) None of these

138. Heteroecious nature of stem rust of wheat was shown by:

(A) De Barry (1865) (B) Eikson (1894)

(C) Stakman (1930) (D) None of these

139. Origin of new physiologic forms of *Puccinia graminis tritici* by hybridization was shown by:

(A) Stakman (1930) (B) Erikson (1894)

(C) De Barry (1865) (D) None of these

140. Physiologic specialization in stem rust of wheat was shown by:

(A) Stakman (1930) (B) Erikson (1894)

(C) De Barry (1865) (D) None of these

141. Biologic form in cereal rusts were distinguished by:

(A) Stakman (1914) (B) Erikson (1894)

(C) Robert Hook (1665) (D) None of these

142. Function of Pycnia was discovered by:

(A) Craigie (1927-31) (B) De Barry (1865)

(C) Stakman (1914) (D) None of these

143. Inheritance of rust resistance in wheat in a Mendelian fashion was demonstrated by:

(A) Stakman (1914) (B) Biffen (1905)

(C) De Barry (1865) (D) None of these

144. The sexual process in uredinales was discussed by:

(A) Buller (1950) (B) K.C. Mehta (1940)

(C) Williams *et al.* (1966) (D) None of these

145. The first artificial culture of cedar rust of apple fungus was done by:

(A) Hotson and cutter (1951) (B) Williams *et al.* (1966-67)

(C) J.C. Arthur (1929) (D) None of these

146. Origin of new physiologic forms of *Puccinia graminis* through heterkaryosis was demonstrated by:

(A) Nelson, Wilcoxson and Christensen (B) HJotson and Cutter

(C) Williams *et al.* (D) None of these

147. The first successful axenic culture of black stem rust from uridiospores was done by:

(A) Neloson *et al.* (B) Williams *et al.*

(C) Hotson and Cutter (D) K.C. Mehta

148. Optimum temperature for the germination of Urediospores of *Puccinia graminis* is:

(A) 20°C (B) 24°C

(C) 30°C (D) None of these

149. Optimum temperature range for the appresorium formation of *Puccinia graminis* is:

(A) 10-15°C (B) 16 - 27°C

(C) 28-32°C (D) None of these

150. Optimum temperature required for penetration of *P. graminis* is:

(A) 20°C (B) 25°C

(C) 29°C (D) 32°C

151. The optimum temperature for germination of urediospore of *Puccinia recondita* is:

(A) 20°C (B) 25°C

(C) 29°C (D) 32°C

152. The optimum temperature range for germination of urediospore of *Puccinia striformis* is:

(A) 18-22°C (B) 25-28°C

(C) 9-13°C (D) None of these

153. Swollen shoot virus in cocoa is transmitted by:

(A) Mealy bugs (B) Aphids

(C) Whitefly (D) Jassids

154. Leaf curl of chilli is transmitted by:

(A) *Thrips tabaci* (B) *Aphis gossypii*

(C) *Bemesia tabaci* (D) *Myzus persicae*

155. The region in India known to be epidemic to wart disease of potato:
(A) Darjeeling (B) Assam
(C) Nilgiris (D) Maharashtra

156. The antibiotic produced by *Trichoderma viride* is:
(A) Mystatin (B) Validacin
(C) Blasticin (D) Viridin

157. Coenocytic mycelium is found in:
(A) Basidiomycetes (B) Oomycetes
(C) Ascomycetes (D) Deuteromycetes

158. Diploid thallus is found in the members of:
(A) Oomycetes (B) Ascomycetes
(C) Deuteromycetes (D) Basidiomycetes

159. Horizontal and vertical resistance was described by:
(A) E.C. Stakman (B) J.C. walker
(C) J.G. Dickson (D) J.E. Vanderplank

160. Gamma particles are found in the zoospore of:
(A) *Blastocladiella emersoni* (B) *Pythium debaryanum*
(C) *Albugo candida* (D) *Phytophthora infestans*

161. Many plants pathogenic fungi are able to penetrate the cuticles of their host plants by elaborating specialized cells known as:
(A) Appresorium (B) Penetration peg
(C) Germ tube (D) Haustorium

162. Programmed cell death following infection by a plant pathogen is called:
(A) Devitalization (B) Necrosis
(C) Death of the cells (D) Apoptosis

163. The following major gene controls resistance to the old multi-site fungicide dichlofluanid in field isolates of *Botrytis cineria*:
(A) R gene (B) R1 gene
(C) A1 gene (D) Dic 1 gene

164. *Ampelomyces quisqualis* is:
(A) Hyperparasite (B) Saprophyte
(C) Parasite (D) Facultative parasite

165. The smut infection occurs by:

(A) Dikaryotic hyphae
(B) Teliospore
(C) Basidia
(D) Basidiospore

166. The casual agent of onion smut is:

(A) *Uromyces phaseoli*
(B) *Urocystis cepulae*
(C) *Ustilago avenae*
(D) None of these

167. Eucarpic hyphae constricted present in the order of:

(A) Saprolegniales
(B) Legnidiales
(C) Leptomitales
(D) Peronosporales

168. *Tuberculina maxima* is:

(A) Saprophyte
(B) Hyperparasite
(C) Parasite
(D) None of these

169. *Tilletiopsis* sp. parasitizes the cucumber powdery mildew fungus:

(A) *Sphaerotheca fuligena*
(B) *Alternaria* spp
(C) *Athelia bombacina*
(D) None of these

170. A device to maintain a continuous culture of microorganism is:

(A) Colorimeter
(B) Thermostat
(C) Photostat
(D) Chesmostat

171. Which of the following is edible?

(A) *Albugo candida*
(B) *Taphrina maculans*
(C) *Alternaria solani*
(D) *Agaricus bisporus*

172. Red bread mould is:

(A) *Neurospora*
(B) *Alternaria*
(C) *Aspergillus*
(D) *Penicillium*

173. The zoospores are not produced in the asexual phase of which of the following fungus:

(A) *Phytophthora infestans*
(B) *Peronospora pisi*
(C) *Albugo candida*
(D) *Pythium aphanidermatum*

174. *Neurospora crassa* was selected for genetic experiments because:

(A) It belongs to Ascomycetes
(B) It is saprotroph
(C) It forms ascospores
(D) The ascospores segregate in 1:1 ratio

175. Blast of rice is caused due to:

(A) *Magnaporthe grisea*
(B) *Helminthosporium oryzae*
(C) *Xanthomonas oryzae*
(D) None of these

176. *Pythium ultimum* is the casual organism of:

(A) Late blight of potato
(B) White rust of crucifers
(C) Damping off
(D) None of these

177. Which of the following stage of rust fungi is called zero stage:

(A) Aecial
(B) Uredial
(C) Telial
(D) Pycnial

178. Which of the following growth regulators at 40 ppm is most suitable for the control of fruit drop in mango?

(A) 1 AA
(B) NAA
(C) 2, 4-D
(D) IBA

179. Late blight of potato can be controlled by:

(A) B.H.C. dust 5 per cent
(B) Bordeaux mixture 4:4:50
(C) DDT
(D) All of the above

180. The haplodization in parasexual reaction take place by any of the following methods:

(A) Meiosis
(B) Mitosis
(C) Loss of genomic material
(D) Non disjunctional phenomenon

181. The erection of the genus *Sclerophthora* is based on:

(A) Occurrence on a new host
(B) Morphology of the entire thallus
(C) The asexual stage resembles the sexual stage of *Sclerospora* and the asexual stage that of *Phytophthora*
(D) None of these

182. *Synchytrium endobioticum* is:

(A) Obligate parasite
(B) Facultative parasite
(C) Saprotroph
(D) Facultative saprotroph

183. In which fungus the lignifications of mycelia were first reported:

(A) *Rhizopus stolonifer*
(B) *Saccharomyces cerevisiae*
(C) *Colletotrichum lagenarium*
(D) *Melampsora lini*

184. Coffee rust is caused by:

(A) *Uromyces* (B) *Cronartium ribicola*

(C) *Hemeleia vastatrix* (D) *Ravenelia embelica*

185. Guava wilt is caused due to:

(A) *Fusarium solani* (B) *Macrophomina phoseoli*

(C) *Cephalosporium* spp *or Fusarium* spp (D) *Gliocladium roseum*

186. Two celled conidia found in:

(A) *Ramularia* (B) *Botrytis*

(C) *Rhynchosporum* (D) None of these

187. Three to many celled conidia found in:

(A) *Ramularia* (B) *Fusicladium*

(C) *Cephalosporium* (D) None of these

188. Filiform, one to many celled conidia found in:

(A) *Colletotrichum* (B) *Marssonia*

(C) *Cylindrosporium* (D) None of these

189. One celled conidia found in:

(A) *Phymatotrichum* (B) *Cylindrocladium*

(C) *Pyricularia* (D) None of these

190. Most damaging disease of papaya:

(A) Collar rot (B) Root rot

(C) Papaya mosaic (D) Papaya leaf curl

191. Lace bug/mealy bug (*Planoccocus breviseps*) in pineapple transmits:

(A) Tristeza (B) Pine apple wilt

(C) Panama wilt (D) Bunchy top

192. Organic phosphate solubliser is:

(A) VAM (B) Thiobacillus

(C) Aspergillus (D) None of these

193. Apothesia is formed in:

(A) *Aspergillus nidulans* (B) *Scelrotinia sclerotiorum*

(C) *Phoma betae* (D) *Alternaria solani*

194. Cleistothesia is found in:

(A) *Aspergillus nidulans* (B) *Scelrotinia sclerotiorum*

(C) *Phoma betae* (D) *Alternoria solani*

195. The cause of brown rot of stone fruits is:

(A) *Monilinia fructicola* (B) *Sclerotinia sclerotiorum*

(C) *Pseudopeziza trifolii* (D) None of these

196. The casual agent of lettuce drop is:

(A) *Monilinia fructicola* (B) *Sclerotinia sclerotiorum*

(C) *Diplocarpon rosea* (D) None of these

197. Black spot of quince and pepper is caused by:

(A) *Diplocarpon maculatum* (B) *Sclerotinia sclerotiorum*

(C) *Diplocarpon rosea* (D) None of these

198. Black spot of rose is caused by:

(A) *Diplocarpom maculatum* (B) *Sclerotinia sclerotiorum*

(C) *Diplocarpon rosea* (D) None of these

199. The alternation of generation is found in:

(A) *Alternaria alternata* (B) *Allomyces arbuscula*

(C) *Tilletia foetida* (D) *Neovossia indica*

200. Which of following disease occur on rabi cereals in India:

(A) Rust (B) Smut

(C) Leaf blight (D) All of the above

201. Most damaging disease of maize in India is:

(A) Stalk rot (B) *Helminthosporium* blight

(C) Downy mildew (D) All of the above

202. Pisidic acid (malic citric) is produced by:

(A) The root of sugarcane (B) The root of sorghum

(C) The root of wheat (D) The root of pigeon pea

203. Green island is formed in the following phathosystem:

(A) Soft rot of crucifers

(B) Early blight of potato

(C) Pinto bean leaves infected by *Uromyces phaseoli*

(D) Loose smut of wheat

204. Which of the following disease of soybean is caused by *Peronospora* sp:

(A) Leaf spot (B) Downy mildew

(C) Root rot (D) Pod blight

205. Black arm of cotton is caused by:

(A) Bacteria (B) Virus

(C) Actinomycetes (D) Fungi

206. Which one of the following genera produces only one ascus in its cleistonecium:

(A) *Erysiphe* (B) *Microsphaera*

(C) *Podosphaera* (D) *Uncinula*

207. Which of the following a disease shows both local and systemic infection:

(A) White rust of mustard (B) Early blight of potato

(C) Leaf spot of turmeric (D) Black stem rust of wheat

208. The disease early blight of potato symptoms shows:

(A) Local systemic infection

(B) Curling of leaf

(C) Spots covered with a deep greenish blue growth of the fungus

(D) None of these

209. Which of the following fungi is polymorphic:

(A) *Puccinia graminis tritici* (B) *Olpidium brassicae*

(C) *Cerospora apii* (D) *Phyllosticta solitaria*

210. The carpogenic and myceliogenic germinations are found in:

(A) *Mucor mucedo* (B) *Sclerotinia sclerotiorum*

(C) *Sclerotium rolfsii* (D) None of these

211. Which of the following fungus is an omnivorous and destructive parasite of many plants:

(A) *Mucor mucedo* (B) *Sclerotinia sclerotiorum*

(C) *Sclerotium rolfsii* (D) None of these

212. The thread like ascospores is found in:

(A) *Claviceps purpurea* (B) *Sclerotinia sclerotiorum*

(C) *Sclerotium rolfsii* (D) None of these

213. The following chemical induces differentiation of infection structures in the wheat stem rust fungus:

(A) Vitamin A (B) Lipid

(C) Vitamin C (D) Acrolein

214. Mark the aboitic elicitors:

(A) Glucon

(B) Salicylic acid

(C) Jasomic acid

(D) 2, 2, dichloro-3, 3-dimethylcyclopropane carboxylic acid

215. The carpogenic and myceliogenic germinations are found in:

(A) *Mucor mucedo*

(B) *Sclerotinia sclerotiorum*

(C) *Sclerotium rolfsii*

(D) None of these

216. Which of the following fungus is an omnivorous and destructive parasite of many plants:

(A) *Mucor mucedo*

(B) *Sclerotinia sclerotiorum*

(C) *Sclerotium rolfsii*

(D) None of these

217. The thread like ascospores is found in:

(A) *Claviceps purpurea*

(B) *Sclerotinia sclerotiorum*

(C) *Sclerotium rolfsii*

(D) None of these

218. The following chemical induces differentiation of infection structures in the wheat stem rust fungus:

(A) Vitamin A

(B) Lipid

(C) Vitamin C

(D) Acrolein

219. It is known that barely plants are more resistant than wheat to microorganisms because it contains:

(A) Thionin

(B) Lignin

(C) Callose

(D) Extension

220. The black brittle rhizomorphs of *Armillaria mellea* commonly called:

(A) Shoe-strings

(B) Network

(C) Cortex

(D) None of these

221. Match the discovery of J.C. Walker:

(A) Role of phenols in disease resistance

(B) Viral nucleoprotein

(C) Pectolytic enzymes

(D) Mycoplasma discovery

222. Damping off disease occurs on:

(A) Adult plant

(B) Seedlings

(C) Old plant

(D) All of these

223. The fungus which shows the phenomenon of bioluminescence:

(A) *Armillaria mellea*
(B) *Ustilago tritici*
(C) *Phytophthora infestans*
(D) *Sphaerotheca fuliginea*

224. Resistance of tomato fruits to *Botrytis cineria* is correlated with:

(A) Presence of hairs
(B) Colour of fruits
(C) Thickness of cuticles
(D) All of the above

225. Smut of bajra perpetuates in:

(A) Seed
(B) Insects
(C) Soil
(D) Water

226. What is the reserved food material in fungi:

(A) Starch
(B) Glycogen
(C) Glucose
(D) Oil bodi

227. Black scurf of potato is caused by:

(A) Bacteria
(B) Fungi
(C) Virus
(D) Nematode

228. Imports of plants into India is allowed by the permission of:

(A) Director General ICAR
(B) Plant Protection Adviser, Govt. of India
(C) Ministry of Agriculture, Govt. of India
(D) All of the above

229. Main source of primary inoculum of downy mildew of bajra is:

(A) Mycelium
(B) Conidia
(C) Oospore
(D) Both mycelium and conidia

230. The downy mildew fungus *Sclerospora graminicola* is primarily:

(A) Soil-borne
(B) Seed borne
(C) Insects
(D) None of these

231. Formation of sporangia *Sclerospora graminicola* is favored by a temperature range of:

(A) 10-12°C
(B) 15-25°C
(C) 28-32°C
(D) 35-40°C

232. The transformation of entire vegetative thallus into a reproductive structure is called:

(A) Holocarpic
(B) Acarpic
(C) Eucarpic
(D) Syncarpic

233. In which region of India, the incidence of mango malformation is much higher:

(A) North – East
(B) North – West
(C) South – East
(D) South – West

234. Which smut is externally seed-borne in nature?

(A) *Ustilago avenae*
(B) *Ustilago tritici*
(C) *Entyloma oryzae*
(D) None of these

235. Phenylamide fungicides are specially developed for the management of disease caused by:

(A) Hyphomycetes
(B) Oomycetes
(C) Zygomycetes
(D) Pyrenomycetes

236. Diplanetism is found in:

(A) *Aspergillus*
(B) *Saprolegnia*
(C) *Rhizopus*
(D) *Mucor*

237. Monoplanetism is found in:

(A) *Pythiopsis*
(B) *Isoachlya*
(C) *Rhizopus*
(D) *Mucor*

238. Multiline involves the concept of:

(A) Horizontal resistance
(B) Specific resistance
(C) Vertical resistance
(D) All of the above

239. Sporulation in *Septoria* is enhanced under:

(A) Near UV light
(B) Green light
(C) Red light
(D) Yellow light

240. *Botrytis cineria* enters host through:

(A) Hydathodes
(B) Lenticels
(C) Nectaries
(D) None of these

241. Appresorium of *Puccinia graminis tritici* enters host through:

(A) Substomatal cavity
(B) Lenticels
(C) Nectaries
(D) None of these

242. *Phytophthora infestans* almost always enters plants through:

(A) Hydathodes
(B) Directly
(C) Lenticels
(D) Wound

243. Members of Sphaeropsidales entering through:

(A) Stomata
(B) Hydathodes
(C) Lenticels
(D) Directly

244. Leaf infecting fungi-imperfecti generally enters through:

(A) Wound (B) Stomata

(C) Directly (D) None of these

245. *Streptomyces scabies* the cause of the common scab of potato, enters young developing tubers through:

(A) Stomata or Lenticels (B) Hydathodes

(C) Nectaries (D) None of these

246. *Oospora pustulans* which causes skin scab of potato tubers enters through:

(A) Hydathodes (B) Lenticels

(C) Nectaries (D) Directly

247. De Barry (1886) was the first to study critically the nature of direct penetration by:

(A) *Sclerotinia sclerotiorum* (B) *Alternaria alternata*

(C) *Erwinia amylovora* (D) None of these

248. *Cephalosporium gramineum* causes infection of wheat seeding enters through:

(A) Wound (B) Hydathodes

(C) Lenticels (D) Directly

249. *Plasmodiophora brassicae* enters through:

(A) Wound (B) Root hairs

(C) Lenticels (D) None of these

250. *Taphrina deformans* enters host through the:

(A) Directly (B) Unfolding leaf buds

(C) Lenticels (D) Stomata

251. Sporangia are forcefully expelled in:

(A) *Phytophthora* (B) *Alternaria*

(C) *Pythium* (D) *Sclerospora*

252. Identify the defense related enzyme:

(A) Lipoxygenase (B) β-1, 3- glucanase

(C) Phenyl ammonia lyase (PAL) (D) All of the above

253. Conidia generally germinate:

(A) Through zoospore (B) Directly

(C) Through formation of germ tubes (D) Through appresorium

254. Sporangia of *Pythium* generally germinate:

(A) Directly (B) Zoospore

(C) Appresorium (D) None of these

255. The disease prevalent in and confined to a particular country or district year to year is termed:

(A) Sporadic (B) Pandemic

(C) Endemic (D) Epidemic

256. The detailed study on physiologic specialization in rust of cereals in USA was made by:

(A) A.E. Diamond (B) E.C. Stakman

(C) J.G. Horsfall (D) G.W. Padwick

257. Elapsed time between infection and appearance of disease symptoms is:

(A) Latent period (B) Latent infection

(C) Infection period (D) Dormant period

258. Sexual stage in the life cycle of the fungus is:

(A) Anamorphic stage (B) Perfect stage

(C) Both anamorphic and perfect stage (D) None of these

259. The thallus bearing more than one reproductive structure called:

(A) Polycentric (B) Monocentric

(C) Eucarpic (D) Holocarpic

260. The thallus bearing a single reproductive structure called:

(A) Polycentric (B) Monocentric

(C) Eucarpic (D) Holocarpic

261. Reproductive structure formed certain parts on the thallus called:

(A) Polycentric (B) Monocentric

(C) Eucarpic (D) Holocarpic

262. Whole thallus enters the host cell called:

(A) Holocarpic (B) Epibiotic

(C) Endobiotic (D) None of these

263. Stolen present in:

(A) *Rhizopus* (B) *Mucor*

(C) *Alternaria* (D) None of these

264. Light green, yellow patches on the plant surface called:

(A) Mosaic (B) Mottle

(C) Curling (D) Wilting

265. The process by which an organism undergoes self dissolution is:

(A) Atomize (B) Attenuation

(C) Autolysis (D) Autotrophism

266. Resting spores of potato wart disease organism can survive at least up to:

(A) 30-40 years (B) 20 years

(C) 10 years (D) 5 years

267. Resting spores of club root of crucifers' disease organism can survive at least up to:

(A) 10 years (B) 15 years

(C) 20 years (D) 25 years

268. Optimum temperature ranges required for club root of crucifers fungus infection:

(A) 17-25°C (B) 28-32°C

(C) 35-40°C (D) None of these

269. *Phytophthora infestans* germinates by a:

(A) Directly (B) Germ tube

(C) Through sporangium (D) None of these

270. The maximum sporangial production of *Phytophthora colocasiae* occurs at:

(A) 11°C (B) 16°C

(C) 21°C (D) 26°C

271. The average size of sporangia of *Phytophthora palmivora* in microns is:

(A) 35-50 (B) 20-25

(C) 40-60 (D) None of these

272. Tab toxin is produced by:

(A) *Alternaria* (B) *Pseudomonas*

(C) *Talaromyces* (D) *Helminthosporium*

273. Ten toxin is produced by:

(A) *Alternaria* (B) *Pseudomonas*

(C) *Talaromyces* (D) *Helminthosporium*

274. Endo PAL of *Erwinia carotovora* induces in potato:

(A) Maceration (B) Electrolytic loss

(C) Necrosis (D) All of the above

275. Match *Mycotypha*:

(A) *Cokeromyces* (B) *Cunninghamella*

(C) *Rhizopus* (D) None of these

276. Match *Sclerotinia*:

(A) *Fusarium* (B) *Botrytis*

(C) *Whetzelinia* (D) None of these

277. Cytokinins are important for symptoms production in case of:

(A) *Uromyces* spp (B) *Rhodococcus faciens*

(C) *Taphrina deformans* (D) All of the above

278. Auxins are important in case of disease caused by:

(A) *Pseudomonas savastanoi* (B) *Agrobacterium tumefaciens*

(C) *Albugo candida* (D) All of the above

279. Prevost proved the cause of bunt of wheat as fungus in:

(A) 1886 (B) 1807

(C) 1883 (D) 1798

280. Tillet published a paper on bunt or stinking smut of wheat in:

(A) 1729 (B) 1755

(C) 1846 (D) 1833

281. Who discovered the method of artificial culture of microorganism?

(A) Anton De bary (B) Millerdet

(C) Brefeld (D) Prevost

282. Who first time reported heterothallism in *mucor*:

(A) Millerdet (1882) (B) Blakeslee (1904)

(C) Biffen (1905) (D) Orton (1900)

283. Who discovered the heterokaryosis?

(A) Burgeff (1912) (B) Hansen E L Smith (1923)

(C) Biffen (1905) (D) None of these

284. American plant pathologist T.J. Burrill for the first time found that fire blight of pear and apple was caused by a bacterium in:

(A) 1878 (B) 1876

(C) 1880 (D) 1882

285. Who recognized the significance of the role of pectic enzyme on pathogenesis of some fungi on plant?

(A) Tanaka (1933) (B) William Brown (1915)

(C) L.R. Jones (1905) (D) Anton de Barry (1886)

286. The wild fire toxin was first toxin to be isolated in pure form by:

(A) Kurosawa (1926) (B) Wooley *et al.* (1952)

(C) Gaumann (1926) (D) Yabuta (1936)

287. Rigid cleistothecial appendages with dichotomously branched tips are characteristics of:

(A) *Unicinula* (B) *Podosphaera* and *Microsphaera*

(C) *Erysiphe* (D) *Laveillula*

288. Rigid with curled tips as in:

(A) *Uncinula* and *Pleochaeta* (B) *Microsphaera* and *Podosphaera*

(C) *Phyllactinia* (D) None of these

289. Rigid spear like with bulbous base and pointed tips are characteristics of:

(A) *Phyllactinia* (B) *Microsphaera*

(C) *Uncinula* (D) None of these

290. Single ascus in cleistothecium is characteristics of:

(A) *Erysiphe* (B) *Phyllactinia*

(C) *Sphaerotheca* and *Podosphaera* (D) *Unicnula*

291. Which one of the following is a sexual spore?

(A) Zoospore (B) Zygospore

(C) Sporangia (D) Conidia

292. The sporangiophores dichotomously branched at acute angle and taper to curved pointed tips in the genus:

(A) *Peronospora* (B) *Plasmopara*

(C) *Sclerospora* (D) *Basidiophora*

293. The sporangiophore is club shaped with a swollen head over which the sporangia are borne on minute sterigmata in the genus:

(A) *Sclerospora* (B) *Plasmopara*

(C) *Basidiophora* (D) None of these

294. The sporangiophore is a long, stout hypha with many upright branches near the end bearing sporangia at the tips in the genus:

(A) *Sclerospora* (B) *Plasmopara*

(C) *Basidiophora* (D) None of these

295. The branches and their sub divisions occur typically at right angle and are irregularly spaced in the genus:

(A) *Sclerospora* (B) *Plasmopara*

(C) *Basidiophora* (D) None of these

296. The sporangia of Pseudoperonospora germinate typically by:

(A) Directly (B) Germ tube

(C) Zoospore (D) None of these

297. The genus *Peronospora* always germinates by:

(A) Directly (B) Germ tube

(C) Zoospore (D) None of these

298. The disease responsible for the well known Bengal famine is:

(A) *Helminthosporium* (B) Blast

(C) Tungro (D) Sheath blight

299. A substance inhibiting the growth of certain microorganism in plants in response to number of chemical, biological and physical stimuli is:

(A) Phytoalexin (B) Phytotoxin

(C) Elicitor (D) Prohibitin

300. Which one of the following is the host specific toxin:

(A) Victorin (B) Tab toxin

(C) Ten toxin (D) All the above

301. The enzyme responsible for splitting 1-4 glycosidic bonds between adjacent uronic acid units are:

(A) Pectic lyases (B) Pectin methylesterases

(C) Pectin polymerases (D) All of the above

302. Blister blight of tea is caused by:

(A) *Colletotrichum virescens* (B) *Phoma lingam*

(C) *Cephaleuros virescens* (D) *Exobasidium vexans*

303. Which of the following fungus is a hyperparasite of the uredia of groundnut rust?

(A) *Darluca filum* (B) *Gliocladium virens*

(C) *Nectria cinnabarina* (D) *Trichoderma viride*

304. *Gaeumannomyces graminis* reduces the uptake of:

(A) K^+ (B) Ca^{++}

(C) PO_4^- (D) All of the above

305. E.J. Butler authored:

(A) Taxonomy of higher plant
(B) Fungi delight of curiosity
(C) Fundamentals of mycology
(D) Fungi and disease in plant

306. B.B. Mundkur authored:

(A) Fungi and plant diseases
(B) Fungi and diseases in plant
(C) Fundamental of mycology
(D) None of these

307. Surface sterilization of leaves is achieved by:

(A) 100 ppm streptomycin
(B) 100 ppm tetracyclin
(C) 1 per cent sodium hypochloride
(D) 5 per cent sodium hypochloride

308. Most successful resistant cultivars in India are developed and released for cultivation for:

(A) Grain mold of sorghum
(B) Pigeon pea wilt
(C) White rust of mustard
(D) Blight of sunflower

309. The antheridium of *Albugo candida* is clavate, paragynous and contains:

(A) 6-12 nuclei
(B) 15-21 nuclei
(C) 25-31 nuclei
(D) None of these

310. A violent reaction in which the invaded tissues upon attacked by pathogen, die promptly preventing further spread of infection is:

(A) Hypersensitivity
(B) Hyperparasitism
(C) Hypertrophy
(D) Hyperplasia

311. Which one is a typical soil borne pathogen?

(A) *Puccinia arachidis*
(B) *Ustilago nuda*
(C) *Fusarium udum*
(D) *Erysiphe chicoracearum*

312. Match *Fusarium*:

(A) *Cochliobolus*
(B) *Thanethophorus*
(C) *Nectria*
(D) *Setosphaeria*

313. Match *Heminthosporum*:

(A) *Cochliobolus*
(B) *Thanetophorus*
(C) *Nectria*
(D) *Setosphaeria*

314. Match *Rhizoctonia*:

(A) *Cochlioblus*
(B) *Thanetophorus*
(C) *Nectria*
(D) *Setosphaeria*

315. Match *Exserohilum*:

(A) *Cochliobolus* (B) *Thanetophorus*

(C) *Nectria* (D) *Setosphaeria*

316. Match Anamorph:

(A) Asexual stage (B) Sexual stage

(C) Physiologic race (D) The whole fungus

317. Match Holomorph:

(A) The whole fungus (B) Sexual stage

(C) Asexual stage (D) Physiologic race

318. Match the discovery of R. Harting:

(A) Viroid (B) Hybridoma technology

(C) Forest pathology (D) Mycoplasma discovery

319. Match *Curvularia*:

(A) *Cochliobolus* (B) *Thanetophorus*

(C) *Nectria* (D) *Setosphaeria*

320. Which of the following classes of fungi have biflagellate zoospores where the two flagella are unequal?

(A) Chytridiomycets (B) Plasmodiophoromycetes

(C) Oomycetes (D) Hyphochytridiomycetes

321. Which of the following fungus is a mycoparasite of sclerotia:

(A) *Fusarium moniliforme* (B) *Dactylium fusariodes*

(C) *Nectria cinnabarina* (D) None of these

322. Myceliogenic and carpogenic gramination shows in:

(A) *Claviceps microsephala* (B) *Mucor mucedo*

(C) *Alternaria triticina* (D) None of the these

323. Alternation of generation is found in:

(A) *Aspergillus* (B) *Allomyces*

(C) *Agaricus* (D) *Albugo*

324. Three types of life cycle have been discovered in:

(A) *Allomyces* (B) *Aspergillus*

(C) *Agaricus* (D) None of these

325. Species of the subgenus *Euallomyces* exhibit a definite:

(A) Alternation of generation (B) Carpogenic germination

(C) Multiplication (D) None of these

326. Biological specialization is found in:

(A) *Alternaria* (B) *Helminthosporium*

(C) *Pleurotus* (D) *Puccinia*

327. *Mucor* belongs to:

(A) Ascomycetes (B) Zygomycetes

(C) Hyphomycetes (D) Basidiomycetes

328. *Penicilium* is produced by:

(A) *Penicillium citrinum* (B) *Penicillium chrysogenum*

(C) *P. oxalicum* (D) *P. roquefortii*

329. Which of the following fungi is hallucinogenic:

(A) *Amanita* (B) *Psilocybe*

(C) *Russula* (D) *Coprinus*

330. Which of the following fungi is VAM:

(A) *Glomus* (B) *Armillaria*

(C) *Pleurotus* (D) *Rhizoctonia*

331. Which of the following fungi is used for genetic studies?

(A) *Peziza* (B) *Sordaria*

(C) *Neurospora* (D) *Pilobolus*

332. Abnormal multiplication of the host cells is called:

(A) Hyperplasia (B) Hypertrophy

(C) Hypersensitivity (D) None of these

333. Enlargement of the host cell is called:

(A) Hyperplasia (B) Hypertrophy

(C) Hypersensitivity (D) None of these

334. Dimorphic fungi is:

(A) *Mucor* (B) *Alternaria*

(C) *Helminthosporium* (D) None of theses

335. Small sporangia designated as:

(A) Sporangiophore (B) Sporangiola

(C) Sporocladia (D) None of these

336. An organism feeding on dead tissues of living hosts is termed as:

(A) Authotroph (B) Necrotroph

(C) Perthophyte (D) Parasite

337. The initial invasion of a host by a pathogen called:

(A) Infection (B) Invasion

(C) Penetration (D) None of these

338. Name of group of fungi in which spore germinate even in the absence of water to cause infection:

(A) Pernosporaceae (B) Ustilaginaceae

(C) Plasmodiophoraceae (D) Erysiphaceae

339. Which of the following genera of zygomycetes bears true conidia in which the sporangial and spore walls are inseparably fused?

(A) *Cunninghamella* (B) *Cokeromyces*

(C) *Chaetocladium* (D) *Thamnidium*

340. Caducous asexual propagules produced on the inflated tip of genus:

(A) *Cunnighamella* and *Mycotypha* (B) *Actionmucor* and *Phizopus*

(C) *Chanephorar* and *Cokeromyces* (D) None of these

341. Dr. A.F. Blakeslee discovered in 1904 the phenomenon of sexual compatibility in the order of:

(A) Peronosporales (B) Mucorales

(C) Leptomitales (D) None of these

342. Spitzenkorper found near the:

(A) Middle hyphae (B) Hyphal apex

(C) Spore wall (D) None of these

343. Which of the following genera has chiastobasidia?

(A) *Agaricus* (B) *Ustilago*

(C) *Macromyces* (D) *Exidia*

344. Who discovered the concept of phytoalexins?

(A) K.O. Muller (B) L.O. Kunkel

(C) Flor (D) Cromwell

345. Gelationous fungi are:

(A) *Exidia* (B) Rust

(C) *Alternaria* (D) None of these

346. Ear cockle of wheat is caused by:

(A) Fungus (B) Nematode

(C) Virus (D) Bacterium

347. ***Pyricularia oryzae*** **got introduced in India during 1918 from:**

(A) Jawa (B) Sri Lanka

(C) Japan (D) Thialan

348. Downy mildew of maize caused by ***Sclerospora phillipinensis*** **was introduced in India during 1912 from:**

(A) Jawa (B) Sri Lanka

(C) Japan (D) Thailan

349. The first plant disease simulator to be developed was:

(A) Potato late blight (B) Blight of crucifers

(C) Early blight of potato (D) None of these

350. EPIDEM forecasting simulation model have been used for:

(A) Southern maize leaf blight (B) Apple scab

(C) Tomato early blight (D) None of these

351. Match simulation model for apple scab:

(A) PIDEM (B) EPIMAY

(C) EPIVEN (D) BARSIM – I

352. Match simulation model for southern maize leaf blight:

(A) EPIDEM (B) EPIMAY

(C) EPIVEN (D) BARSIM-I

353. Match simulation model for barley leaf rust:

(A) EPIDEM (B) EPIMAY

(C) EPIVEN (D) BARSIM-I

354. Onion smut caused ***Urocystis cepulae*** **introduced in India from Europe during:**

(A) 1950 (B) 1958

(C) 1960 (D) 1965

355. Match ***Drechslera*****:**

(A) Pyrenophora (B) Physalospora

(C) Pleospora (D) None of these

356. Match ***Diplodia*****:**

(A) Pyrenophora (B) Physalospora

(C) Pleospora (D) Noen of these

357. Match *Gibberella pulicaris*:

(A) *Fusarium sambucinum* (B) *F. moniliforme*

(C) *F. xylarioides* (D) *F. graminearum*

358. Match *Erysiphe betae*:

(A) Sphaceloma necator (B) Sphacelia typhina

(C) Oidium erysiphodies (D) None of these

359. Match *Mycosphaerella graminicola*:

(A) *Septoria tritici* (B) *Cladosporium iridis*

(C) *Ascochyta pinodes* (D) None of these

360. Match *Phoma lingam*:

(A) *Septoria tritici* (B) *Pleospora betae*

(C) *Leptoshaeria maculans* (D) None of these

361. Who established the mechanism of *Fusarium* wilt of cotton:

(A) T.S. Sadasivam (B) S.N. Das Gupta

(C) R.N. Tondon (D) None of these

362. *Fusarium oxysporum* f. sp. *ciceri* which causes wilt disease of chickpea harborus in:

(A) Cortex (B) Xylem

(C) Phloem (D) Stomata

363. Which of character is found in Ascomycetes:

(A) Mycellium devoid of cross wall

(B) Sexual reproduction lacking

(C) Spore are produced in asci

(D) All the plants are multicellular

364. Which type of nutrition does not occur in fungi:

(A) Parasitism (B) Saprophytism

(C) Cymbiotic relation (D) Photoautotrophism

365. Pink disease of mango is caused by:

(A) *Pestaliopsis theae* (B) *Corticium salmonicolor*

(C) *Diplodia cajani* (D) *Exobasidium vexans*

366. Grey blight of tea is caused by:

(A) *Pestaliopsis theae* (B) *Corticium salmonicolor*

(C) *Diplodia cajani* (D) *Exobasidium vexans*

367. Stem canker of Arhar is caused by:

(A) *Diplodia cajani*
(B) *Xanthonobas campestris*
(C) *Pseudomonas solanacearum*
(D) *Exobasidium vexans*

368. Which of the following occurs in yeast?

(A) Hypha
(B) Hyphae
(C) Budding
(D) Mycelium

369. Which of the group includes heterotrophic plants:

(A) Algae
(B) Fungi
(C) Brayophyta
(D) Pteridophyta

370. Which of the following effect is caused to the hosts when a parasitic fungus causes infection?

(A) The loss their chlorophyll
(B) They show healthy growth
(C) They are destroyed by the decay
(D) They show pronounced symptoms

371. Which of the following disease is caused by parastitic fungi?

(A) Tuberculosis
(B) Cholera
(C) Rust
(D) Pneumonia

372. Which is not saprophytic fungus:

(A) Yeast
(B) *Penicillium*
(C) *Puccinia graminis*
(D) *Mucor*

373. Which of the following is not related with other three?

(A) Hypha
(B) Hyphae
(C) Mycelium
(D) Chloroplast

374. Mycelium on its outer sides bounded by cell wall which is made up of:

(A) Chitin in most fungi
(B) Chitin in all fungi
(C) Cellulose in all fungi
(D) Protein is some fungi

375. Which of the following lacks mycelium:

(A) *Mucor*
(B) *Penicillium*
(C) *Saccharomyces*
(D) *Puccinia graminis*

376. Which of them germinate?

(A) Ascogenous hyphae
(B) Ascogonium
(C) Ascospores
(D) Ascus

377. Which of the following does not occur in Ascomycetes?

(A) Unicellular plant
(B) Conidia
(C) Septate mycelium
(D) Coenocytic mycelium

378. Which of the condition is met in heterothallic species of *Mucor* at the time of sexual reproduction?

(A) Fusions take place in between the mycelium obtained by the germination of two spores of different strains
(B) Mycelium do not fuse
(C) Mycelium obtained by the germination of one spore fuse
(D) None of these

379. Saccharomyces has no relation with:

(A) Sugar
(B) Host
(C) Enzyme
(D) Fermentation

380. Which of the following fungus causes disease of animals:

(A) *Rhizopus nigricans*
(B) *Saccharomyces*
(C) *Penicillium*
(D) *Puccinnia*

381. The person related with the discovery of antibiotic substance penicillin obtained from *Penicillium notatum* is:

(A) Blackeslee
(B) Pasteur
(C) Flaming
(D) Robert Brown

382. Which of them is used as tool in genetic studies:

(A) *Neurospora*
(B) *Saccharomyces*
(C) *Penicillium*
(D) *Rhizopus*

383. Which of the place will be most suitable for the growth of *Rhizopus nigricans*?

(A) Wounded plant
(B) Dry and dead plant
(C) Healthy plant
(D) Injured plant

384. Which of the asexual spore is most common?

(A) Zygospore
(B) Conidiophore
(C) Oospore
(D) Sporangiospore

385. Culture of powdery mildew fungus is maintained on:

(A) PDA
(B) Sand
(C) Seed
(D) Host plant

386. Which oil is used for oil immersion lens:

(A) Coconut oil (B) Cedar wood
(C) Linseed oil (D) Mustard oil

387. Laminar flow unit is used to obtain:

(A) Desired temperature (B) Desired pH
(C) Germ free air (D) Desired relative humidity

388. In loose smut of wheat the fungus is associated with seed as:

(A) Intra embryo infection (B) Contaminant
(C) Externally seed borne (D) Inert matter

389. Entry of club root pathogen in root is mainly through:

(A) Wound (B) Hydathodes
(C) Lenticels (D) Mechanical means

390. The disease club root of crucifers is reduced by:

(A) Low K and P (B) High K and P
(C) Low N and Ca (D) None of these

391. *Venturai inaequalis* in living leaf is located at:

(A) Epidermal layer (B) Parenchyma
(C) Cuticular layer (D) Sub cuticular layer

392. Which of the following enzyme can degrade plant waxes:

(A) Macerating (B) Pectic
(C) Cellulytic (D) None of these

393. Phenols are known to effect pectic enzyme by:

(A) Inactivation (B) Activation
(C) Stagnation (D) Retardation

394. Alternaric acid is a:

(A) Vivotoxin (B) Phytotoxin
(C) Pathotoxin (D) None of these

395. Chickpea blight fungus *Ascochyta* is:

(A) Soil borne (B) Seed borne
(C) Air borne (D) Both (B) and (C)

396. Chickpea blight caused by *Ascochyta rabiei*. The pathogen survives as:

(A) Conidia (B) Conidiophore
(C) Pycnidia (D) None of these

397. Mostly the propagules of the fungus *Ascochyta rabiei* are present:
(A) On or in the seed coat
(B) In the soil
(C) In the seedling
(D) None of these

398. Perfect stage of *Ascochyta rabiei* the cause of chickpea blight is:
(A) *Didymella* sp.
(B) *Penicillium* sp.
(C) *Sclerotinia*
(D) *Thanetophorus* sp.

399. Match *Alternaria alternata*:
(A) *Alternaria tenuis*
(B) *Macrosporium solani*
(C) *Trichoconis padwickii*
(D) None of these

400. Match *Fusarium solani*:
(A) *Cladosporium fulvum*
(B) *Fusarium javanicum*
(C) *Fusarium redolens*
(D) None of these

401. Match *Helminthosporium solani*:
(A) *Fomes annosus*
(B) *Helminthosporium atrovirens*
(C) *Helminthosporium satvium*
(D) None of these

402. Match *Colletotrichum coccodes*:
(A) *Colletotrichum musae*
(B) *Colletotrichum falcatum*
(C) *Colletotrichum atramentianum*
(D) None of these

403. *Saprolegnia parasitica* causes disease in:
(A) Plants
(B) Fish
(C) Insects
(D) Reptiles

404. Match *Colletotrichum musae*:
(A) *Gloeosporium musae*
(B) *Colletotrichum falcatum*
(C) *Colletotrichum atramentianum*
(D) None of these

405. Heterogametangic reproduction found in:
(A) Oomycetes
(B) Hyphomycetes
(C) Zygomycetes
(D) None of these

406. Which of the following order have more than one egg in the oogonium:
(A) Legnidiales
(B) Leptomitales
(C) Peronosporales
(D) Saprolegniales

407. Which of the following genera of Ascomycetes is autogamous:
(A) *Neurospora*
(B) *Schizosaccharomyces*
(C) *Sphaerotheca*
(D) *Talaromyces*

408. Resolution of a microscope is limited by:

(A) Eye piece (B) Object

(C) Wave length of light (D) Condenser

409. Slime molds are characterized by the presence of:

(A) Capillitium (B) Elaters

(C) Pseudoelaters (D) None of these

410. In which of the following fungi zoospore are formed inside vesicles:

(A) *Phytophthora* (B) *Penicillium*

(C) *Sclerophthora* (D) *Pythiogeton*

411. The study of Plant Pathology in India was initialized by:

(A) Butler (B) Pavgi

(C) Mundkur (D) Sadsivan

412. Stem rot of chickpea is caused by:

(A) *Fusarium oxysporum* (B) *Sclerotinia sclerotiorum*

(C) *Botrytis cinerea* (D) *Peronospora parasitica*

413. Generic name of *Sclerotinia* is:

(A) *Whetzelinia* (B) *Sclerotia*

(C) *Cercospora* (D) None of these

414. Bunchy top disease of banana perpetuates through:

(A) Fallen Leaves (B) Suckers

(C) Fallen fruit (D) Stems

415. The sclerotia are actually the ergot of commerce is associated:

(A) *Fomes* sp. (B) *Ustilago* sp.

(C) *Claviceps* sp. (D) *Agrobacterium* sp.

416. Amoeboid naked, multinucleate body in Myxomycotina is known as:

(A) Rhizomycelium (B) Plasmodium

(C) Promycelium (D) Meiosporangium

417. Chlamydospore is a modified:

(A) Pycnidium (B) Zygospore

(C) Oospore (D) Hyphal cell

418. Naked asci are present in:

(A) Apothecium (B) Cleistothecium

(C) Perithecium (D) None of these

419. Conidia having muriform septation with long beak and colour belong to:

(A) Stemphyllium (B) Helminthosporium

(C) Alternaria (D) Curvularia

420. Peritrichus flagella are seen in:

(A) *Erwinia* (B) *Streptomyces*

(C) *Clavibactor* (D) *Pseudomonas*

421. High sterol content is characteristic of:

(A) Prokaryotes (B) Eukaryotes

(C) Both A and B (D) None of the

422. Rust of pea is caused by:

(A) *Uromyces fabae* (B) *Puccinnia purpurea*

(C) *Puccinia striformis* (D) *Uromyces appendiculatus*

423. Powdery mildew of pea can be controlled by:

(A) Ridomil (B) Sulfex

(C) Thiram (D) Vitavax

424. The maximum germination of conidia of *Erysiphe polygony* occurs at:

(A) 20-24ºC (B) 30-35ºC

(C) 35-40ºC (D) None of these

425. Flattening or cohering of organs such as stems, flowers and roots is known as:

(A) Callus (B) Intumescences

(C) Fasciations (D) Fasciculation

426. Premature development of shoot from bud is known as:

(A) Proliferation (B) Prolepsis

(C) Restoration (D) Rosetting

427. Vertical resistance in plants is the resistance against:

(A) Many races (B) One race

(C) A few races (D) None of these

428. Plant parasitic fungi and bacteria that pass a part of their life cycle in soil called:

(A) Soil inhabitation (B) Soil transient

(C) Soil invaders (D) None of these

429. Plant disease was written by:

(A) B.B. Mundkur (B) E.J. Butler

(C) Ashok Agrawal (D) R.S. Singh

430. Downy mildew of lettuce is caused by:

(A) *Bremia lactucae* (B) *Sclerospora graminicola*

(C) *Sclerospora sacchari* (D) None of these

431. Quantitative assessment of spores in air is done through:

(A) Barometer (B) Plenimeter

(C) Thermometer (D) Hirst spore trap

432. Cobb scale for the assessment of wheat rust is based on:

(A) Descriptive key (B) Standard area diagram

(C) Incidence severity relationship (D) Inoculum disease intensity relationship

433. Maceration of host tissue in damping off is brought about by:

(A) *Taphrina* (B) *Erysiphe*

(C) *Uncinula* (D) *Nectria*

434. Downy mildew of maize is caused by:

(A) *Sclerospora sacchari* (B) *Peronospora pisi*

(C) *Sclerospora graminicola* (D) *Peronospora parasitica*

435. The casual agent of downy mildew of pea is:

(A) *Sclerospora sacchari* (B) *Peronospora pisi*

(C) *Sclerospora graminicola* (D) *Peronospora parasitica*

436. Downy mildew of crucifers is caused by:

(A) *Sclerospora sacchari* (B) *Peronospora pisi*

(C) *Sclerospora graminicola* (D) *Peronospora parasitica*

437. Stem gall of coriander is caused by:

(A) *Protomyces macrosporus* (B) *Taphrina maculans*

(C) *Taphrina deformans* (D) *Exoascus* sp.

438. Powdery mildew of apple is caused by:

(A) *Podoshaera leucotricha* (B) *Uncinulla necator*

(C) *Erysiphe cichoracearum* (D) *Sphaerotheca fuligenea*

439. Appendages having basal swellings (bulbous) are in genus:

(A) *Phyllactinia* (B) *Erysiphe*

(C) *Uncinula* (D) *Microsphaera*

440. Appendages having hypha like are in the genus:

(A) *Phyllactinia* (B) *Erysiphe*

(C) *Uncinula* (D) *Microsphaera*

441. Appendages having without bulbous base are in the genus:

(A) *Microsphaera* and *Uncinula* (B) *Phyllactinia* and *Erysiphe*

(C) *Sphaerotheca* and *Podosphaera* (D) None of these

442. *Albugo candida* causes:

(A) Red rust (B) Brown rust

(C) Yellow rust (D) White rust

443. Optimum temperature for late blight development is:

(A) 11-30°C (B) 16-18°C

(C) 20-21°C (D) 24-26°C

444. Association of curling, puckering and thickening in leaves of patches is attributed to:

(A) *Taphrina pruni* (B) *Taphrina cerasi*

(C) *Taphrina deformans* (D) None of these

445. Brown stripe downy mildew of maize is caused by:

(A) *Sclerophthora rayssiae* (B) *Sclerospora sorghii*

(C) *Plasmopara viticola* (D) *Sclerospora sacchari*

446. Two types of appendages are formed on the surface of cleistothecia of:

(A) *Sclerospora graminicola* (B) *Taphrina maculans*

(C) *Podoshaera leucotricha* (D) None of these

447. Conidia of *Podospaera leucotricha* germinate at an optimum temperature:

(A) 19°C and 22-25°C (B) 28 - 30°C

(C) 10°C and 14-18°C (D) 30-35°C

448. What is the best method for the control of ergot of bajra:

(A) Fungicide spray (B) Biological control

(C) Use of clean seed (D) Cultural method

449. Mature smut spore represent:

(A) Haplophase (B) Diplophase

(C) Dikaryophase (D) None of these

450. *Nectria cinnabarina* cause infection of fruit trees by penetrating through:

(A) Wounds (B) Cuticle

(C) Lenticels (D) Corticells

451. False smut of rice is cause by:

(A) *Ustilaginoidea virens* (B) *Ustilago scitaminea*

(C) *Ustilago hordei* (D) *Ustilago tritici*

452. Wheat rust life cycle in India was given by:

(A) E.J. Butler (B) K.C. Mehta

(C) R. Prasad (D) None of these

453. Mycology is the study of:

(A) Fungi (B) Virus

(C) Viroid (D) Bacteria

454. Micro organism was first observed by:

(A) Antony von Leeuwenhoek (B) Stanley

(C) Louis Pasteur (D) T.J. Burrill

455. Soil pH adjustment to 7.2 is suggested for managing:

(A) Wart disease of potato (B) Club root of cabbage

(C) Root rot of pea (D) Root rot of papaya

456. Defoliation disease of tomato is causes by:

(A) *Septoria* sp. (B) *Cercospora* sp.

(C) *Alternaria* sp. (D) *Curvularia* sp.

457. Germinating sporangia of *Peronospora destructor* gives rise to:

(A) Zoospores (B) Zygospores

(C) Oospores (D) Germ tubes

458. The pathogen causing false smut of rice grows:

(A) On the leaves (B) On the stem

(C) In the ovary of individual kernels (D) In the xylem

459. Match *Ustilaginoidea virens*:

(A) *Claviceps oryzae-sativae* (B) *Peronospora parasitica*

(C) *Peronospora pisi* (D) *Sclerospora graminicola*

460. The size of spore of *Ustilago nuida tritici* is:

(A) 1-3 μ (B) 5-9 μ

(C) 11-15 μ (D) 17-20 μ

461. Primary infection of karnal bunt of occurs through:

(A) Teliospores

(B) Sickle shaped primarily sporidia

(C) Alantoid filiform secondary sporidia

(D) Chlamydospores

462. The term vertical resistance and horizontal resistance were coined by:

(A) Robinson (1971)
(B) Van der Plank (1963)
(C) Nelson (1973)
(D) Russell (1978)

463. Vertical resistance is also known as:

(A) Major gene resistance
(B) Non specific resistance
(C) Race specific resistance
(D) All of the above

464. The class ends with:

(A) Mycotina
(B) Oomycetes
(C) Mycetes
(D) Zygomycetes

465. Horizontal resistance is also referred to as:

(A) Minor gene resistance
(B) Polygenic resistance
(C) General resistance
(D) All of the above

466. Conidia and sporangia are the spores produced by:

(A) Fungi
(B) Bacteria
(C) Virus
(D) Viroids

467. Gene for gene hypothesis was produced by:

(A) Van der Plank (1963)
(B) Robinson (1971)
(C) Flor (1956)
(D) Russell (1978)

468. The term commensalism shows:

(A) Living together of two or more species of organism for mutual benefit

(B) Living two organism together which are harmful to each other

(C) Living together of the two species one of which is benefited by the association where as the other is neither beneficial non harmful

(D) None of these

469. Concept of gene for gene hypothesis was developed in:

(A) Pearl millet
(B) Linseed
(C) Sorghum
(D) Pea

470. In cotton, example of durable resistance includes resistance to:

(A) Bacterial blight (B) Jassid

(C) Leaf curl (D) All of the above

471. Vertical resistance can be exploited through:

(A) Development of multiline

(B) Gene pyramiding, Gene deployment

(C) Development of varieties with individual major gene

(D) All of the above

472. *Pythium* belongs to class:

(A) Oomycetes (B) Chytridimcetes

(C) Zygomycetes (D) Hyjohomycetes

473. *Albugo* is a:

(A) Facultative parasite (B) Obligate parasite

(C) Facultative saprophyte (D) Predator

474. Ripe rot in chillies is induced by:

(A) Anthracnose (B) Drought

(C) Ca deficeincy (D) Bright of sun shine

475. Growth and development in fungi from spore to spore is called:

(A) Life Cycle (B) Pathogenesis

(C) Disease cycle (D) Incubation period

476. Wart disease of potato is restricted to:

(A) Kulu (B) Shimla

(C) Darjelling (D) Kashmir

477. Frontier quarantine station in West Bengal is located at:

(A) Kolkata (B) Singoor

(C) Siliguri (D) Sukhiapokari

478. Resistance to anthracnose disease of beans is:

(A) Monogenic (B) Polygenic

(C) Oligogenic (D) Modifiergenic

479. Resistance to cotton wilt disease is:

(A) Monogenic (B) Polygenic

(C) Oligogenic (D) Modifiergenic

480. *Lycopersicon chilense* contains gene for breeding resistance against:

(A) *Septoria* leaf spot
(B) Spotted wilt
(C) Bacterial canker
(D) Anthracnose

481. Match Trichoderma:

(A) *Hypocrea*
(B) *Microsphaera*
(C) *Uncinula*
(D) *Phyllactinia*

482. Antifungal activity of glyodine is directed to:

(A) Cell wall
(B) Cytoplasm
(C) Enzyme system
(D) Nucleic acid

483. *Mucor* is included in:

(A) Zygomycetes
(B) Oomycetes
(C) Hyphomycetes
(D) Chytridiomycetes

484. *Sclerotinia sclerotiorum* produces ascospores:

(A) Apothecium
(B) Perithecium
(C) Naked asci
(D) Cleistothecium

485. In India, foci of infection of stem rust of wheat are found in:

(A) Himalayas
(B) Southern hills
(C) Central plains
(D) Northern hills

486. What is the best temperature for the growth of *Puccinia striformis*:

(A) 11°C
(B) 21°C
(C) 31°C
(D) None of these

487. What is the source of primary infection of *Uromyces ciceris-arietini* on chick pea in north Indian plains:

(A) Seed of chickpea
(B) Other leguminous plants on the plains
(C) Soil
(D) Other leguminous plants on the hills

488. Diplodia produces spores in:

(A) Acervuli
(B) Pycnidia
(C) Oospores
(D) Conidiophores

489. F value denotes:

(A) Variance of ratio
(B) Sum of square
(C) Interaction
(D) Mean square

490. The fish odour in hill bunt affected wheat plants is due to a volatile compound called:

(A) Tetramethylamine (B) Pentachloroethane

(C) Trimethylamine (D) Chlorinated benzoic acid

491. Which fungus produces two kinds of conidia namely the α and β:

(A) *Phomopsis* (B) *Phoma*

(C) *Fusarium* (D) *Macrophomina*

492. Which of the following are poisonous mushroom?

(A) Toad stool (B) Puff fungi

(C) Oyster mushroom (D) Button mushroom

493. Secondary spread of ergot of bajra is through:

(A) Seed (B) Insect

(C) Both (A) and (B) (D) Air

494. Alternate host or black or stem rust of wheat:

(A) Barberry (B) Mahonia

(C) Thalictrum (D) All of the above

495. Solar energy treatment for the control of loose smut of wheat was given:

(A) Mitra (B) Butler

(C) Luthra and Sattar (D) Thind

496. In plains of India which wheat rust appears first:

(A) Yellow rust (B) Black rust

(C) Orange rust (D) All of the above

497. Cultivars of pigeon pea resistant to wilt have high content of:

(A) Cysteine (B) Nitrogen

(C) Silicon (D) Cellulose

498. Zoospore with unequal flagella is called:

(A) Heterokont (B) Anisokont

(C) Isokont (D) None of these

499. Destructive Insect Pest Act came into effect in India during the year:

(A) 1960 (B) 1950

(C) 1938 (D) 1914

500. Foolish disease of rice is the other name of:

(A) Root rot (B) Bakane

(C) Sheath rot (D) Stem rot

501. The eye spot of tobacco is caused by:

(A) *Helminthosporium nicotiana* (B) *Alternaria nicotiana*
(C) *Cercospora nicotiana* (D) *Drechslera nicotiana*

502. Organism which kills host tissues in advance of penetration and then lives saprophytically is known as:

(A) Perthotrophs (B) Saprophytes
(C) Biotrophs (D) Hemibiotrophs

503. Catechuic acid is present in the:

(A) Red onion skin (B) Bitter gourd skin
(C) Ginger leaves (D) Garlic leaves

504. Hypersensitivity reaction was first observed due to virus by:

(A) Mayer (B) Stakman
(C) Holmes (D) Klement

505. Pyricularin by *Pyricularia oryzae* is a typical example of:

(A) Phytotoxin (B) Pathotoxin
(C) Vivotoxin (D) None of these

506. The casual organism of Dutch elm disease is:

(A) *Alternaria tenuis* (B) *Cercospora* sp.
(C) *Ceratocystis ulmi* (D) None of these

507. Which fungus produces conidia in acervuli?

(A) *Septoria* (B) *Pyricularia*
(C) *Helminthosporium* (D) *Gloeosporium*

508. Yeast is used in bakery because they:

(A) Produce CO_2 (B) Give aromatic flavor
(C) Produce alcohol (D) Produce protein

509. *Venturia inaequalis* overwinters in:

(A) Air (B) Soil
(C) Dead leaves of apples (D) In plant debris of other host

510. *Botrytis* can cause:

(A) Leaf spot (B) Stem lesions
(C) Damping off (D) All of the above

511. *Botrytis* overwinters in the soil, decaying plants debris as:

(A) Zoosporangia (B) Planogametangia
(C) Zygospores (D) Mycelium and sclerotia

512. Some species of *Botrytis* occasionally produce a *Botryotinia* perfect stage in which ascospores are produced in:

(A) An apothecium (B) Perithecium

(C) Cleistothecium (D) Oogonium

513. Powdery scab of potato is caused by:

(A) Virus (B) MLO

(C) Fungus (D) Bacterium

514. When monomers of cutin enter into the cells of a pathogen producing cutinases, the production of cutinase enzyme is:

(A) Decreased (B) Increased

(C) Stopped (D) None of these

515. Glucan-cellulose is the chief constituent of cell wall of:

(A) Oomycetes (B) Ascomycetes

(C) Chytridiomycetes (D) Zygomycetes

516. The sexual form of a fungus is called:

(A) Holomorph (B) Anamorph

(C) Teleomorph (D) None of these

517. Posteriorly uniflagellate zoospore with flagella of whiplash type is found in:

(A) *Synchytrium* (B) *Phytophthora*

(C) *Pythium* (D) *Saprolegnia*

518. Zoospore is not produced by the genus of:

(A) *Sclerospora* (B) *Pseudoperonospora*

(C) *Peronospora* (D) *Plasmopara*

519. The tips of the sporangiophore branches inflate into bulbous apophyses bearing sterigmata in:

(A) *Bremia* (B) *Bremielella*

(C) *Plasmopara* (D) *Basidiophora*

520. The ascospores of *Claviceps purpurea* are:

(A) Thread like (B) Oval

(C) Spherical (D) Two celled

521. Sporangiophores are dichotomously branched at acute angle and taper to gracefully curved pointed tips on which sporangia are borne in the genus:

(A) *Basidiophora* (B) *Pseudoperonospora*

(C) *Plasmopara* (D) None of these

522. Plasmogamy in rust fungi occurs in:

(A) Receptive hypha
(B) Aeciospores
(C) Spherical
(D) Promycelium

523. The conidia are globose to ovoid and under the microscope resemble glass beads in the genus:

(A) *Aspergillus*
(B) *Penicillium*
(C) *Elaphomyces*
(D) *Ascosphaera*

524. Match *Penicillium*:

(A) *Talaromyces*
(B) *Aspergillus*
(C) *Elaphomyces*
(D) *Ascosphaera*

525. Ascospores dextrinoid in:

(A) *Aspergillus*
(B) *Ascosphaera*
(C) *Ascoscleroderma*
(D) *Microascales*

526. Asci borne in spore ball formed inside a trophocyst in the genus:

(A) *Talaromyces*
(B) *Aspergillus*
(C) *Ascosphaera*
(D) *Ascoscleroderma*

527. Capitate paraphyses are production in:

(A) *Melampsora lini*
(B) *Uromyces fabae*
(C) *Puccinia recondita*
(D) *Puccinia striformis*

528. Which of the following represent perfect stage of rust:

(A) Pycniospore
(B) Teliospore
(C) Aeciospore
(D) Unediospore

529. The pycnidia are produced by members of:

(A) Hyphomycetes
(B) Coelomycetes
(C) Blastomycetes
(D) None of these

530. How many sporidia, a spore of *Neovossia indica* produces on the promycelium:

(A) Numerous
(B) 4
(C) 8
(D) 16

531. In hot air oven the sterilization of glass wares at 160°C is obtained in:

(A) 60 minute
(B) 90 minutes
(C) 120 minute
(D) 150 minutes

532. Which of the following biochemical test are related to bacteria:

(A) Nitrate reduction
(B) H_2S production
(C) Gelatin liquification
(D) All of the above

533. In the meantime the mycelial mat which has produced the conidiophores, continues to develop in:

(A) *Claviceps purpurea*
(B) *Colletotrichum falcatum*
(C) *Helminthosporium sativum*
(D) None of these

534. Ultra voilet light alters DNA activity of microorganisms by dimerization of:

(A) Adenine
(B) Guanine
(C) Cytosine
(D) Thiamine

535. Match $\log_e [1/(1-x)]$:

(A) Simple interest disease
(B) Compound interest disease
(C) Logarithmic disease
(D) None of these

536. Milo root rot is caused by:

(A) *Rhizobium leguminosarum*
(B) *Alternaria alternata*
(C) *Periconia circinata*
(D) *Curvularia lunata*

537. Match $\log_e (x/(1-x))$:

(A) Simple interest disease
(B) Compound interest disease
(C) Logarithmic disease
(D) None of these

538. Match facultative parasites:

(A) *Pythium* and *Rhizoctonia*
(B) *Puccinia graminis tritici*
(C) *Erysiphe polygoni*
(D) *Claviceps oryzae stativae*

539. Match *Claviceps oryzae sativae*:

(A) *Ustilaginoidea virens*
(B) *Erysiphe polygoni*
(C) *Periconia circinata*
(D) None of these

540. Wheat rust epidemic came in Madhya Pradesh in:

(A) 1942-43
(B) 1946-47
(C) 1953-54
(D) 1963-64

541. Match *Endothia*:

(A) *Cryphonecteria*
(B) *Helminthosporium*
(C) *Albugo*
(D) *Exobasidium*

542. The role of Plant Pathology in the scientific and social development of the world had been discussed by:

(A) Stakman (1958)
(B) Padwick (1956)
(C) Ricker (1964)
(D) Klinkowaski (1970)

543. Catastrophic plant diseases were summarized by:

(A) Stakman (1958)
(B) Padwick (1956)
(C) Ricker (1964)
(D) Klinkowaski (1970)

544. The hyphae of *Taphrina maculans* are largely confined to:

(A) Epidermal cells
(B) Xylem
(C) Phloem
(D) Cortical cells

545. *Alternaria triticina* was first demonstrated by:

(A) Prasad and Prabhu
(B) L.M. Joshi
(C) B.B. Mundkur
(D) G. Rangaswami

546. Which of the following causes systematic infection?

(A) *Helminthosporium sativum*
(B) *Helminthosporium oryzae*
(C) *Helminthosporium gramineum*
(D) None of these

547. Stripe disease of barely is caused by:

(A) *Helminthosporium sativum*
(B) *Helminthosporium oryzae*
(C) *Helminthosporium gramineum*
(D) *Helminthosporium sigmoideum*

548. Maximum stripe disease of barely occurs when seeds are planted:

(A) 4-5 cum deep
(B) 2-3 cum deep
(C) 6-7 cum deep
(D) 8-10 cum deep

549. The epidermis of host is ruptured by the telial stage of:

(A) *Puccinia recondita*
(B) *Puccinia striformis*
(C) *Melampsora lini*
(D) *Puccinia graminis tritici*

550. The infection of loose smut of wheat occurs at the time of:

(A) Germination of seed
(B) Flowering stage
(C) Vegetative stage
(D) Emerging stage

551. Powdery mildew can be effectively controlled by:

(A) Metalaxyl
(B) Karathane
(C) Mancozeb
(D) Hinosan

552. Downy mildew can be effectively controlled by:

(A) Metalaxyl (B) Karathane

(C) Mancozeb (D) Hinosan

553. Sammelzellen in usually formed with in the infected host in case of:

(A) *Pythium debaryanum* (B) *Peronospora parasitica*

(C) *Phytophthora infestans* (D) *Physoderma zea maydis*

554. In *Synchytrium endobioticum* zygote infection in host cause:

(A) Hypertrophy (B) Hyperplasia

(C) Necrosis of cell (D) All of these

555. In *Synchytrium endobioticum* zoospore infection is mainly responsible for:

(A) Hypertrophy (B) Hyperplasia

(C) Necrosis of cells (D) All of these

556. In the plains of India, *Phytophthora infestans* survives during off season in:

(A) Soil (B) Plant debris

(C) Seed tubers in cold storage (D) Left over tubers

557. Anthracnose refers to symptoms produced by:

(A) *Colletotrichum* (B) *Phoma*

(C) *Botriodiplodia* (D) *Cercospora*

558. Powdery mildew of sheeshum is caused by:

(A) *Uncinula* (B) *Podosphaera*

(C) *Sphaerotheca* (D) *Phyllactinia*

559. Powdery mildew of grape is caused by:

(A) *Uncinula* (B) *Podosphaera*

(C) *Sphaerotheca* (D) *Phyllactinia*

560. Powdery mildew of gooseberries is caused by:

(A) *Uncinula* (B) *Podoshaera*

(C) *Sphaerotheca* (D) *Phyllactinia*

561. Powdery mildew of lilac is caused by:

(A) *Uncinula* (B) *Microsphaera*

(C) *Sphaerotheca* (D) *Phyllactinia*

562. *Plasmodiophora brassicae* is most severe at pH:

(A) 8.0 (B) 7.0

(C) 7.5 (D) 5.7

563. Which forms cushion like hyphae before penetrating hyphae into the host cells:

(A) *Rhizoctonia solani* (B) *Erysiphe polygoni*

(C) *Venturia inaequalis* (D) *Fusarium oxysporum*

564. The axenic culture of *Puccinia graminis tritici* was first reported by:

(A) W.M. Cutter (B) M.D. Coffey

(C) H.H. Hotson (D) Williams and coworkers

565. Rishitin is a phytoalexin produced by:

(A) Potato (B) Pepper

(C) Clover (D) Bean

566. Elicitors are determinates of pathogens:

(A) Avirulence (B) Virulence

(C) Susceptibility (D) None of these

567. A disease occurring every year in a locality in moderate to severe form is known as:

(A) Epidemic (B) Endemic

(C) Sporadic (D) Pandemic

568. The time spent between infection and appearance of disease symptoms is:

(A) Latent infection (B) Latent period

(C) Infection period (D) Dormant period

569. The process by which an organism undergoes self dissolution is:

(A) Automise (B) Autolysis

(C) Attenuation (D) Autotrophism

570. Rigid cleistothecial appendages with coiled tips are characteristics of:

(A) *Leveillula* (B) *Uncinula*

(C) *Phyllactinia* (D) *Erysiphe*

571. The sporangiophores are dichotomously branched at acute angle and taper to curved pointed tips in the genus:

(A) *Plasmopara* (B) *Sclerospora*

(C) *Basidiophora* (D) *Peronospora*

572. A substrate inhibiting the growth of certain microorganisms in plants in response to a number of chemicals, biological and physical stimuli is:

(A) Phytoalexins (B) Phytotoxin

(C) Elicitors (D) None of these

573. Which one of the following is the most specific toxin?

(A) Tentoxin (B) Taboxin

(C) Alfatoxin (D) Vivotoxin

574. A violent reaction in which the invaded tissue upon attack by pathogen, dies promptly preventing further spread of infection is:

(A) Hyperplasia (B) Hypertrophy

(C) Hypersensitivity (D) Hyperparasitism

575. A polymer of esterified fatty acids linked by peroxide and other linkage is:

(A) Cutin (B) Cytokinin

(C) Chitin (D) Calose

576. Degree of pathogenicity of an isolate or a race of the pathogen is termed as:

(A) Virulence (B) Infectivity

(C) Pathogenesis (D) Aggressiveness

577. Physiological races of the pathogen were first reported by:

(A) E.C. Stakman (B) Blackeslee

(C) Biffen (D) Smith

578. Genetically identical populations of individuals are called:

(A) Serotype (B) Physiotype

(C) Biotype (D) Pathotype

579. Establishment of the pathogen inside the host following penetration is termed:

(A) Pathogenesis (B) Infection

(C) Inocutation (D) Invasion

580. Zoospores are biflagellate in genus:

(A) *Pythium* (B) *Peronospora*

(C) *Synchytrium* (D) *Peronosclerospora*

581. Leaf shredding symptoms are observed in:

(A) Leaf blight of wheat (B) Downy mildew of bajra

(C) Rust of jowar (D) Downy mildew of jowar

582. Clistothecial appendages are hypha like in genus:

(A) *Uncinula* (B) *Erysiphe*

(C) *Microsphaera* (D) *Phyllactinia*

583. Alternate host of *Puccinia penniseti* is:

(A) *Solanum melongena*
(B) *Bromus japonicus*
(C) *Oxalis corniculata*
(D) *Thalictrum polyganum*

584. When aecial stage in rust life cycle is omitted then that life cycle is known as:

(A) Eu type
(B) Brachy type
(C) Opsistype
(D) Endotype

585. Ascocarp is not produced in the genus:

(A) *Taphrina*
(B) *Erysiphe*
(C) *Leveillula*
(D) *Claviceps*

586. Teleutospores remain united in solid balls in the gems:

(A) *Tilletia*
(B) *Ustilago*
(C) *Taphrina*
(D) *Tolyposporium*

587. Basidiospores are sessile in the genus:

(A) *Sphacelotheca*
(B) *Uromyces*
(C) *Puccinia*
(D) *Melampsora*

588. To control monocyclic disease it is important to:

(A) Reduce the primary inoculums
(B) Reduce the secondary spread
(C) Check both
(D) None of these

589. Tyloses formed in xylem vessels of plants during wilting by most of the vascular pathogens are:

(A) Overgrowth of xylem vessels themselves
(B) Overgrowth of phloem cell
(C) Gum deposits
(D) Overgrowth of protoplast of adjacent living parenchymatous cells protruding into xylem vessels through pits:

590. Red scaled onion varieties are resistant to onion smudge due to the presence of:

(A) Maleic acid
(B) Citric acid
(C) Catechol acid and catechuic acid
(D) Chlorogenic acid

591. When uredial stage in rust life cycle is omitted then that life cycle is known as:

(A) Eu type (B) Opsistype

(C) Brachy type (D) Endotype

592. When aecial and uredial stage in rust life cycle is omitted and teliospores not resting then that life cycle are known as:

(A) Lepto type (B) Hemi type

(C) Micro type (D) None of these

593. The resting teliospores in rust life cycle are known as:

(A) Lepto type (B) Hemi type

(C) Micro type (D) None of these

594. When karyogamy in teliospore in rust life cycle is followed then that life cycle is known as:

(A) Eu type (B) Lepto type

(C) Hemi type (D) None of these

595. When karyogamy in aeciospores in rust life cycle is followed then that life cycle is known as:

(A) Endotype (B) Hemi type

(C) Eu type (D) Microtype

596. When incomplete life cycle is found in many tropical rust then that life cycle is known as:

(A) Endotype (B) Hemi type

(C) Eu type (D) Microtype

597. Match the book "Out line of the History of Plant Pathology":

(A) Whetzel (1918) (B) Large (1940)

(C) Orlob (1971) (D) None of these

598. Match the book "The Advance of Fungi":

(A) Whetzel (1918) (B) Large (1940)

(C) Orleb (1971) (D) None of these

599. The book entitled "The Disease of Cultivated Crops: Their Causes and Their Control" is written by:

(A) Alexopolus (B) J.G. Kuhn

(C) Burnett (D) None of these

600. In 1956, who published an authoritative book entitled "Principles of Fungicidal Action":

(A) J.G. Horsfall (B) J.G. Kuhn

(C) Burnett (D) G. Rangaswami

601. Who published a very useful book "Principles of Plant Disease Management" in 1982:

(A) Burnett (B) J.G. Kuhn

(C) Fry (D) Owen

602. Vidyasekaran (1998) is known for his contribution in:

(A) Molecular biology of pathogenesis and induced systemic resistance

(B) Extracellular enzyme

(C) Tissue culture in Plant Pathology

(D) None of these

603. Who was the first to elucidate the biochemical mechanism of mutation?

(A) K.K. Reddi (B) Mundry

(C) K.M. Smith (D) None of these

604. Who was the first to review the work on Indian Forest Mycology and Plant Pathology in India?

(A) Das Gupta (B) Mehta

(C) Bagchee (D) Harting

605. Who identified the casual organism of red rust of tea in Assam caused by *Cephaleurous virescens*?

(A) G.B. Bisbay (1931) (B) D.D. Cunninham (1989)

(C) Mehta (1963) (D) None of these

606. The major contribution of Puskharnath is:

(A) Seed Pathology (B) Ascomycologist

(C) Potato disease (D) None of these

607. Book 'Fungicide in Plant Disease Control' is written by:

(A) G. Rangaswami (B) S.P. Kapoor

(C) V.P. Bhinde (D) Y.L. Nene

608. K.S. Bilgrami is known as for his contribution in:

(A) Mycotoxin (B) Diseases of oil seed

(C) Mycorrhiza (D) None of these

609. Who established the school of Mycology and Plant Pathology at Madras University:

(A) K.J. Narasimhan (B) J.C. Luthra

(C) T.S. Sandasivan (D) C.V. Subramanian

610. The first Indian mycologist is:

(A) E.J. Butler (B) T.S. Sadasivan

(C) V.V. Subramanian (D) C.M. Manoharachary

611. Who has received the role of plant disease clinics in disease diagnosis?

(A) Barnes (1994) (B) Agrios (1997)

(C) Nene (1987) (D) Williams (1955)

612. Who have discussed the importance of computer based diagnosis of plant disease:

(A) G.N. Agrios (1997) (B) Barries (1994)

(C) Nene (1987) (D) Ariena and Bruggen (1991)

613. The term 'appressorium' was introduced by:

(A) Frank (1883) (B) Hasselbring (1906)

(C) Peries (1962) (D) None of these

614. *Plasmodiophora brassicae*, the cause of finger and the toe disease of crucifers, often enters through:

(A) Root hairs (B) Lenticels

(C) Stomata (D) Trichomes

615. *Uromyces pisi* is the rust fungi which enter through:

(A) Lenticels (B) Stomata

(C) Unfolding buds (D) None of these

616. The peach leaf curl fungus *Taphrina deformans* enters its host through the:

(A) Stomata (B) Unfolding leaf buds

(C) Lenticels (D) None of these

617. The *Fusarium oxysporum* f. sp. *pisi* enters through:

(A) Lenticels (B) Cotylendonary bundles

(C) Root hairs (D) Trichomes

618. Match Gaeumannomyces:

(A) *Ophiobolus* (B) *Corticium*

(C) *Thielaviopsis* (D) None of these

619. Match *Pyricularia oryzae*:

(A) *Pyricularia grisea*
(B) *Pyricularia polymorpha*
(C) *Pyricularia rosea*
(D) None of these

620. Enzyme Phospholipase B is produced by:

(A) *Thielaviopsis*
(B) *Helminthosporium*
(C) *Pyricularia*
(D) None of these

621. Book 'Toxins in Plant Pathogenesis' is written by:

(A) Strobel (1974)
(B) JM Daly and BJ Deverall (1983)
(C) Ludwig (1960)
(D) Wheeler and Luke (1963)

622. Fusaric acid was first isolated as the metabolic product of *Fusarium heterosporum* Nees, non-specific parasite, by:

(A) E.F. Smith (1952)
(B) Diamond (1953)
(C) Yabuta *et al.* (1934)
(D) Gaumann (1954)

623. Lycomarasmin is a wilt toxin of the tomato wilt pathogen *Fusarium oxysporum* f. sp. *lycopersici* Sacc described by:

(A) Yabuta *et al.* (1934)
(B) Diamond (1953)
(C) Clauson-kass *et al.* (1944)
(D) Sanwal (1961)

624. Tab toxinine is toxic to cells because it inactivates the enzyme:

(A) Glutamine
(B) α-glactosidase
(C) Glucosidase
(D) None of these

625. T-toxin is produced by race T of:

(A) *Helminthosporium (Cochliobolus) heterostrophus*
(B) *Colletotrichum fuscum*
(C) *Endothia parasitica*
(D) None of these

626. Diporthin toxin is produced by:

(A) *Alternaria solani*
(B) *Endothia parasitica*
(C) *Phyllosticta maydis*
(D) *Rhizopus* sp.

627. Match Hydrogen cyanide (HCN):

(A) Pathotoxin
(B) Vivotoxin
(C) Phytotoxin
(D) Phaseolo toxin

628. The green island formation in pinto bean leaves is infected with:

(A) *Uromyces phaseoli* (B) *Pseudomonas solanacearum*

(C) *Ustilago zeae* (D) None of these

629. Phytoalexin glyceollin is produced in Soybean plants infected with the fungus:

(A) *Phytophthora megasperma* f. sp. *glycinea* (B) *Phytophthora infestans*

(C) *Monilinia fructicola* (D) None of these

630. Which phytoalexin was isolated from the parsnip root discs inoculated with *Ceratocystis fimbriata*?

(A) Capsidiol (B) Gossypol

(C) Rishitin (D) Xanthotoxin

631. Defense genes quiescent in:

(A) Healthy plant (B) Disease plant

(C) Rooted plant (D) None of these

632. Adaptation in fungi has been reviewed by:

(A) Pantecorvo (1963) (B) Sermonti (1954)

(C) Buxton (1960) (D) None of these

633. Hybridization is encouraged by incompatible mechanism which prevents or reduces:

(A) Inbreeding (B) Hybridization

(C) Mutation (D) None of these

634. This size of urediniospores of *Puccinia graminis tritici* is:

(A) 32 x 20 µm (B) 27 x 17 µm

(C) 28 x 20 µm (D) 24 x 17 µm

635. In *Puccinia striformis,* how many physiological races are available in India?

(A) 10 (B) 14

(C) 16 (D) 18

636. In *Puccinia recondita,* how many physiological races are available in India?

(A) 20 (B) 22

(C) 25 (D) 28

637. In *Puccinia graminis tritici,* how many physiological races are available in India?

(A) 28 (B) 30

(C) 32 (D) 34

638. Match "Fungal Spores":

(A) Ingold (1971)
(B) Muskett (1960)
(C) Broadbent (1960)
(D) None of these

639. The first example of an important plant disease epidemic in India was the:

(A) Brown spot of rice
(B) Stem rust of wheat
(C) Damping off of cotton
(D) Coffe rust

640. *Armillaria mellea* can be eliminated by:

(A) CS_2
(B) PCNB
(C) Formaldehyde
(D) None of these

641. Match *Cryphonectria*:

(A) *Endothia*
(B) *Agrobacterium*
(C) *Peniophora*
(D) *Heterobasidion*

642. Best example of splash dispersal:

(A) Coffee rust
(B) Angular leaf spot of cotton
(C) Red rust of mango
(D) None of these

643. The hyphovirulent strains carry virus like:

(A) ssRNA
(B) dsRNA
(C) ssDNA
(D) dsDNA

644. In wheat rust resistance is completely suppressed by a:

(A) Dominant inhibitor gene
(B) Recessive inhibitor gene
(C) Dominant gene
(D) None of these

645. The first classical example of hybridization is provided by:

(A) Muller (1920)
(B) Howard (1904)
(C) Vavilov (1950)
(D) William A Orton (1907)

646. Who showed that temperature plays an important part in the onset of wilt disease of cabbage :

(A) L.R. Jones
(B) Leppik
(C) Zhukovsky
(D) None of these

647. Wilt disease of cabbage is caused by:

(A) *Fusarium oxysporum* f. sp. *lini*
(B) *Fusarium oxysporum* f. sp. *conglutinanas*
(C) *Fusarium oxysporum* f. sp. *lycopersici*
(D) *Fusarium oxysporum* f. sp. *solani*

648. In which of the following wheat varieties is resistant against all reces (*Puccinia garminis tritici*) but high degree of sterility which prevented progress in hybridization:

(A) Khapli emmer
(B) Karned
(C) Kota
(D) Imillo

649. Match Phylloticta:

(A) *Rhizoctonia*
(B) *Ascochyta*
(C) *Verticillium*
(D) None of these

650. Root rot of gram is caused by:

(A) *Rhizoctonia bataticola*
(B) *Verticillium lecani*
(C) *Fusarium undum*
(D) *Cercospora cruenta*

651. Root rot of groundnut is caused by:

(A) *Rhizoctonia bataticola*
(B) *Verticillium leani*
(C) *Fusarium udum*
(D) *Rhizoctonia destrens*

652. Paragynous antheridia always present in the genus:

(A) *Pythium*
(B) *Basidiophora*
(C) *Helminthosporium*
(D) *Phytophthora*

653. The book "Phytophthora Disease in India" written by:

(A) Mishra and Hall
(B) Mehrotra and Aggarwal
(C) Ram Nath
(D) None of these

654. Leaf spot on pea is caused by:

(A) *Alternaria alternata*
(B) *Curvularia lunata*
(C) *Mycosphaerella pinodes*
(D) None of these

655. In which of the following, high amounts of nutrients seem to increase the severity of rice blast caused by *Magnaporthe grisea*:

(A) Phosphorus
(B) Potassium
(C) Nitrogen
(D) Calcium

656. The epidemic rate 'r' for potato late blight is:

(A) 1.6 units per day
(B) 0.3-0.5 units per day
(C) 0.15 units per day
(D) 0.02 units per day

657. Cedar apple rust completes a reproductive cycle in:

(A) One year
(B) Two year
(C) Three year
(D) Four year

658. Match *Chalara elegans*:

(A) *Thieleviopsis basicola*
(B) *Phytophthora nicotianae*
(C) *Colletotrichum lagenarium*
(D) None of these

659. Match *Leucostoma*:

(A) *Cystospora*
(B) *Phytophthora*
(C) *Colletotrichum*
(D) None of these

660. Non-motile asexual spores formed in terminal sporangia is the main characteristic of the order:

(A) Peronosporales
(B) Mucorales
(C) Saprolegniales
(D) Plasmodiophorales

661. Match *Epichloe*:

(A) *Acremonium*
(B) *Balansia*
(C) *Atkinsonella*
(D) *Myriogoenospora*

662. Match *Guignardia*:

(A) *Spilocaea*
(B) *Rhabdocline*
(C) *Phyllosticta*
(D) None of these

663. Ascocarps are black, spherical, discoid or elongate and are produced in stomata are main characters of the order:

(A) Phytismales
(B) Helotiales
(C) Peronosporales
(D) Mucorales

664. The pathogens *Plasmopara viticola* overwinters as oospores in:

(A) Dead leaf lesion and shoots
(B) Seeds
(C) Roots
(D) None of these

665. Basidiorarp lacking, basidia produced on surface of parasitized tissue is the main characters of the genus:

(A) *Thanatephorus*
(B) *Armillaria*
(C) *Cronartium*
(D) *Exobasidium*

666. Basidiocarp is web like, inconspicuous, basidia without cross walls, with four prominent sterigmata present in the genus:

(A) *Thanatephorus*
(B) *Armillaria*
(C) *Cronartium*
(D) *Exobasidium*

667. Crown wart of alfalfa disease is caused by the genus:

(A) *Synchytrium*
(B) *Physoderma*
(C) *Urophlyctis*
(D) *Olpidium*

668. Development of disease can be mathematically expressed by an equation propounded by Vander Plank (1963) is:

(A) X= Xoert (B) X=Xgert

(C) X=Xeort (D) None of these

669. Zygospore is the overwintering or resting stage of the fungus, when it germinate and produces a:

(A) Zygosporangium (B) Oogonium

(C) Sporangiophore (D) Conidia

670. Match *Blumeriella* sp:

(A) *Higginsia* sp. (B) *Cochliobolus* sp.

(C) *Microcyclus* sp. (D) None of these

671. Match the genus *Microcyclus*:

(A) *Dothidella* (B) *Higginsia*

(C) *Elytroderma* (D) *Lophodermium*

672. Germ tubes of the pathogen *Glomerella* sp. Penetrates uninjured tissue:

(A) Wound (B) Directly

(C) Lenticels (D) None of these

673. *Fusarium oxysporum* survive in the soil for long time as only:

(A) Microconidia (B) Macroconidia

(C) Chlamydospores (D) Acervulus

674. *Verticillium dahliae* overwinters in the soil as microsclerotia, which can survive up to:

(A) 15 Years (B) 18 Years

(C) 20 Years (D) 25 Years

675. In dying or dead trees which type of spores produces by the mycelium *Ophiostoma ulmi*:

(A) Sporothrix-type (B) Graphium-type

(C) Both (A) and (B) (D) None of these

676. The fungus *Rhizoctonia*, the optimum temperature for infection is about:

(A) 15-18°C (B) 20-25°C

(C) 28-30°C (D) None of these

677. The book entitled "The Advance of Fungi" is written by:

(A) Burnett (B) De Barry

(C) E.C. Large (D) Brodie

678. Protein degrading enzymes (proteolytic enzymes) degrade proteins to form:

(A) Starch (B) Polypeptide

(C) Glucose (D) Amylase

679. Starch is hydrolyzed by enzyme, amylase to produce________used by the pathogens as nutrients:

(A) Starch (B) Glucose

(C) Lipid (D) Proteins

680. If there is no known differential interaction among genotypes of the host and genotypes of the pathogen called:

(A) General resistance (B) Specific resistance

(C) Dilatory resistance (D) None of these

681. Who was reported the casual organism of linseed wilt as *Fusarium lini* (*F. oxysporum* f. sp. *lini*):

(A) H.L. Bolley (1901) (B) W.A. Orton (1900)

(C) R.H. Biffen (1905) (D) E.C. Stakman (1918)

682. The cob length may be normal but only lower parts is converted into a leaf mass and upper portion bears grains is the symptom characteristics of :

(A) Downey mildew of pearl millet (B) Downy mildew of pea

(C) Downey mildew of maize (D) None of these

683. Shredding of leaves is the most characteristics symptoms of the disease:

(A) Downey mildew of maize (B) Downey mildew of pea

(C) Downey mildew of sorghum (D) Powdery mildew of apple

684. Which of the following disease is common in the Indo-Gangetic plain:

(A) Downey mildew of grape vine (B) Downey mildew of pea

(C) Downey mildew of sorghum (D) None of these

685. Match the genus Physalospora:

(A) *Botryosphaeria* (B) *Venturia*

(C) *Monilinia* (D) *Diporthe*

686. Rhizomorphs producing fungi is:

(A) *Alternaria* (B) *Colletotrichum*

(C) *Armillaria* (D) None of these

687. *Heterobasidion annosum* causing:

(A) Seed rot (B) Seedling rot

(C) Root rot and butt rot of forest tree (D) None of these

688. Match *Fumes*:

(A) *Inonotus* (B) *Ganoderma*

(C) *Peniophora* (D) *Heterobasidion*

689. *Alternaria alternata* survives in barley and wheat seeds for:

(A) 2 years (B) 5 years

(C) 8 years (D) 10 years

690. Who introduced the term hypersensitivity?

(A) E.C. Stakman (B) H.M. Ward

(C) Smith (D) None of these

691. The possibility of *Phytophthora palmivora* the cause of bud rot of palm survives during the dry month in:

(A) Leaf (B) Stem

(C) Root (D) All of the above

692. Pleomorphic life cycle present in:

(A) Smut (B) Bunt

(C) Rust (D) None of these

693. Which of the following major contribution of Sadasivan (1961):

(A) Physiology of the wilt disease

(B) Disease cycle of downy mildew

(C) Toxin theory

(D) None of these

694. Who wrote a book "The Physiology of the Wild Diseases":

(A) Sadasivan (1961) (B) Ludwing (1952)

(C) Gaumann (1951) (D) Talboy (1958)

695. Leaf falling in rubber plant due to cause by:

(A) *Phytophthora capsici* (B) *Phytophthora palmivora*

(C) *Macrophomina phaseolina* (D) None of these

696. Match *Paecilomyces*:

(A) *Eurotium* (B) *Setosphaeria*

(C) *Byssochlamys* (D) *Hypocrea*

697. What is the mode of development of conidia of *Bipolaris*, causing leaf spot on grasses:

(A) Porosporus conidia produced apically and laterally from determinate

(B) Phialidic blastic dry

(C) Thallic mode of conidiogenesis

(D) None of these

698. Match *Alternaria*:

(A) *Aethelium* (B) *Lewia*

(C) *Hypocrea* (D) None of these

699. The book "Principles of Seed Pathology" is written by:

(A) Agarwal and Sinclair (B) Singh and Trivedi

(C) Neergaard (D) Nene and Agarwal

700. The book entitled "Basic Plant Pathology Methods" is written by:

(A) Dhingra and Sinclair (B) Tuite

(C) Boudoin *et al.* (D) None of these

701. The book entitled "Plant Pathological Method-Fungi and Bacteria" is written by:

(A) Dhingra and Sinclair (B) Tuite

(C) Boudoin *et al.* (D) None of these

702. How many histones or histone like protein present in eukaryotes:

(A) 8 (B) 5

(C) 10 (D) 15

703. Match chlorosis:

(A) Hypoplasia (B) Hyperplasia

(C) Hypertrophy (D) None of these

704. A more general and more flexible plant disease simulator for stripe rust of wheat called:

(A) EPIDEMIC (B) CERCOS

(C) MYCOS (D) EPICORN

705. Match *Geotrichum candidum* (Sour rot of fruit and vegetable):

(A) Saccharomycetales (B) *Eurotium*

(C) Byssochlamys (D) *Cochliobolus*

706. Match *Exerohilum*:

(A) *Setosphaeria* (B) *Eurotium*

(C) *Byssochlamys* (D) *Cochliobolus*

707. Match *Chalara*:

(A) *Setosphaeria* (B) *Ceratocystis*

(C) *Byssochlamys* (D) *Cochliobolus*

708. Match *Rosellinia*:

(A) *Dematophora* (B) *Ceratocytis*

(C) *Byssochlamys* (D) *Cochliobolus*

709. Who coined the word "microbe":

(A) Se Dillot (1878) (B) Clayton 1934)

(C) Hershey (1952) (D) None of these

710. Size of eukaryotic flagellum up to:

(A) 200 μ long and 2000 A° in diameter

(B) 4-5 μ long 120 A° in diameter

(C) 500 μ long and 4000 A° in diameter

(D) None of these

711. The pili help in adhesion of bacterial cells during conjugation and to the water surface for better aeration. When they are bring about adhesion of bacterial cells with red blood cells, the phenomena is called:

(A) Hemaglutination (B) Slide agglutination

(C) Ambisexualis (D) None of these

712. *Geotrichum citri-aurantii* cause past harvest disease of citrus fruit invades through:

(A) Wound (B) Stomata

(C) Hydethodes (D) None of these

713. C.V. Subramanian (1961) is known for his contribution in:

(A) Discomycetes (B) Oomycetes

(C) Hyphomycetes (D) Pyrenomycetes

714. Who discovered parasexuality in fungi?

(A) Smith (B) Fischer

(C) Ervin (D) Pontecorvo

715. The tip of the branches is expanded into cup shaped apophyses with four sterigmata each bearing the sporangia along their margins in:

(A) Bremia (B) Bremiella

(C) Basidiophora (D) Peronospora

716. Bunchy top of groundnut was reported in India in:

(A) 1940 (B) 1955

(C) 1966 (D) 1977

717. Amylase enzyme act on:

(A) Starch (B) Protein

(C) Lipid (D) None of these

Fill in the Blanks with Correct Answer

1. In 2000________has discussed 'Plant Pathology and Indian Agriculture Past, Present and Future'.

   Nagarajan

2. A monograph entitled '**Further Studies on Cereal Rust in India**' was published by________in 1940

   K.C. Mehta

3. According to ________the evolution of the plant disease clinic arose from a perceived need for organized systematic professional efforts to assist in the identification of plant disease problem.

   Barnes (1994)

4. *Erwinia amylovora* enters through ________of apple and pear flowers.

   Nectaries

5. Degradation of RNA by *Phytophthora infestans*, the late blight pathogen, was demonstrated by

   Page (1959)

6. Lycomarasmin and alternaric acid is best example of________

   Phytotoxins

7. Piricularin toxin produced by *Pyricularia grisea* is a typical________

   Pathotoxin

8. Several disease resistant genes have been cloned and characterized by

   Baker et al. (1997)

9. Changes in the pathogenicity of *Fusarium oxysporum* f. sp. *pisi* is brought about by________________

   Exudates

10. The effect of environmental factor of plant disease have been reviewed by________________

    Cohoun (1973)

11. Autonomous dispersal by plant pathogens have been discussed by.

    Muskett (1960)

12. Autonomous transmission takes place by active growth of

    Hyphae or hyphal strands

13. The rate of autonomous transmission in *Phymatotrichum omnivorous*, root rot of cotton is estimated at________ per season and ________per month in alfalfa crop.

    5 to 30 feet; 2 to 8 feet

14. Wind dispersal pathogen is termed ________given by Gaumann (1950).

    Anemochory

15. Monogenic resistance is governed by ________ gene.

    One

16. Corn with Cms-T cytoplasm is more susceptible to yellow leaf blight cuased by

    ***Phyllosticta* sp.**

17. Inhibitor genes affecting recessive genes for rust resistance are also known to be presented in________

    Maize

18. Book '**Phytophthora Disease Worldwide**' is written by

    Erwin and Ribeiro (1996)

19. Sporagia inflated, lobulate branched or unbranched, zoosopores 7-12 µm, oogonia spherical, terminal, anthridia monoclinous or diclinous intercalary or terminal dome shaped oogonium, oospores aplerotic is the main characteristics of the genus

    Pythium aphamnidermatum

20. The epidemic rate for wheat stem rust is________

    0.3 to 0.6 units per day

21. Perosyacetyl nitrate (PAN) may injure plants in concentrations as low as

    0.01 to 0.02 ppm

22. Rust of oak is caused by________

    Cronatrium fusiforme

23. Motile gametes of equal or unequal size fuse to form ________

    Meiosporangia

24. Fragments of hyphae and hard masses of mycelium are known as.

    Sclerotia

25. ________produces a plasmodium or plasmodiums like structure.

    Maxomycota

26. A naked amorphous plasmodium body is the main characteristics of________

    Slime molds

27. Some Ascomycetes produces conidiophores on a cushion shaped stroma of mycelium and the whole structure is called as________

    Sporodochium

28. When environmental conditions or nutrition becomes unfavorable the fungus may produce________ containing one or few asci.

    Cleistothecia

29. Older plants infected with *Verticillium* are usually________, and their vascular tissues show characteristics.

    Stunting, Discoloration

30. ________are signals, presumably chemical signals generated by an organism to bring about metabolic shifts in another organism.

    Elicitors

31. Lipids are degraded by enzyme________

    Lipase

32. The first symptoms appear as creamy droplets of a sticky liquid exuding from young florets of infected heads by ________

    Claviceps purpurea

33. The conidia of *Claviceps purpurea* exude from the young flarets creamy droplets droplets known as the ________stage.

    Honeydew

34. When the fungus *Ophiostoma ulmi* reaches the large xylem vessels of the spring wood, it produces________

    Sporothrix-type spores

35. In dying or dead trees, the mycelium produces mostly

    Graphium-type spores

36. Sclerotia also produce occasional conidia on sporodochia, these conidia, however, seems to be________

    Sterile

37. ________ first observed loose smut hyphae in wheat seed embryo.

    Maddox (1896)

38. Teliospores of *Ustilago nuda* germinate on the stigma/ovary surface and form dikaryotic promyceila (hyphae) which penetrates the ________

    Ovary wall

39. *Drechelera* sp. survives in oats, barley, paddy and wheat for________years

    10

40. Teliospores of *Ustilago zeae* carried on seed surfaces of maize as well as sporidia may send out germ tubes that directly penetrate young

    Epidermal cells

41. Yellowing due to lack of light is called.

    Etiolation

42. Facultative parasites are basically________

    Saprophytic

43. The soil borne obligates parasites are few are few such as.

    Plasmodiophora brassicae* and *Synchytrium endobioticum

44. Most of the characters of a simple interest disease are found in.

    Slow epiphytotics

45. Host cell necrosis restricts the spread of the parasite beyond the necrotic zone even without causing________

    Latter's death

46. The species of the *Physodrma* are obligate parasites on vascular plants producing thick-walled and thin-walled slipper shaped________

    Sporangia

47. *Psilocybe maxicana* is________

    Hallucinogens

48. Loss of pathogenic virulence in culture is sometimes called________

    Attenuation

49. ________is the ability of a pathogen to cause the disease.

    Pathogencity

50. ________is a measure of degree of pathegenicity of an isolate or race of a pathogen.

    Virulence

51. ________is the capacity of the pathogen/parasite to invade and grow in its host plant and to reproduce on or in it.

    Aggressiveness

52. When a progeny of the pathogen exhibited characteristics that are different from those present in the parental individual, it is called________

    Variant

53. ________is a population of individuals which are genetically identical produced by variant?

    Biotype

54. ________is the strain of a pathogen virulent toward specific resistant gene.

    Pathotype

55. Elicitors are signals, presumably chemical signals, generated by an organism to bring about________ in another organism.

    Metabolic shifts

56. Varksha-Ayurveda, a book by Surapal in ancient India is the ________in which much light has been thrown on plant.

    First book

57. In 1675, the Dutch worker ________developed the first microscope and in 1683 used it to describe bacteria seen with this.

    Leeuwenhoek

58. The French scientist Prevost proved in 1807 that ________ are caused by microorganisms.

    Diseases

59. During 1830-1845 ________of potato spread fast in England, Ireland and continental Europe.

    Late blight

60. Tulasne brothers of France (R.L. Tulasne and C. Tulasne) produced illustrated descriptions of________causing fungi.

    Rust and smut

61. In 1878, the downy mildew of grape vine was introduced into Europe/France from________

    America

62. The first Indian University established in 1857 at Kolkata, Mumbai and Madras emphasized ________of fungi.

    Taxonomy

63. In 1830, the late blight of potato was introduced into Europe from________

    South America

64. In 1940, the bunchy top of banana was introduced in India from________

    Sri Lanka

65. In 1953, the wart disease of potato was introduced into India from________

    Netherland

66. In 1879, the leaf rust of coffee was introduced into India from________

    Sri Lanka

67. In 1940, the fire blight of apple and pear was introduced into India from________

    England

68. During 1850-1875 A.D. Cunningham and A. Barclay started ________of fungi in India itself.

    Identification

69. ________was first Indian scientist who collected and identified fungi in this country.

    KR Kirtikar

70. T.S. Sadasivan's school contributed to the concept of ________ And worked out the mechanism of wilting in cotton due to *Fusarium vasinfectum*

    Vivotoxin

71. Typical plant diseases are caused by inanimate (non-living) or________causes.

    Animate (living)

72. A________is a compact, often hard mass of dormant fungus mycelium.

    Sclerotium

73. ________diseases are characterized by drying of the entire plant.

    Wilt

74. In dieback or whither tip diseases plant organs especially stem or branches dry from the________

    Tip backwards

75. Certain infectious pathogens, such as viruses are neither seed nor can be grown on________

    Artificial media

76. ________parasites are basically saprophytic in their feeding mechanism.

    Facultative

77. The infection by________parasites never kills the host.

    Obligate

78. The chain of events leading to disease development is referred to as_______

    Pathogenesis

79. Sum total of all symptoms characterizing a disease called_______

    Disease syndrome

80. When the pathogens survive on only one gene or species of plants it is called_______

    Homogenous infection chain

81. Most of the wilt causing species of *Fusarium* are _______

    Host specific

82. A living organism harboring a parasite is called_______

    Host

83. When a pathogen present in the host does not produce symptoms, the host is regarded as its_______

    Carrier

84. A mutually beneficial association of two or more different kind of organism are called as _______

    Symbiosis

85. Club rot of crucifers is caused by_______

    Plasmodiophora brassicae

86. Mature particles of a plant viruses is generally called_______

    Virion

87. Whole infective virus particles are_______

    Nucleotide

88. In red rot of sugarcane, causal fungus is *Colletotrichum falcatum* and its primary infection comes from_______

    Diseased setts

89. The stage in the growth and development of an organism is referred to as_______

    Life cycle

90. The primary incolum of rice spot (*Helminthosporium oryzae*) and blast (*Pyricularia) oryzae*) is believed to come from_______

    Wind borne spores

91. ________fungi over-summer in hills from where come the primary inoculums through wind or insects.

    Cereal rusts

92. The role of alternate host is not as important as of________hosts.

    Collateral

93. Generally, the soil mycoflora reduce saprophytic activity in the process of________survival of fungi.

    Saprophytic

94. ________are organisms which kill the host tissues in advance of penetration and then lives saprophytically.

    Necrotroph or perthotrophs

95. ________is the sexual stage of the life cycle of fungus.

    Perfect stage

96. When the pathogen rapidly grows in the host tissues is called________

    Invasion

97. The establishment of a parasite within the host plant is called as________

    Infection

98. The degree of infectivity is called as________

    Inoculum potential

99. Fungi have declining saprophytic phase limiting their survival in soil________

    Root inhabiting

100. The.________ are the main structures for dormant survival.

    Spores

101. ________is extreme degree of susceptibility in which rapid death of the cells in the vicinity of the invading pathogen occurs.

    Hypersensitivity

102. ________ is the descendants of the single isolation in pure culture; an isolates.

    Strain

103. ________cannot be infected by a given pathogen.

    Immune

104. ________is the inability of a plant to resist the effect of a pathogen or any other damaging factor.

    Susceptibility

105. Fungal species like *Pythium, Phytophthora* etc. form zoospores provided with ________for locomotion in film of water.

 Flagella

106. The growth of________ fungi in soil is either extremely slow or non-existent.

 Pathogenic

107. ________is the completely effective against some races of as pathogens but not against others.

 Vertical resistance

108. ________is the partial resistance, equally effective against all races of a pathogen.

 Horizontal resistance

109. ________is the science of the cause of disease.

 Aetiology

110. The________ can carry the fungal propagules on their body and help in entry through punctures for their own entry.

 Nematode

111. Dissemination of red rot fungus and wilt organism in active or dormant state may be________by agency of water.

 Transported

112. The uredospores of *Puccnia graminis* are disseminated by________

 Wind

113. H.V. toxin produced by the fungus________

 Helminthosporium victoriae

114. Direct penetration by penetration peg (fungi) produced by________

 Appresoria

115. Fungal species like *Pythium, Fusarium, Rhizoctonia* are saprophytes and ________

 Soil inhabitant

116. ________is the partly responsible for several bacterial galls of plants.

 Cytokinin

117. Only uredia and telia are formed on the________

 Wheat

118. ________are repeating spores.

 Uredospores

119. The uredospores or *Puccinia graminis va.r tritici* have been detected as high as 1400 feet above________wheat fields.

 Infected

120. In Australia, the introduction of *Puccinia graminis* (Rust causing fungus) has been suspected due to wind borne spores from________and elsewhere.

 India

121. The pathogens embodying enzyme based success have to produce adequate________to produce needed quantum of enzymes for success.

 Hyphae

122. Polycyclic fungus also called________

 Ployetic (Multiyear)

123. Pectic substance degraded by________

 Pectinease or pectolytic enzymes

124. Fusicocin produced by________causes twig blight in almond and peach.

 Fusicoccum amygdoli

125. Rhizomorphic fungus________can produce sufficient amount of hyphae to produce desired amount of enzymes for success.

 Armillaria mellea

126. The________fungi are usually more aggressive and endoparasitic nematodes also attack roots.

 Vascular wilt

127. Pathogens for direct penetration have to neutralize the natural________to establish in the host plant system.

 Barriers

128. The glands in leaf hairs of chickpea contain________which is antifungal and provides resistance against rust fungus.

 Maleic acid

129. The tendency of treatments and conditions acting before inoculum or introduction of biotic and abiotic pathogens is referred to as________

 Predisposition

130. Uredia and telia are formed on the________

 Wheat

131. Pycnia and aecia developed on the alternate host________

 Barberry

132. *Puccinia striformis* is highly susceptible to________

High temperature

133. The manifestation of resistant reaction of the host at a particular stage of growth is termed as ________

Latent infection

134. ________are cell wall component of the pathogen which are capable of inducing phytoalexin synthesis.

Elicitors

135. ________is defined as the time between inoculation and the start of sporulation by the resulting infection.

Latent period

136. The phenomenon of temporarily inhibiting the growth of the fungus is called________

Fungistatis

137. ________deals with breaks out and spread of disease in a population.

Epidemiology

138. ________disease usually occurs widely but periodically in a destructive form.

Eptiphytotic

139. The period of incubation and ________is short in epiphytotics.

Sporulation

140. Stem rust of wheat (*Puccinia graminis tritici*) serves as an example of________ nature disease spread.

Compound interest

141. In simple nature of disease spread ________infection rarely occurs during the same season.

Secondary

142. ________is the best method for the control of slow epidemic.

Crop sanitation

143. Longer the distance from the source of survival of the pathogen, the longer will be time required for build up epiphytotic in a________crop.

Susceptible

144. Continuous use of susceptible variety occupying large contiguous area help in buildup of inoculums and augments chance of________

Epiphytotics

145. High birth rate coupled with low rate of mortality enhance the chance of occurrence of________at a high level.

Epiphytotics

146. Disease severity is the product of inoculums potential and________

Disease potential

147. Inoculum and disease potential vis-à-vis environment favorably or unfavorably influence the development of ________

Epidemics

148. ________is the amount of increase of disease per unit of time in a plant population.

Epidemic rate

149. When it is constantly present in a moderate or serve from and it is confined to a particular country or district referred to as ________.

Endemic disease

150. Wart disease of potato (Synchytrium endobioticum) is________in Darjeeling.

Endemic

151. The most important disease of banana in India is________

Bunchy top

152. Bean rust caused by________

Uromyces phaseoli

153. Bean rust is an ________long cycle rust fungus.

Autoecious

154. Casual agent of leaf rust of coffee is________

Hemeillia vastratix

155. ________occurs at very irregular interval and locations and in relatively few instances.

Sporadic disease

156. ________occurs all over the world and result in mass mortality.

Pandemic disease

157. Bordeaux mixture consist of________and________

CaCo3, lime, water

158. Apple scab is caused by________

Venturia inaequalis

159. ________is the most important fungus associated with guava wilt.

     Gliocladium

160. ________or fungus that complete its entire life cycle on the same host or single host.

     Autoecious rust

161. As a system, epidemic exists in a relatively very small compartment of a much bigger system, the ________

     Agroecosystem

162. A system is regarded as an interlocking complex of process characterized by many reciprocal ________pathways.

     Cause effect

163. ________model could simulate, imitate or mimic an epidemic on a computer provided complete information about the epidemic is fed to the computer.

     Epidem

164. Harmful effect of infection on host physiology comprise the chief caused of.________expression and loss.

     Symptoms

165. The most common symptoms of plant diseases are those caused by ________ of tissues.

     Disintegration.

166. It is believed that________ enzymes mainly help in penetration of infection thread by pressure.

     Cuticular

167. Viruses require________host.

     Living

168. Fungi, bacteria and nematodes embody pectic or pectionolytic enzymes which degrade the components of ________and cell wall.

     Middle lamella

169. ________DNA is not degraded, but cytoplasmic DNA can be degraded by pathogen.

     Chromosomal

170. Viruses are ________under microscopic.

     Not-visible

171. Viruses can be ________by electron microscopic.

     Photographed

172. Usually a single kind of virus infects only________type of host organism.

One

173. ________enzyme separates the amino groups from nucleotide and makes them available to the parasite.

Deaminases

174. Viruses are provided with ________

Nuecleic acid

175. Indole acetic acid (IAA) is produced either by the pathogen in small quantities or by________the plant to produce more of it.

Induce

176. Due to the effect of IAA on oxidative enzyme system of the plant there may occur an abnormal increase in ________ of the tissues.

Respiration

178. Viruses can be culture on ________media.

Living

179. The fungi causing late blight of potato, smut of maize, panama diseases of banana, downy mildew of maize and pearl millet and nematodes causing root knot induce to produce________

IAA

180. The________were discovered by Japanese, while investigating bakanae (foolish seedling) disease of rice caused by *Gibberella fujikuroi*.

Gibberellin

181. The loss from disease is owing to reduction in ________of the plant.

Reproductivity

182. Bacteriopages are those viruses that attack________

Bacteria

183. Bacteriophages possess either DNA or RNA which is present in________

Head

184. Unit of bacterial measurement is ________

Micron

185. As mitochondria are not found in bacterial cell, respiration take place probably in________

Plasma membrane

186. Bacterial cell wall is composed of ________

Diaminopimelic acid

187. The injury or wounds caused by pathogens to plant roots reduce their effective number and proportionally the water________

Uptake

188. The fungi or bacteria causing damping off, stalk rot and canker can enter the ________vessels.

Xylem

189. Bacteria and fungi capable of producing polysaccharides that can induce wilt symptoms ________

In vitro

190. Tyloses are outgrowth of________adjacent to xylem and appear as peg like structure.

Parenchyma

191. ________is a fruiting body of fungus *Claviceps purpurea* which is a parasite on rye and other grasses.

Ergot

192. ________reproduction is the most common process for increasing the number of bacteria.

Vegetative

193. Bacteria that obtain their carbon from organic compounds synthesized by other organisms are________

Heterotrophs

194. The toxin hypothesis with experimental evidence was first provided by metabolites of *Heliminthosporium victoriae*, a fungus which causes________of oats.

Victoria blight

195. A________toxic is a metabolic product of a pathogenic microorganism, selectively toxic only to susceptible host of the pathogen.

Host specific

196. In cereal rusts the ________penetrates through the stoma.

Infection tube

197. The species of ________invade only juvenile tissues in which secondary thickening has not yet developed.

Pythium

198. The ________and post infectionally produced metabolites resistance.

   Phytoalexins

199. ________leaves are known to excrete chemicals that provide resistance to attack of *Botrytis cineria*.

   Tomato

200. ________leaves resistance to *Cercospora* leaf spots possess toxic substance that inhibits the germination of conidia.

   Cowpea

201. Gum deposition in cells along the border of________ often serves as a protective mechanical barrier.

   Diseased tissues

202. Cell wall of most bacteria on its outer side is covered by a gelatinous ________

   Capsule

203. *Bacilli* when present in large numbers with their linear arrangement in one row or chain appear________

   Filamentous

204. ________is the common black mold of breads

   Rhizopus nigricans

205. Black color of the mold is due to ________

   Sporangiospores

206. Most of the common methods of reproduction in *Mucor* is________

   Spore

207. Sexual reproduction in *Mucor* occurs only in between the plant of different________

   Strain

208. Zygospores are produced in ________species of *Mucor*.

   Heterothallic

209. Fungi are well known for their ________role since historical time.

   Harmful

210. The cytoplasmic defense reactions weaken, localize and eliminate the________ bacteria or the endomycorrihal fungi.

   Endophytic nodule

211. Compatible (susceptible) soybean *Phytophthora* produce very little________a pterocarpanoid phytoalexin.

 Glyceollin

212. Elicitors are cell wall components of the pathogens capable of inducing ________synthesis.

 Phytoalexin

213. Chemically different crop residues in rotation help in environmental modification in favor of________and against pathogen.

 Host

214. Fungi are important agent for ________

 Decay

215. Cell wall of the mycelium of most fungi is made up of ________

 Chitin

216. Reserve food material of fungi is in the form of ________

 Glycogen

217. Mycorrhiza is a close association between the root of higher plants and

 Fungi

218. Conidia are produced in ________

 Penicilium

219. CO_2 is given off during the fermentation which make yeast useful for

 Baking

220. Fugal hyphae are usually highly branched and together are known as________

 Mycelium

221. Growth of the hyphae is restricted to the ________of filaments.

 Tips

222. Proteins formed in older parts of the mycelium are known as________

 Cytoplasmic streaming

223. The most common type of reproductive unit of fungi ________

 Spores

224. Most striking morphological feature of yeast is________

 Their unicellular nature

225. Most of the space of yeast cell is occupied by ________

Vacuole

226. Vegetative reproduction in yeast occurs by________

Budding

227. Most common methods of yeast multiplication is ________

Budding

228. Asci are produced in yeast under ________

Unfavorable

229. Mycelium which is not divided by septa is known as ________Mycelium.

Coenocytic

230. The outgrowths of fungal mycelia are known as ________

Haustoria

231. During asexual reproduction spores like cells are cut off from the tips of hyphae which are known as________

Conidia

232. Fungi are more akin to ________than to higher plants.

Algae

233. Chemically RH-124 is ________

4-n-butyl 1, 2, 4-triazole

234. Aureofungin antibiotics produced in submerged culture by________

Streptoverticillium cinnamomeous* var. *tericola

235. Variation in the colors of algae is due to ________

Pigment

236. Agar-agar is a substance obtained from the cell wall of a ________

Algae

237. Green algae have ________

Whiplash

238. Flagellum is an organelle for ________

Locomotion

239. The most important criteria that are taken into consideration for classification of algae is________

Pigment

240. Because of a ________outer wall layer, sporogyra is silky to touch. --

Gelatinous

241. Plant body of algae is known as________

Thallus

242. Algae grow more commonly in ________

Water

243. Small and free floating algae are known as ________

Phytoplankton

244. Algae differ from fungi by the ________of chlorophyll.

Presence

245. Most common food for marine animals is ________

Algae

246. Horizontal resistance governed by ________

Many genes

247. Vertical resistance governed by ________

Single gene

248. Horizontal resistance reduce the ________of pathogens.

Infection rate

249. Vertical resistance delay in the ________

Start of epidemic

250. In 1925, Hilson used orgonomercuriales for the first time in India for the control of ________through seed treatment.

Sorghum smut

251. ________is a condition, in which genetically different nuclei are associated in the same protoplast or the same mycelium.

Heterokaryosis

252. Heterokaryosis is the main source of variation in the________which lacking of sexual.

Imperfect fungi

253. The term heterokaryosis is proposed by.________in 1932, who reported the first time in *Botrytis cineria*.

Hansen and Smith

254. _______ is a major factor in natural variability and sexuality.

Heterokaryosis

255. In 1962, reported that heterokaryons were formed by hyphal anastomosis and nuclear migration in_______

Cochliobolus sativus

256. Puhalla (1969) described a method of including diploid formation in_______on agar medium.

Ustilago maydis

257. De Barry (1886) in his last work, poineered research on the_______

Physiology of parasitism

258. Parasexualism process first time discovered in the fungus_______ in 1952 by Pontecorvo and Roper and Glasgow.

Aspergillus nidulans

259. Fusion between genetically different hyphae called _______—

Anastomosis

260. Fusion between haploid nuclei to form diploid nuclei called_______ -

Diplodization

261. Two compatible bacterial cell come in contact and interchange of genetic material take place of both cells is altered called_______

Conjugation

262. In 1905, Biffen showed that resistance to yellow rust in_______ was genetic and governed by a_______

Wheat, Recessive gene

263. In 1911_______________demonstrated that there is genetic varilability within a pathogen species that is, there are different pathogen races.

Barrus

264. In 1914, Stakman described neumerous physiological races of_______on wheat.

Puccinia graminis tritici

265. In 1905_______reported the involvement of cytolytic enzymes produced by bacteria in several soft rot diseases of vegetables.

LR Jones

266. Vertical resistance controlled by _______is strong but race specific.

Few major genes

267. Horizontal resistance is________

Field resistance

268. Group of biotype of a pathogen is known as________

Formae specialis (fsp)

269. In case of horizontal resistance, reproduction rate is________

Not zero

270. Slow down rate of spread in ________

Horizontal resistance

271. Horizontal resistance is governed by________ is weaker but effective against all races of pathogen species.

Many minor genes

272. Epidemics are common in case of ________

Vertical resistance

273. If one or more resistance gene are not matched by the pathogen with appropriate virulence gene________

Resistance reaction

274. ________would result only when the pathogen is able to match the entire resistance gene present in the host with appropriate virulence genes. —

Susceptible reaction

275. Red scales of onion contain________ the phenolic compounds which impart resistance to attack by *Colletotrichum circinans.*

Protocatechuic acid and Catechol

276. In seedling disease of radish and lettuce caused by.

Rhizoctonia solani

277. Phytoalexins are not present in________

Healthy plants

278. Polygenic resistance is governed by________

Many genes

279. The gene for gene relationship between a host and its pathogen was postulated by________

Flor in 1956

280. Copper fungicide was first used by________in1807 for seed treatment.

Prevost

281. Downy mildew of onion is caused by ________

Peronospora destructor

282. White rust of crucifers is caused by________and control by________

Albugo candida and 03 per cent copper oxychloride preparation

283. Leaf spot of ginger is caused by ________

Phyllosticta zingiberi

284. Organic mercuriales contains 18.8 per cent Hg in form of the________it was introduced in 1915 by the Bayer Company of Germany.

Chlorophenol mercury

285. Systemic fungicide is not transported downwards through the________

Phloem tissues

286. Antibiotics are transported in________with adequate amount.

All places

287. Toxin seems to act directly on________ and interfere the permeability of its membrane and with its function.

Protoplast

288. Polysaccharides seems to play a role only in the________________in which they passively interfere with the translocation of water in the plants.

Vascular diseases

289. Parasitism and disease resistance run parallel and are________

Inseparable

290. Enzymes, toxin, growth regulators, polysaccharides are the important________ of the pathogen.

Chemical weapons

291. Death of the cells, tissues, organs or entire plants is referred to as

Necrosis

292. ________are large protein molecules which catalyze and interrelated reactions in the living cells.

Enzyme

293. Dry of the entire plants is referred to as ________

Wilt

294. ________is a organism derived the food materials from dead organic matter.

Saprophyte

295. ________are extremely poisonous substances and effective at very low concentration produced by microorganism.

 Toxins

296. ________enzyme which help in breakdown of fatty acid.

 Cutinase

297. *Puccinia graminis tritici* produce pectic enzyme during________

 Spore germination

298. The degradation of starch is brought about by the action of enzymes called________

 Amylases

299. Alternaric acid produced by *Alternaria solani* is the best example of________

 Phytotoxin

300. Phytotoxins are ________toxin.

 Non host specific

301. Piricularin toxin produced by ________are typical example of phytotoxin.

 Piricularia oryzae

302. Ceratoulmin toxin is produce by ________in the Dutch elm disease.

 Ceratocystis ulmi

303. The inoculums produce from primary infection is called________

 Secondary inoculums

304. Viruses, Viroides, plant mycoplasma and fastidious bacteria reproduced only________

 Inside cells

305. ________induces phytoalexin formation in some tissues and stimulates synthesis or activity of several enzymes that may play a role of increasing plant resistance to infection.

 Ethylene

306. Large polysaccharide molecules released by pathogen in the xylem, may be sufficient to causes a mechanical blockage of vascular bundles and initiates.

 Wilting

307. Aflatoxin produce by ________

 Aspergillus flavus

308. ________is used for reducing the danger of infection.

 Laminar Flowhood

309. ________is an instrument used for counting population of bacteria based on the principles of turbidity determination.

    Spectrophotometer or colorimeter

310. For laboratory glassware's a temperature of________for two hours is sufficient to kill all bacteria and spores.

    160°C

311. ________are widely used as surface sterilization agent.

    Hypochloride

312. The most lethal wave length of ultra violet light is________

    265nm

313. In lyophilization technique, the cultures are frozen at________and then dehydrated by vaccume.

    -70°C

314. Preservation of microbes culture in liquid nitrogen methods at temperature of.________

    -196°C

315. The migration of charged particles under the influence of an electric field is called________

    Electrophoresis

316. Vesicular arbuscular mycorrhizal (VAM) fungi belonging to the family________

    Endogonaceae (Zygomycetes)

317. The term________was introduced in 1904 by the German Scientist Hiltner.

    Rhizosphere

318. Paraffin prevents dehydration of ________

    Medium

319. The bacteria that loss the crystal violet and counter strained by safranin (appear red) are referred to as________

    Gram negative

320. Giemsa stain used for________

    Fungal nuclei

321. ________may be used by itself as a mounting medium.

    Lactic acid

322. ________is a stain commonly used for making semi permanent microscopic preparation of fungi.

Lactophenol cotton blue

323. Use glyseel a cellulose based sealant for permanent or________

Semi permanent slides

324. Oil-immersion lenses produce as________

Good image

325. ________provide greater light for illuminating the specimens.

Condenser

326. *Chaetomium globosum* produce antibiotic________

Cochliodinol

327. ________is a microorganism that adversely affects another.

Antagonist

328. *Penicillium nigricans* produce antibiotics *i.e.* ________which control *Pythium* sp. and *Rhizoctonia solani*.

Griseofulvin

329. Libration of an antibiotics or other chemical substance produced/induced by one microorganism and harmful to the other pathogens called________

Antibiosis

330. Siderophore are low molecular weight compound it has bindable capacity it binds the________

Fe

331. ________is a parasite parasitic on another parasite.

Hyperparasite

332. Take all disease of wheat and barley caused by________

Gaeumannomyces graminis

333. Any plant that can attack by a given pathogen called ________

Suscept

334. Loose smut of wheat is caused by ________and transmitted through seed.

Ustilago nuda tritici

335. The ________Enhance uptake fo phosphorus and other nutrients.

Mycorrhiza

336. Soil borne pathogens develop well and cause severe disease in some soils known as________

Conducive soil

337. Soil, in which certain disease are suppressed is referred to as ________

Suppressive soil

338. ________have been shown to parasitize/and or inhibit the pathogenic fungi *Pythium* sp. and *Gaeumannomyces tritici.*

Streptomyces and Pseudomonas

339. Mycophagous nematode *Aphelenchus avenae* parasitizes________

Rhizoctonia and Fusarium

340. Chestnut blight caused by the fungus ________

Endothia parasitica

341. In 1851.________, first time gave the possibility that not exudates effect a pathogen.

Wight

342. ________was first to name a vesicular arbuscular mycerrhiza (VAM)/fungus.

Dangeared (1900)

343. When the mycorrhiza is made up of a fungal sheath as well as some intracellular hyphae it is known as________

Ectomycorrhiza

344. Loose weft of hyphae present outside the host is referred to as________

Endomycorrhiza

345. VAM fungi are able to bind the soil and improve the________of soil.

Structure

346. Continuous cultivation (monoculture) of wheat and cucumber leads to reduction of ________, respectively.

Take all disease of wheat (Gaeumannomyces graminis) and Rhizoctonia damping off of cucumber

347. Resistance nucleic capsid gene involved in transgenic development for the control of________

Tomato spotted wilt virus disease in tomato

348. Resistance osmotin gene involved in transgenic development for the control of________caused by *Phytophthora infestans.*

Late blight of potato

349. Arabidopsis plant is resistant to________

Chilling

350. ________plant used in genetic engineering because the presence of very small genome structure.

Arabidopsis thaliana

351. The soil surrounding the roots called ________

Rhizosphere

352. Microflora found on root surface called________

Rhizoplane

353. The term phyllosphere was introduced by ________in 1955.

Last

354. *Cercospora* and *Cladosporium* are high________

Sugar pathogen

355. Microbial intracellular granules which on staining with certain basic dyes (i.e. Polychrome methylene blue, toluidine blue) exhibit________

Metachromasy

356. First time cultured of rust fungus in artificial media by________

Hutson and Cutter

357. Anthracnose of bean is governed by________

Three gene (several genes)

358. The temperature at which all bacterial cells in a solution after on exposure of 10 minutes are killed is known as________

Thermal death point or thermal death rate

359. ________is defined as the time required killing all cells of a bacterial species at a given temperature.

Thermal death time

360. Mycologists have found various species of *Allomyces* and *Blastocladiella* to be valuable research tools in the study of________

Morphogenesis

361. Mode of action of cyclohexamide is due to its interference with________

RNA synthesis

362. The term vertical and horizontal resistance were coined by________

Von der Plank (1963)

363. ________resistance refers to resistance of a host to all the prevalent races of a pathogen.

Horizontal

364. Long lasting resistance is referred to as ________

Durable resistance

365. A host pathogen interaction which leads to death of infested tissues is called________

Hypersensitivity

366. In India striga is a parasitic weed of________

Sorghum

367. If two plasmids are present, one of which cannot promote conjugation, the one which can promote conjugation may bring about the simultaneous transfer of the non conjugational plasmid, this is known as________

Mobilisation

368. The integration of F (fertility) and chromosome (by crossing over) occurs in about once per________at each generation.

105 cells

369. Discovery and use of streptomycin by________

Selman Waksman (1952)

370. In 1980, Paul Berg and Frederick Sanger worked on________

Genetic engeering

371. The Ovulariopsis type with clavate conidia and conidia are not in chains as in________

Phyllactinia

372. The oidiopsis type in which the conidiophores are branched and typically arises from a stroma as in________

Leveillula

373. *Venturia inaequalis* fungus is________

Heterothallic

374. Genus *Pernosclerospora* germinated by________

Germ tube

375. ________has written a history of physiological Plant Pathology.

Fuchs (1976)

376. _______has discussed modern trends in plant pathological research.

Ubrizsy (1964)

377. Kottaivazhai is a serious malady in_______variety of banana.

Poovan

378. Penicilline antibiotic is produced by_______

Penicillium notatum, P chrysogenum

379. Phenolic compound umbliferon is produced in response to injury in _______

Sweet potato

380. Teleomorphic stages of Chalara fungus is_______

Ceratocystis

381. A.L. toxin is produced by_______

Alternaria lycopersicae

382. Gummosis in citrus is caused by_______

Phytophthora palmivora and P parasitica

383. Downy mildew of poppy is caused by_______

Peronospora arborescens

384. Thread blight of ginger is caused by_______

Penicillium filamentosa

True or False

1. Eukaryotes have two types of ribosomes. (**True**)
2. Wilt of cotton (*Fusarium oxysporum* f. sp. *vasinfectum* and tomato (*F. oxysporum* f. sp. *lycopersici*) is favored by acidic soil. (**True**)
3. Alcohal is formed as byproduct during the respiration of yeast. (**True**)
4. Sugarcane red rot cannot spread through drainage and irrigation water. (**False**)
5. Elicitors are considered as determinates of pathogen avirulence. (**True**)
6. The club root pathogen and the fungi causing vascular wilts, produce spores within the host tissues. (**True**)
7. Most of the lignin in the world is degraded and utilized by a group of basidiomycetes called white rot fungi. (**True**)
8. Salicylic acid and isonicotinic acid are true systemic acquired resistance activated. (**True**)
9. Bunchy top is caused by fungus transmitted by aphid. (**False**)

10. Chitinase genes provide fungal resistance. (**True**)
11. The epidemic rate (r) for polycyclic disease is much greater than the rate of epidemic increase (r_m) for monocyclic diseases. (**True**)
12. Pathogenic causing polyetic epidemic are present for more than one year in the infected plant before they produce effective inoculums. (**True**)
13. Foot and stem rot of papaya (*Pythium aphanidermatum* and *Phytophthora parasitica*) spread through drainage and irrigation water. (**True**)
14. Cucurbitaceous weeds aid in survival and spread of the pathogens causing powdery mildew and mosaic in them. (**True**)
15. Sporadic disease occurs at very irregular intervals and locations. (**True**)
16. Pandemic disease occurs all over the world and result in mass mortality.(**True**)
17. Association of a specific pathogen is not essential with infectious disease. (**False**)
18. The virus is not maintained on its original host, but on artificial media. (**False**)
19. The facultative saprophytes live on dead organic matter. (**True**)
20. Usually the obligate parasites are not host specific. (**False**)
21. Spores are reproductive bodies consisting of one or few cells. (**True**)
22. During the parasitic phase of fungi assume various positions in relation to the plant cells and tissues. (**True**)
23. An epiphytotic disease is one which occurs widely but periodically in a destructive form. (**True**)
24. Biotrophs can grow only in association with the living cells, being unable to feed on dead cells. (**True**)
25. Spore dissemination in almost all fungi is passive, although the initial discharge of spores in some fungi is forcible. (**True**)
26. The means of survival are the first link in disease cycle. (**True**)
27. Germination of spores brings about antagonistic elimination of the pathogens. (**True**)
28. *Trichoderma* and *Gliocladium* are used as bio-control agents against several plant pathogenic fungi. (**True**)
29. *Melanospora*, of which its anamorphs *Philalophora* and *Gonatobotrys* parasitize the mycelium of many fungi including the important plant pathogens Ophiostoma, *Ceratocystis*, *Fusarium* and *Verticillium*. (**True**)
30. The movement of propagules of animate pathogens through wind is not common for soil or seed borne pathogens. (**False**)

31. *Calviceps purpurea* causing ergot of rye, is poisonous to human and animals. (**True**)
32. The green ear is a common disease of pearl millet and occurs in several countries including India, Israel, Iran, Fijji, Japan, China, U.S.A., European and African countries. (**True**)
33. Autonomous dispersal of pathogens is accomplished through soil, seed and plant organs during normal agronomic operations. (**True**)
34. No fungal pathogen first grows on the surface of the host before causing penetration. (**False**)
35. *Setosphaera* (anamorph is *Exserohilum*) cause leaf spots on cereals and grasses. (**True**)
36. *Pleospora* anamorph is *Phoma*, causing black and foot rot of cabbage. (**False**)
37. *Leptosphaeria*, anamorph is *Stemphylium* causing black mold rot of tomato. (**True**)
38. *Magnaporthe grisea* causing the very important rice blast disease, its anamorph is *Pyricularia oryzae*. (**True**)
39. Slim molds are saprophytic. (**True**)
40. *Pythium* species occur in water and soil. They live on dead plant, animal materials as saprophytes or as parasites fibrous roots of plants. (**True**)
41. *Rhizoztonia solani* (attacking radish/tomato roots) on coming in contact with root surface, forms infection cushions and appressoria and form these no multiple infection occur through infection peg. (**False**)
42. Many fungal pathogens first grow one the surface of the host before causing infection. (**True**)
43. In *Colletotrichum* species even the hyphae of the fungus within the host cells do not form apperssoria. (**False**)
44. In indirect infection, the germs tubes of the infection threads enter the host through wound and or through natural openings. (**True**)
45. *Phytophthora infestans*, pathogen overwinters as mycelium in infected potato tubers. (**True**)
46. *Phytophthora infestans* zoospores do not germinate and penetrate the tubers through lenticels or through wound. (**False**)
47. The infection occurring through wounds is known as trauma infection. (**True**)
48. Most of the pathogens do not use wounds as portal of entry into the host. (**False**)
49. In many host specific pathogens like rust and powdery mildew entry of fungi through stomata is a rule to infect a non cereal host. (**False**)

50. The entry of the pathogen into the host plant always insures infection and disease development. (**False**)
51. The downy mildew fungi do not produce sporangia on sporangiophores. (**False**)
52. The oospore of the downy mildew usually germinates by germ tubes but in few causes they may produce a sporangium that release zoospores. (**True**)
53. In most downy mildews, the pathogens routinely causes systemic shoot infection of its host when it is carried in the seed bulb or when infection takes place at the seeding or young plant stage. (**True**)
54. *Olpidium* is a vector of at least six plant viruses, including tobacco necrosis virus and lettuce big vein virus. (**True**)
55. *Rhizopus* and *Mucor*, both are common bread mold fungi. (**True**)
56. The plant pathogenic zygomycetes are not weak parasites. (**False**)
57. The mycelium of *Rhizopus* grows on surface, it produces stolens. (**True**)
58. Sexually produced spore of *Rhizopus* spp is called zygospores and is the overwintering or resting stage of the fungus. (**True**)
59. The perfect stage is usually the overwintering stage. (**True**)
60. The zygospores help the fungus to survive periods of starvation and adverse temperature and moisture. (**True**)
61. Usually nine ascospores are found in per asucs. (**False**)
62. The powdery mildews, ascocarp is completely closed spherical container called perithecium. (**False**)
63. Many species of *Alternaria* are mostly saprophytic. (**True**)
64. Plant pathogenic species of *Alternaria* overwinter as mycelium or spores in infected plant debris and in or on seed. (**True**)
65. Black rot of grape is caused by *Guignardia bidwelli*. (**True**)
66. *Rhabdocline* fungus infects leaves of Douglas fir. (**True**)
67. *Botrysphaeria dothidea*, cause canker of poplar. (**False**)
68. *Cryptodiaporthe populea*, causing canker of apple. (**False**)
69. The fungus *Leucostoma*, previously known as *Valsa* is most commonly found in its anamorph stage, *Cystospora*. (**True**)
70. The time between penetration by pathogen and appearance of symptoms is known as incubation period. (**True**)
71. The pathogens unable to produce mycelium are ectobiotic as the thallus is entirely contained within a host cell. (**False**)

72. *Diplocarpon*, causes black spot of rose is *D. rosae* and leaf scorch of strawberry is *D. earliana*. (**True**)
73. *Coryneum* (*Stigminia*) causing blight, shot hole of fruit spot of stone fruits especially peach and apricot. (**True**)
74. Bitter rot of grapes is caused by *Melanconium* sp. (**True**)
75. Bitter rot of apple is caused by *Glomerella* sp. (**True**)
76. The fungus *Glomerella* over season in the diseased stem, leaves and fruits as mycelium or spore. (**True**)
77. Sycamore anthracnose disease is caused by *Gnomonia* sp. (**True**)
78. Fungi after growing for some time in the host tissues produce branched spore bearing hyphae when conditions are favorable to cause primary infection. (**False**)
79. *Calviceps purpurea* causes ergot of cereals and grasses. (**True**)
80. *Physalospora obtusa* causes black rot, frogeye leaf spot and canker of apple. (**True**)
81. *Venturia inaequalis* produce at first light olive colored, irregular spots on the under surface of sepals or young leaves of flower buds. (**True**)
82. Genetic response of the plant to the pathogen is an innate plant factor determining success of infection. (**True**)
83. Both conidia and ascospores of *Monilinia* cannot cause blossom infection. (**False**)
84. The conidia of *Monilinia* are windblown or may be carried to floral parts by rainwater splashes or insects. (**True**)
85. Brown rot of stone fruits caused by *Monilinia fructicola* can be controlled best by completely controlling the blossom blight phase of the disease. (**True**)
86. Damping off of seedling due to *Botrytis* occurs primarily in cold frames, where the humidity is high. (**True**)
87. Some species of *Botrytis* do not cause leaf spots on gladiolus onion and tulip. (**False**)
88. Predisposition is the action of set environment, prior to penetration and infection which makes the plant vulnerable to attack by pathogen. (**True**)
89. Ontogenic disposition plays no role in the infection by many pathogens. (**False**)
90. Storage organs such as onion bulbs can be protected from *Botrytis* infection by keeping them at 32 to 50°C for 2 to 4 days to remove excess moisture and then keeping the mat 3°C in dry environment. (**True**)
91. *Ophiostoma* causes the vascular wilt of elm trees. (**True**)

92. Most of the wilt causing *Fusarium* is *Fusarium oxysporum*. (**True**)
93. *Fusarium* wilt is not one of the most prevalent and damaging disease of tomato. (**False**)
94. *Verticilium* wilts occur worldwide but are most important in temperate regions. (**True**)
95. *Verticilium* is not main cause of the potato early dying disease. (**False**)
96. Temperature predisposes infection chiefly by producing injury, a real predisposing agent. (**True**)
97. Humidity is major predisposing factor in as much as its effect on the plant is concerned. (**False**)
98. Dutch elm disease is the result of an usual partnership between a fungus and an insects. (**True**)
99. Aflatoxins are produced by *Aspergillus flavus* and several other species of *Aspergillus*. (**True**)
100. Southern corn rust is caused by *Puccinia polyspora*. (**True**)
101. *Hemileia* cause rust on roses and yellow rust on raspberry. (**False**)
102. *Fragmidium*, cause the devastating coffee leaves rust. (**False**)
103. Rust of cotton caused by *Puccinia stakmanii*. (**True**)
104. *Peridermium harknessii*, cause western gall rust in pine. (**True**)
105. *Gymnoconia*, cause orange rust of blackberry and raspberry. (**True**)
106. *Phakopsora pachyrhizi*, causing the potentially catastrophic rust of peach. (**False**)
107. *Tranzschelia* cause rust of soybean. (**False**)
108. Dark treatment before inoculation decreases susceptibility of potato to infection by *Macrophomina phaseolina* and *Fusarium coeruleum*. (**False**)
109. Resistance of cotton to wilt in zinc amended soil has been assigned to its vigorous growth. (**True**)
110. The plant exudates are rich in sugars, amino acids and other nutrients unsuitable for parasitic growth. (**False**)
111. *Urocystis cepulae* cause onion smut. (**True**)
112. *Neovossia barclayana* cause kernel smut of rice. (**True**)
113. Leaf smut of rice is caused by the fungus *Entyloma oryzae*. (**True**)
114. The presence of sugars in wrinkled seeded pea varieties enhances their susceptibility to infection of *Pythium*. (**True**)
115. The fungus *Ustilago maydis* overwinters as teliospores in crop debris and in the soil. (**True**)

116. Galls in developed plant are seem always to be the result of local infections by *Ustilago maydis* and systematic infection occur occasionally in very young seedling of corn. (**True**)
117. *Sclerotium sepivorum* causes the white rot disease of onion and garlic. (**True**)
118. The sporophores of most wood rotting fungi such as *Inonotus* and *Phellinus* are formed annually while *Heterobasidion* are perennial. (**True**)
119. Ectomycorrhizal fungi usually produce a tightly interwoven fungus mantle around the outside of the feeder roots and only grow around the cortical cells and form haring net. (**True**)
120. Arbuscules (haustoria) is formed by the endomycorrhiza. (**True**)
121. Mycorrhizae apparently improve plant growth by absorbing and accumulating certain nutrients especially calcium. (**False**)
122. Any phenomenon, structural or functional, having two or more separate component and some interaction between these components represents a system. (**True**)
123. The activity of pectic enzymes may not be aggravated by synergistic action of other enzymes and metabolites. (**False**)
124. The foolish seedling disease of rice entails abnormal elongation of stem to excessive elongation of internodes. (**True**)
125. Gibberellins are considered as normal constituents of green plants and are not found in microorganism. (**True**)
126. The best known cytokinin is kinetin. (**True**)
127. Many seed borne pathogens are asymptomatic. (**True**)
128. *Sclerospora graminicola*, which causes green ear or downy mildew of pear millets, is carried internally as hyphae in the scutellum. (**True**)
129. Barely stripe mosaic virus and *Ustilago tritici* are internally seed borne pathogens. (**True**)
130. Black root rot of tobacco (*Thielaviopsis basicola*) is favored by wet soil or high relative humidity. (**True**)
131. The endosperm is the common site of infection by seed borne fungi. (**True**)
132. The pathogen or its growth regulators, toxin and enzymes alter cell wall permeability of roots. (**True**)
133. The decreased suction tension due to increased transpiration may result in collapse of vessels or formation of tyloses. (**False**)
134. Rust and mildew fungi are known to cause accumulation of carbohydrates and minerals away from infection sites. (**False**)

135. Chlamydospores of *Ustilago nuda* and *U. tritici* in barley and wheat germinate on style surfaces and the hyphae pass through pollen tubes into the integuments and embryo. (**True**)

136. *Claviceps fusiformis* (ergot of pearl millet), *Claviceps purpurea* (ergot of rye), *Sclerotinia sclerotiorum* (rots of crucifers) and legumes are mixed with seeds during threshing. (**True**)

137. The teliospores of *Melampsora lini* are carried in infected chaff mixed with linseed seeds. (**True**)

138. Higher nitrogen fertilization reduces the incidence of corn kernels infection from *Fusarium moniliforme*. (**True**)

139. Embryos are most susceptible to the loose smut fungus within 4 days of heading. (**True**)

140. The pentose pathway of respiration dominates in diseased plants. (**True**)

141. There is evidence that the Krebs's cycle is activated in infected plants due to decrease in concentration of coenzymes A. (**False**)

142. Leaf blights, spots and identical disease pathogens do not obstruct photosynthesis by adversely affecting chloroplasts. (**True**)

143. In blight affected potato tuber growth ceases when 75 per cent foliage is killed. (**True**)

144. *Ustilago hordei* promycelia from chlamydospores penetrate seedling of barley without promycelim fusion.(**True**)

145. Four types of infection processed by *Ustilago hordei* and *U. avenae* may occur in a single host plant. (**True**)

146. Teliospores of *Ustilago hordei* (covered smut of barley) are carried as a seed surface contaminant. Teliospores give rise to sporidia which penetrate seedling through coleoptiles. (**True**)

147. In 1954, Ernet Gaumann had boldly asserted that a microorganism cannot be pathogenic unless it is toxigenic. (**True**)

148. A toxin represents microbial metabolites excreted (exotoxin) or released by lysed cell (endotoxin) which in very low concentration is directly toxic to cell of suscept. (**True**)

149. The concentration of glucocides in roots, stems, leaves flowers and seeds of different plants vary from 1-5 mg/gram of fresh weight. (**True**)

150. Fungal invasion of oat roots triggers release of avenacins from protoplasts simultaneously with leakage of electrolytes and nutrients. (**True**)

151. Root rot of cotton (*Phymatotrichum omnivorum*) is favored by water deficiency. (**False**)

152. Stalk rot of maize (*Fusarium moniliforme*) infection is favored by wet soil or high relative humidity. (**False**)

153. Wet soil or high relative humidity favors the occurrence of safflower root rot (*Phtytophthora cryptogea*). (**False**)

154. Wilt of pea (*Fusarium oxysporum* f. sp. *pisi* race 1) incidence is favored by wet alkaline soil. (**True**)

155. Club root of cabbage (*Plasmodiophora brassicae*) infection is favored by acid soil. (**True**)

156. Faculative parasites are basically obligate in their feeding mechanism. (**False**)

157. The non specialized facultative parasites (*e.g. Pythium, Rhizoctoia*) can pass their entire life in soil. (**True**)

158. Specialized facultative parasites (*Armillaria mellea, Ophiobolus graminis, Phymatotrichum omnivorum*) can pass their life in the soil in the absence of the host buy depend more on the residue or their host plant. (**True)**

159. Many downy mildew fungi such as *Sclerospora* as *Pernospora* produce oospore in tissues of pods and there may go with the seed if the latter is not cleaned. **(True)**

160. Individual hypae of the rhizomorphic fungus *Armilaria mellea* do not succeed in penetration but when they have developed the rhizomorphs infection occurs easily. **(True)**

161. Epidemics of plant disease occur due to aggressiveness, rapid rate of reproduction and rapid long distance dispersal of the causal agent. (**True**)

162. Most of the characters of simple interest diseases are found in rapid epiphytotics. (False)

163. Crop sanitation is best method of control of slow epidemics. (**False**)

164. Disintegration of tissues occurs by the action of enzymes of the pathogen. (**True**)

165. More specialized pathogens such as *Puccinia graminis tritici* produce pectic enzymes during spore germination. (**True**)

166. May wood rooting fungi degrade lignin but cannot utilize it. (**True**)

167. Necrosis caused by many obligate parasites and pathogens including viruses is not due to enzymatic action of their own but due to indirect effect. (**True**)

168. Chromosomal DNA is not degraded but the cytoplasmic DNA can be degraded by fungal and bacterial enzyme. (**True**)

169. The tumors cells contain more than normal quantity of IAA. (**True)**

170. The production of gel and gums are as a result of enzymatic action of the pathogen. (**True)**

171. The level of cytokinins is low in diseased plant cells than in healthy cells. (**True**)
172. Basidiospores of *Puccinia graminis* are not infected in the alternate host, *Berberis* spp by direct penetration of the epidermis. (**False**)
173. *Puccinia recondita* cannot penetrate the stomata a in the light as well as in dark. (**False**)
174. Necrotic or hypersensitive reactions are both, structural as well as physiological. (**True**)
175. Phytoalexin production is stimulated by organisms which are nonpathogenic or avirulent on the particular host. (**True**)
176. Necrobiosis does not lead to hypersensitive reaction and is not related to phytoalexin production. (**False**)
177. Sexual reproduction in *Phytophthora infestans* is common in nature. (**False**)
178. The binding of the proteins or proteins polymerization is hydrophobic *i.e.* the surface reject water in favour of proteins. (**True**)
179. Suppressiveness is related to inoculum density of the pathogen. (**True**)
180. High potassium (K) reduces incidence of disease. (**True**)
181. Excess potassium predisposes tomato plants to wilt and citrus to *Phytophthora* root and trunk rot. (**True**)
182. Optimum temperature and moisture are needed for germination of spore and entry of germ tube in the host. (**True**)
183. The disease escaping varieties have true genetic resistance to pathogen. (**False**)
184. Monogenic resistance is stable in different environmental conditions but easily susceptible to new race of the pathogen. (**True**)
185. Milo root rot is caused by a powerful toxin produced by fungus (*Periconia circinata*) growing in the root zone. (True)
186. *Penicillum expansum* which causes the storage rot of apple enters through stomata or lenticels. (**True**)
187. *Oospora pustulans* which causes the skin scab of potato tubers attacks directly. (**False**)
188. Suberin deposition in potato tuber cells enables better resistance to penetration of pathogens. (**True**)
189. Dispersal of inoculums by insects and other animals including man has been discussed by Broadbent. (**True**)
190. Oligogenic resistance is determined by single gene. (**False**)
191. Polygenic resistance is governed by one gene with a 3:1 Mendelian ratio. (**True**)

192. Polygenic resistance involves many genes. (**True**)
193. Resistance to smudge disease of onion is controlled by single dominant gene. (**True**)
194. When a variety is resistance to some race, resistance is called horizontal. (**True**)
195. Horizontal resistance has been called field resistance against *Phytophthora infestans*. (**False**)
196. Horizontal resistance is monogenic in inheritance and is effective against a race of pathogen. (**False**)
197. Vertical resistance is an apparent reduction in the initial inoculums, as a result the epidemic is delayed. (**True**)
198. The interface of *Fusarium* wilt resistance in pulse is reported to be controlled by a single pair of genes. (**True**)
199. Rapid pollination reduces ergot infection in peal millet. (**True**)
200. Ribosomes are of 80S and mitochondria occur in the cytoplasm. (**True**)
201. The species of the *Physoderma* are saprophytes of vascular plants. (**False**)
202. Brown leaf spot of maize is caused by *Physoderma maydis*. (**True**)
203. The teliospores of *Puccinia* are one celled, dark and stalked. (**False**)
204. The telisopores of *Uromyces* are stalk less and two celled. (**False**)
205. *Graphiola phoenicis* causes spots on palm. (**True**)
206. Zoospores formed in a vesicle, antheridia, always and paragynous in the genes *Phythium*. (**True**)
207. The species of *Phytophthora* and Phythium have less competitive saprophytic ability. (**False**)
208. *Pythium irregularis, P. spinosum* and *P. ultimum* are more damaging at higher temperature. (**False**)
209. *Pythium myriotylum, P. aphanidermatum, P. arrhenomanes* and related species are damaging at lower temperature. (**False**)
210. *Pythium* spp is most destructive at soil temperature between 24°C and 30°C. (**True**)
211. Seed treatment with spores of *Trichoderma harzianum* and *Penicillium oxalicum* and cells of *Pseudomonas cepacia, P. fluorescens* are found very effective biological control agents. (**True**)
212. Koleroga of mahali disease of areca palms is caused by *Phytophthora palmivora*. (**False**)
213. Bud rot of palm is caused by *Phytophthora palmivora*. (**True**)

214. *Phytophthora palmivora* pathogen is spread by rhinoceros beetle and rains. (**True**)
215. Leaf shredding disease of sorghum is caused by *Sclerospore sorghi*. (**True**)
216. Flag smut of wheat is caused by *Urocystis tritici*. (**True**)
217. Hill bunt or European bunt or stinking smut of wheat is caused by *Tilletia foetida* and *T. carries*. (**True**)
218. The aeciospores are the first dikararyotic spores. (**True**)
219. The occluding plugs in elm infected by *Ceratocystis ulmi* contain both pectin and lignin. (**True**)
220. *Pyricularia grisea* are produced numerous conidia under humid conditions usually during the night. (**True**)
221. *Cerospora archidicola* is potentially more damaging. (**False**)
222. *Penicillium* spp cause considerable damage to citrus fruits in storage and transit. (**True**)
223. Milo disease of sorghum is caused by the fungus *Periconia circinata*. (**True**)
224. Chitin is a component of the cell walls of all fungi except the oomycetous ones. (**True**)
225. Loose smut of wheat is caused by *Ustilago segetum* var. *tritici*. (**True**)
226. The book 'Basic Plant Pathology Method' is written by Tuite (1969). (**False**)
227. 'Plant Pathology Pocket Book' was published by commonwealth mycological institute, England (1968). (**True**)
228. 'Laboratory Exercise in Plant Pathology' edited by Boudoin *et al.* (1988) was published by the American Phytopathological society. (**True**)
229. The Book 'Diseases of Vegetable Crops' is authored by R.S. Singh. (**True**)
230. Budding and fission are characteristics of some yeast and bacteria, although conidia, ascospores (*e.g. Taphrina* sp.) and basidiospores (Ustilaginales) may bud under certain condition. (**True**)
231. In the case of *Erysiphe cichoracearum* (lettuce mildew), cells of resistant osmotic pressure than those of the susceptible ones. (**True**)
232. The defense reaction occurs only in the living cells. (**True**)
233. The phytoalexins is non-specific in its toxicity towards fungi, however, fungal species may be differentially sensitive to it. (**True**)
234. Heterokaryosis plan an important role in homothallic and imperfect fungi. (**True**)
235. *Rhizoctonia* cannot survive under poor aeration, it is very sensitive to oxygen tension. (**True**)
236. Red rust of mango and papaya is caused by alga. (**True**)

237. Mango malformation was first noted by G. Watt (1891) in Darbhanga, Bihar. (**True**)
238. Powdery mildew of grape is caused by fungus *Uncinula necator*. (**True**)
239. Red rot of sugarcane caused by *Colletotrichum falcatum*, the second injury take place through the borer of stem and leaves. (**True**)
240. The concept of inoculums potential was describe as disease producing power of the host environment. (**True**)
241. Polygalacturases (PG) and lyases are chain spliting enzymes. (**True**)
242. Dikaryotic mycelia responded to selection with increases in growth rate, while haploid monokaryotic mycelia did not. (**True**)
243. Wilt of pigeon pea incidence is quite high during the flowering and pod formation stages. (**True**)
244. The silicon content of rice leaves has been implicated in resistance to blast disease of rice. (**True**)
245. Black pod disease of cocoa is caused by the *Phytophthora palmivora*. (**True**)
246. Cocoa canker is caused by the fungus *Phytophthora*. (**True**)
247. Ophiobolin toxin is produced by *Helminthosporium* spp (**True**)
248. Ampelomycin antibiotic is produced by *Ampelomyces quisqualis*. (**True**).
249. Splitting/gumming is associated with peach. (**True**)
250. Internal fruit necrosis is common in aonla. (**True**)
251. Yellow leaf spot is associated with cashew. (**True**)
252. Clump rot of small cardamom is caused by *Pythium* spp (**True**)
253. R.A. Gray has reported a new antibiotic, noformicin, which inhibits both the production of local lesions and systemic infections caused by southern bean mosaic virus and by tobacco mosaic virus in intact plants. (**True**)
254. Teleomorphic stage of *Exerohilium* is *Setosphaera*. (**True**)
255. Amylase enzymes act on lipid. (**False**)
256. Athrospores is formed by fragmentation of hypha. (**True**)
257. Phenolic compound chlorogenic acid and scopoletin are produced in response to injury in sweet potato plant. (**True**)
258. *Colletotrichum furarioides* is control the weed *Aslepias seriacea*. (**True**)
259. Mite (Eriophyidae) is controlled by the fungus *Hirsutella thompsonii*. (**True**)
260. Rhizobi toxin produced in soybean which is associated with target site methionine metabolism. (**True**)

261. Fungal hyphae are capable of indefinite growth under favorable conditions. (**True**)

262. In Tolyposporium the sori consist of permanent spore balls. (**True)**

263. In *Ustilago* contain one celled teliospores, dusty at maturity. (**True**)

264. The powdery mildew fungi produce a mycelium only on the surface of epidermis. (**True**)

265. Irrigation and drainage water do not spread downy mildew of maize. (**False**)

266. Tylosis are found more frequently in healthy plants than in infected plants (**True**)

267. *Alternaria alternata* in sunflower, *A. brassicae* and *A. brassicicola* in rape and cabbage, *A. triticina* in wheat and so many fungi are not found in the seed coat. (**False**)

268. Bean common mosaic virus in bean and urdbean, bean southern mosaic virus in bean, cucumber mosaic virus in cucumber and cowpea mosaic virus in cowpea are carried in both, the embryo and seed coat. (**True**).

269. False smut disease of palm is caused by *Graphiola phoenicis* (**True**).

Match the Following

1. **Match disease and host with the parasite**

Disease Host	*Parasite*
I. Small maize	(A) *Synchytrium endobioticum*
II. Wart, potato	(B) *Gymnosporangium juniper-virginianae*
III. Canker, apple	(C) *Ustlago maydis*
IV. Rust, cedar apple	(D) *Nectria gaglligena*

Answer

I	*II*	*III*	*IV*
C	*A*	*D*	*B*

2. **Match disease and host with the parasite**

Disease Host	*Parasite*
I. Downey mildew, sugarcane	(A) *Puccinia graminis*
II. Bakanae disease, rice	(B) *Sclerospora sacchari*
III. Rust, wheat	(C) *Gibberella fujikuroi*
IV. Leaf blight, cherry	(D) *Gnomonia erythrostoma*

Answer

I	*II*	*III*	*IV*
B	*C*	*A*	*D*

3. Match disease and host with the parasite

Disease Host	*Parasite*
I. Rust, safflower	(A) *Omphalia flavida*
II. Leaf spot, coffee	(B) *Sclerospora sacchari*
III. Downey mildew, maize	(C) *Corynebacterium fascians*
IV. Witches broom, sweet pea	(D) *Puccinia carthami*

Answer

I	*II*	*III*	*IV*
D	*C*	*B*	*A*

4. Match disease and host with the parasite

Disease Host	*Parasite*
I. Club root of crucifers	(A) *Plasmodiophora brassicae*
II. Wart, potato	(B) *Synchytrium endobioticum*
III. Stem rot, papaya	(C) *Pythium aphanidermatum*
IV. Bud rot, palm	(D) *Phytophthora palmivora*

Answer

I	*II*	*III*	*IV*
A	*B*	*C*	*D*

5. Match disease and host with the parasite

Disease Host	*Parasite*
I. Gumosis, citrus	(A) *Plasmopara viticola*
II. White blister, crucifers	(B) *Phytophthora parasitica*
III. Downey mildew, pea	(C) *Pernospora pisi*
IV. Downey mildew, grapevine	(D) *Albugo candida*

Answer

I	*II*	*III*	*IV*
B	*D*	*C*	*A*

6. Match disease and host with the parasite

Disease Host	*Parasite*
I. Kolaraga or Mahali, areca palm	(A) *Sclerospora graminicola*
II. Green ear, pearl millet	(B) *Phytophthora arecae*
III. Downey mildew, maize	(C) *Protomyces macrosporous*
IV. Stem gall, coriander	(D) *Pernospora philipinensis*

Answer

I	*II*	*III*	*IV*
B	*A*	*D*	*C*

7. Match disease and host with the parasite

Disease Host	*Parasite*
I. Leaf curl, peach	(A) *Taphrina maculans*
II. Leaf spot, turmeric	(B) *Penicillium expansum*
III. Soft rot, apple	(C) *Taphrina deformans*
IV. Powdery mildew, pea	(D) *Erysiphe polygoni*

Answer

I	*II*	*III*	*IV*
C	*A*	*B*	*D*

8. Match disease and host with the parasite

Disease Host	*Parasite*
I. Powdery mildew, cucurbits	(A) *Leptosphaeria sacchari*
II. Powdery mildew, grapevine	(B) *Podosphaeria leuctotricha*
III. Powdery mildew, apple	(C) *Uncinula necator*
IV. Ring spot, sugarcane	(D) *Erysiphe cichoracearum*

Answer

I	*II*	*III*	*IV*
D	*C*	*B*	*A*

9. Match disease and host with the parasite

Disease Host	*Parasite*
I. Stem rot, rice	(A) *Sclerotium oryzae*
II. Ergot, cereals	(B) *Claviceps purpurea*
III. Ergot, pearl millet	(C) *Claviceps microcephala*
IV. False smut, rice	(D) *Ustilaginoidea virens*

Answer

I	*II*	*III*	*IV*
A	*B*	*C*	*D*

10. Match disease and host with the parasite

Disease Host	*Parasite*
I. Smut, sugarcane	(A) *Ustilago scitaminea*
II. Smut, oat	(B) *Ustilago avenae*

III. Smut, maize (C) *Ustilago maydis*

IV. Smut, sorghum (D) *Sphacelotheca sorghi*

Answer

I	*II*	*III*	*IV*
A	*B*	*C*	*D*

11. Match disease and host with the parasite

Disease Host	*Parasite*
I. Loose smut, wheat	(A) *Sphacelotheca cruenta*
II. Covered smut, barley	(B) *Ustilago nuda*
III. Loose smut, barley	(C) *Ustilago nuda tritici*
IV. Loose smut, sorghum	(D) *Ustilago hordei*

Answer

I	*II*	*III*	*IV*
C	*D*	*B*	*A*

12. Match disease and host with the parasite

Disease Host	*Parasite*
I. Long smut, sorghum	(A) *Tolyposporium ehrenbergi*
II. Smut, pearl millet	(B) *Entyloma oryzae*
III. Karnal bunt, wheat	(C) *Neovossia indica*
IV. Leaf smut, rice	(D) *Tolyposporium penicilariae*

Answer

I	*II*	*III*	*IV*
A	*D*	*C*	*B*

13. Match disease and host with the parasite

Disease Host	*Parasite*
I. Leaf rust, coffee	(A) *Uromyces fabae*
II. Rust of linseed	(B) *Uromyces ciceris arietini*
III. Rust of chickpea	(C) *Hemileia vastatrix*
IV. Rust of pea and lentil	(D) *Melamspora lini*

Answer

I	*II*	*III*	*IV*
C	*D*	*B*	*A*

14. Match disease and host with the parasite

Disease Host	*Parasite*
I. Sheath blight, rice	(A) *Colletotrichum falcatum*
II. Charcoal rot, soybean	(B) *Colletotrichum gloeosporioides*
III. Red rot, sugarcane	(C) *Macrophomina phaseolina*
IV. Anthracnose, mango	(D) *Rhizoctonia solani*

Answer

I	*II*	*III*	*IV*
D	*C*	*A*	*B*

15. Match disease and host with the parasite

Disease Host	*Parasite*
I. Smut of onion	(A) *Urocystis cepulae*
II. Cedar apple rust	(B) *Gymnosporangium juniperi-virginianae*
III. Rust of soybean	(C) *Phakopsora pachyrhizi*
IV. Rust of beans	(D) *Uromyces appendiculatus*

Answer

I	*II*	*III*	*IV*
A	*B*	*C*	*D*

16. Correctly match in section 'A' to those given under the section 'B'

Section 'A' imperfect stage	*Section 'B' perfect stage*
I. *Cecospora arachidicola*	(A) *Mycosphaerella arachidicola*
II. *Cercospora personata*	(B) *Phaeoisariopsis personata*
III. *Drechslera graminea*	(C) *Pyrenophora graminea*
IV. *Drechslera oryzae*	(D) *Cochliobolus miyabeanus*

Answer

I	*II*	*III*	*IV*
A	*B*	*C*	*D*

17. Correctly match in section 'A' to those given under the section 'B'

Section 'A' imperfect stage	*Section 'B' perfect stage*
I. *Pyricularia oryze*	(A) *Magnaporthe grisea*
II. *Colletotrichum falcatum*	(B) *Glomerell tucumenensis*
III. *Colletotrichum gloeosporioides*	(C) *Glomerella cingulata*
IV. *Colletotrichum lindemuthianum*	(D) *Glomerella cingulata* f. sp. *phaseoli*

Answer

I	*II*	*III*	*IV*
A	*B*	*C*	*D*

18. Correctly match in section 'A' to those given under the section 'B'

Section 'A' imperfect stage	*Section 'B' perfect stage*
I. *Fusarium udum*	(A) *Gibberella indica*
II. *Ascochyta rabiei*	(B) *Didymella rabiei*
III. *Rhizoctonia solani*	(C) *Thanatephorus cucumeris*
IV. *Septoria tritici*	(D) *Leptosphaeria tritici*

Answer

I	*II*	*III*	*IV*
A	*B*	*C*	*D*

19. Correctly match in section 'A' to those given under the section 'B'

Section 'A' imperfect stage	*Section 'B' perfect stage*
I. *Penicillium*	(A) *Talaromyces*
II. *Aspergillus*	(B) *Byssochlamys*
III. *Paecilomyces*	(C) *Eurotium*
IV. *Oidium*	(D) *Erysiphe*

Answer

I	*II*	*III*	*IV*
A	*C*	*B*	*D*

20. Correctly match in section 'A' to those given under the section 'B'

Section 'A' Anamorphic stage	*Section 'B' Teleomorphic stage*
I. *Sporothrix* and *Graphium*	(A) *Ophiostoma*
II. *Trichoderma*	(B) *Hypocrea*
III. *Alterrnaria*	(C) *Lewia*
IV. *Drechslera*	(D) *Pyrenophora*

Answer

I	*II*	*III*	*IV*
A	*B*	*C*	*D*

22. Match the pathogen with the toxin produced by it.

Pathogen	*Toxin*
I. *Alternaria kikuchiana*	(A) H.S. toxin
II. *H. maydis* race T	(B) H.C. Toxin
III. *H. carbonum*	(C) A.K. Toxin/Phytoalternarin
IV. *H. sacchari*	(D) H.M.T. – Toxin

Answer

I	*II*	*III*	*IV*
C	*D*	*B*	*A*

23. Match the pathogen with the toxin produced by it.

	Pathogen		*Toxin*
I.	*Periconia circinata*	(A)	P.C. Toxin
II.	*Phyllosticta maydis*	(B)	Ten toxin
III.	*Rhizopus* spp	(C)	P.M. Toxin
IV.	*Alternaria tenuis*	(D)	Fumaric acid

Answer

I	*II*	*III*	*IV*
A	*C*	*D*	*B*

24. Match the pathogen with the toxin produced by it.

	Pathogen		*Toxin*
I.	*Pseudomonas phaseolicola*	(A)	Phaseolotoxin
II.	*Fusicoccum amygdale*	(B)	Fusicoccin
III.	*Sclerotinia* spp	(C)	Oxalic acid
IV.	*Pyricularia oryzae*	(D)	Pyricularin

Answer

I	*II*	*III*	*IV*
A	*B*	*C*	*D*

25. Match the pathogen with the toxin produced by it.

	Pathogen		*Toxin*
I.	*Helminthosporium* spp	(A)	Syringomycin
II.	*Rhizobium japonicum*	(B)	Rhizobitoxin
III.	*Alternaria mali*	(C)	A.M. Toxin
IV.	*Corynespora cassiicola*	(D)	C.C. Toxin

Answer

I	*II*	*III*	*IV*
A	*B*	*C*	*D*

26. Match the following of appresoria with epidermal cells.

I.	53.0 per cent	(A)	Long cell
II.	16.2 per cent	(B)	Hairs
III.	12.8 per cent	(C)	Motor cell
IV.	1.7 per cent	(D)	Short cell

Answer

I	*II*	*III*	*IV*
C	*A*	*D*	*B*

27. Match the antimicrobial compound with the crops producing it.

Antimicrobial compound	*Crop name*
I. Allyl sulphoxide	(A) Pear and walnut
II. Glucosinolate	(B) Onion and garlic
III. Gaynogenic glucosides	(C) Cabbage
IV. Para-hydroquinone glucoside	(D) Sorghum, lima bean, peach

Answer

I	*II*	*III*	*IV*
B	*C*	*D*	*A*

28. Match the antimicrobial compound with the crops producing it.

Toxin	*Crop*
I. Benzoxazine	(A) Green skin potato tubers
II. Solanin	(B) Potato leaves
III. Chaconin	(C) Tomato
IV. Tomatin	(D) Maize, wheat, rye

Answer

I	*II*	*III*	*IV*
D	*A*	*B*	*C*

29. Match the lignifications agent imparting fungi resistance with the crop.

Fungi	*Crop*
I. *Alternaria japonica*	(A) Wheat
II. *Septoria nodorum, S. tritici*	(B) Radish
III. *Cladosporium cucumerinum*	(C) Sweet potato
IV. *Helicobasidium mompa*	(D) Cucumber

Answer

I	*II*	*III*	*IV*
B	*A*	*D*	*C*

30. Match the phytoalxine with the plant (Host) producing it.

Phytoalexin	*Host*
I. Pisatin	(A) Pea pod
II. Phaseolin	(B) French bean
III. Medicarpin	(C) Alfalfa leaves
IV. Orchinol	(D) Orchid tuber

Answer

I	*II*	*III*	*IV*
A	*B*	*C*	*D*

31. Match the phytoalxine with the plant (Host) producing it.

Phytoalexin	*Host*
I. Rishitin	(A) Potato and tomato
II. Isocoumarin	(B) Carrot
III. Ipomeamarone	(C) Sweet potato root
IV. Glutinosone	(D) Tobacco

Answer

I	*II*	*III*	*IV*
A	*B*	*C*	*D*

32. Match the phytoalxine with the plant (Host) producing it.

Phytoalexin	*Host*
I. Capsidiol	(A) Tobacco and Pepper
II. Phytuberin	(B) Potato
III. Hirinol	(C) Orchid
IV. 6-methoxy mellien	(D) Carrot
V. Avenalumin I, II, III	(E) Oats

Answer

I	*II*	*III*	*IV*
A	*B*	*C*	*D*

33. Match the phytoalexin with the pathogen producing by it.

Phytoalexin	*Pathogen*
I. Ipomeamarone	(A) *Ceratocystis fimbriata*
II. Isocaumerin	(B) *Helminthosporium carbonum*
III. Rishitin	(C) *Phytophthora infestans*
IV. Orchinol	(D) *Rhizoctonia repens*

Answer

I	*II*	*III*	*IV*
A	*B*	*C*	*D*

34. Match the phytoalexin with the pathogen producing by it.

Phytoalexin	*Pathogen*
I. Medicarpin	(A) *Helminthosporium turcicum*
II. Phaseolin	(B) *Sclerotinia frunctigena*
III. Pisatin	(C) *Monilinia fructicola*
IV. Capsidiol	(D) *Phytophthora capsici*

Answer

I	*II*	*III*	*IV*
A	*B*	*C*	*D*

35. Match the varieties of stem rust fungus *Puccinia graminis* with uredospore size (microns)

Fungus	*Size*
I. *Tritici*	(A) 22x16
II. *Secalis*	(B) 32x20
III. *Avenae*	(C) 25x17
IV. *Agrostis*	(D) 28x20

Answer

I	*II*	*III*	*IV*
B	*C*	*D*	*A*

36. Match the host-parasite relationship given below:

Host	*Parasite*
I. *Triticum*	(A) *Helminthosporium victoriae* (Victoria blight of oat)
II. *Avenae*	(B) *Cladosporium fulvum* (Leaf mold of tomato)
III. *Lycopersicon*	(C) *Venturia inaequalis* (Scab of apple)
IV. *Malus*	(D) *Puccinia striformis* (Yellow rust of wheat)

Answer

I	*II*	*III*	*IV*
D	*A*	*B*	*C*

37. Match the predisposing soil temperature with the pathogen:

Temperature	*Pathogen*
I. 30-35°C	(A) *Drechslera graminae* (Barley stripe)
II. 10°C	(B) *Fusarium culmorum* (Wheat root rot)
III. 17-23°C	(C) *Sclerotium rolfsii* (Apple root rot)
IV. 24-28°C	(D) *Thielaviopsis basicola* (Tobacco root rot)

Answer

I	*II*	*III*	*IV*
C	*A*	*D*	*B*

38. Match the disease with the pathogen exhibited by suppressive soils:

Disease	*Pathogen*
I. Wilt of beans	(A) *Fusarium oxysporum* f. sp. *cucumerinum*
II. Wilt of cucurbit	(B) *Streptomyces scabies*

III. Root rot of lettuce	(C) *Fusarium solani* f. sp. *phaseoli*
IV. Common scab of potato	(D) *Olpidium brassicae*

Answer

I	*II*	*III*	*IV*
C	*A*	*D*	*B*

40. Match the disease with their causal agent:

Disease	*Causal agent*
I. Black root rot, cucurbit	(A) *Pseudomonas solanacearum*
II. Damping off vegetables	(B) *Phomopsis sclerotioides*
III. Wilt, vegetable	(C) *Rhizoctonia solani*
IV. Bare patch, wheat	(D) *Pythium ultimum, Pythium* spp

Answer

I	*II*	*III*	*IV*
B	*D*	*A*	*C*

41. Match the decay crops which reduce the pathogen populations:

Crops	*Pathogen*
I. Ryegrass, *Papaver rhoeas, Reseda odorata*	(A) *Plsamodiophora brassicae*
II. Datura stramonium	(B) *Striga asiatica*
III. Sudan grass	(C) *Heterodera avenae*
IV. Maize	(D) *Spongospora subterranea*

Answer

I	*II*	*III*	*IV*
A	*D*	*B*	*C*

42. Match the antagonistic pathogens with each other:

I. Penicillium vermiculation	(A) *Ophiobolus graminis*
II. Didymella exitialis	(B) Conidia of many fungi
III. Gliocladium roseum	(C) *Sclerotinia* of *Sclerothinia*
IV. Coniothyrium minitans	(D) *Rhizoctonia solani*

Answer

I	*II*	*III*	*IV*
D	*A*	*B*	*C*

43. Match the pathogen with organic amendment of soil pathogen:

Pathogen	*Amendment*
I. *Fusarium oxysporum* f. sp. *cubense* (Banana wilt)	(A) Green manure with pea melilotus
II. *Streptomyces scabies* (Potato scab)	(B) Oat straw, Maize stover, lucern hay
III. *Phymatotrichum omnivorum* (Cotton, root rot)	(C) Sugarcane residues
IV. *Thielaviopsis basicola* (Bean, root rot)	(D) Green manure with soybean

Answer

I	*II*	*III*	*IV*
C	*D*	*A*	*B*

44. Match the pathogen with organic amendment of soil pathogen:

Pathogen	*Amendment*
I. *Aphanomyces euteiches* (Pea, root rot)	(A) Barley straw
II. *Verticillium alboatrum* (Potato, wilt)	(B) Wheat straw, saw dust, oil cakes
II. *Rhizoctonia solani* (Poatao, black scarf)	(C) Green manure with rape, pea mixed
IV. *Gaeumannomcyces graminis* (Wheat, take all)	(D) Crucifer tissues

Answer

I	*II*	*III*	*IV*
D	*A*	*B*	*C*

45. Match the antibiotic with antagonistic pathogen producing by it:

Antibiotics	*Producing antagonistic pathogen*
I. Gryseofulvin	(A) *Penicillium nigricans*
II. Pyrollnitrin	(B) *Pseudomonas cepacia*
III. Cochliodinol	(C) *Chaetomium globosum*
IV. Scytalidin	(D) *Scytalidium uridinicola*

Answer

I	*II*	*III*	*IV*
A	*B*	*C*	*D*

46. Match the section 'A' to those given under section 'B':

Section 'A' (Disease/Pathogen)	*Section 'B' (Bio-control agent)*
I. Damping off (*Pythium ultimum*)	(A) *Penicillium cyclopium*
II. Potato blight (*Phytophthora infestans*)	(B) *Pseudomonas fluorescens*
III. Wilt of cereals (*Fusarium* sp. and *Verticillium* sp.)	(C) *Bacillus* sp.
IV. Crown gall of stone fruit (*Agrobacterium tumefacians*)	(D) *Agrobacterium radiobacter*

Answer

I	*II*	*III*	*IV*
B	*A*	*C*	*D*

47. Match the section 'A' to those given under section 'B':

Section 'A' (Bio-control agent)	*Section 'B' (Trade name)*
I. *Agrobacterium radiobacter* K-84	(A) Galltrol-A and No gall
II. *Bacillus* sp.	(B) Quantam-4000
III. *Pseudomonas fluorescens*	(C) Dagger-G
IV. *Gliocladium virens*	(D) Gliogard

Answer

I	*II*	*III*	*IV*
A	*B*	*C*	*D*

48. Match the following some fruits and their maladies:

Malady	*Cause*
I. Mango rust	(A) Algal (*Cephaleuros mycoides*)
II. Panama wilt of banana	(B) *Fusarium oxysporum* f. sp. *cubens*
III. Root knot of grape	(C) *Meloidogyne* spp
IV. Spongy tissue in mango	(D) Ecophysiological

Answer

I	*II*	*III*	*IV*
A	*B*	*C*	*D*

49. Match the following crop disease and their casual organism.

Crop, Disease	*Casual organism*
I. Anthracnose of black pepper	(A) *Colletotrichum necator*
II. Wilt of black pepper	(B) *Pythium* spp

III. Powdery mildew of coriander (C) *Eryiphe polygoni*

IV. Katte/marble disease of cardamom (D) Virus

Answer

I	*II*	*III*	*IV*
A	*B*	*C*	*D*

50. Match the following crop disease and their casual organism.

Crop, Disease	*Casual organism*
I. Wilt of cumin	(A) *Fusarim oxysporum* f. sp. *cumini*
II. Blight of garlic	(B) *Alternaria palandui*
III. Soft rot of ginger	(C) *Pythium aphanidermatum*
IV. Rust of pyrethrum	(D) *Puccinia chrysanthemi*

Answer

I	*II*	*III*	*IV*
A	*B*	*C*	*D*

51. Match the following crop disease and their casual organism:

Crop, Disease	*Casual organism*
I. Brown eye leaf spot of tobacco	(A) *Cercospora nicotianae*
II. Black leg of tobacco	(B) *Phytophthora palmivora*
III. Brown rape of tobacco	(C) *Orobanche cernua*
IV Leaf blotch of turmeric	(D) *Taphrina maculans*

Answer

I	*II*	*III*	*IV*
A	*B*	*C*	*D*

52. Match the following crop disease and their casual organism.

Disease, crop	*Causal organism*
I. Stem bleeding of arecanut	(A) *Ceratostomella paradoxa*
II. Pink disease of cincona	(B) *Pellicularia salmonicolor*
III. Root rot of arecanut, coconut	(C) *Ganoderma lucidum*
IV. Die back of coffee	(D) *Colletotrichum coffeanum*

Answer

I	*II*	*III*	*IV*
A	*B*	*C*	*D*

53. Match the following crop disease and their casual organism.

Disease, crop	*Causal organism*
I. Smut of sugarcane	(A) *Ustilago scitaminea*
II. Gummosis of sugarcane	(B) *Xanthomonas vasculorum*
III. Bunga of sugarcane	(C) *Aeginetia indica* (root parasite)
IV. Red rust (algal red spot) of tea	(D) *Cephaleuros mycoides*

Answer

I	*II*	*III*	*IV*
A	*B*	*C*	*D*

54. Match the following crop disease and their casual organism.

Disease, crop	*Causal organism*
I. Blister blight of tea	(A) *Exobasidium vexans*
II. Bird's eye spot of tea	(B) *Guignardia camelliae*
III. Copper blight of tea	(C) *Cercospora theae*
IV. Charcoal rot of tea	(D) *Ustulina zonata*

Answer

I	*II*	*III*	*IV*
A	*C*	*B*	*D*

55. Match the following crop disease and their casual organism.

Disease, crop	*Causal organism*
I. Storage rot of pineapple	(A) *Ceratostomella paradoxa*
II. Peach leaf curl	(B) *Taphrina deformans*
III. Malformation of mango	(C) *Gibberella fujikuroi*
IV. Grey blight of guava	(D) *Pestalotia psidii*

Answer

I	*II*	*III*	*IV*
A	*B*	*C*	*D*

56. Match the following crop disease and their casual organism:

Disease, crop	*Causal organism*
I. Rust of grapevine	(A) *Colletotrichum gloeosporioides*
II. Rust of fig	(B) *Cerotelium fici*
III. Wither tip of citrus	(C) *Phakospora vitis*
IV. Rind disease of sugarcane	(D) *Pleocyta sacchari*

Answer

I	*II*	*III*	*IV*
C	*B*	*A*	*D*

57. Match the following crop disease and their casual organism.

Disease, crop	*Causal organism*
I. Fruit rot of jackfruit	(A) *Rhizopus articarpi*
II. Inflorescence rot of jackfruit	(B) *Rhizopus articarpi*
III. Charcoal rot of jackfruit	(C) *Ustilana zonata*
IV. Downy mildew of grape	(D) *Plasmopara viticola*

Answer

I	*II*	*III*	*IV*
A	*B*	*C*	*D*

58. Match the following crop disease and their casual organism.

Disease, crop	*Causal organism*
I. Cigar end rot of banana	(A) *Verticillium fructigena*
II. Anthracnose of banana	(B) *Gloeosporium musarum*
III. Leaf spot of banana	(C) *Mycosphaerella musicola*
IV. Black finger of banana	(D) *Macrophoma musae*

Answer

I	*II*	*III*	*IV*
A	*B*	*C*	*D*

59. Match the following crop disease and their casual organism.

Disease, crop	*Causal organism*
I. Black rot, canker and leaf spot of apple and pear	(A) *Botryosphaeria obtusa*
II. Stem brown of apple and pear	(B) *Botryosphaeria ribis*
III. Stem black of apple and pear	(C) *Glomerella cingulata*
IV. Bitter rot of apple and pear	(D) *Coniothecium chomatosporum*

Answer

I	*II*	*III*	*IV*
A	*B*	*D*	*C*

60. Match the following crop disease and their casual organism.

Disease, crop	*Causal organism*
I. Scab of apple and pear	(A) *Venturia inaequalis*
II. Brooks fruit spot of apple	(B) *Mycosphaerella pomi*

III. Powdery mildew of apple and pear	(C) *Podosphaera leucotricha*
IV. Black root rot of apple	(D) *Xylaria mali*

Answer

I	*II*	*III*	*IV*
A	*B*	*C*	*D*

61. Match the following crop disease and their casual organism.

Disease, crop	*Causal organism*
I. Pink disease of mango	(A) *Erythricium salmonicolor*
II. Black mildew of mango	(B) *Botryosphaeria pruni-spinosae*
III. Dieback of peach and apricot	(C) *Meliola mangiferae*
IV. Powdery mildew of peach and apricot	(D) *Sphaerotheca pannosa*

Answer

I	*II*	*III*	*IV*
A	*C*	*B*	*D*

62. Match the following crop disease and their casual organism.

Disease, crop	*Causal organism*
I. Leaf spot of broad bean	(A) *Alternaria brassicae*
II. Rust of broad bean	(B) *Uromyces fabae*
III. Dry root rot of cluster bean	(C) *Macrophomina phaseoli*
IV. Rust of lima bean	(D) *Uromyces appendiculatus*

Answer

I	*II*	*III*	*IV*
A	*B*	*C*	*D*

63. Match the following crop disease and their casual organism.

Disease, crop	*Causal organism*
I. Ashy stem blight of lima bean	(A) *Cercospora cruenta*
II. Leaf spot of lima bean	(B) *Macrophomina phaseoli*
III. Black stalk rot of onion	(C) *Pleospora tarda*
IV. Leaky fruit rot of brinjal	(D) *Pythium aphanidermatum*

Answer

I	*II*	*III*	*IV*
B	*A*	*C*	*D*

64. Match the following crop disease and their casual organism.

Disease, crop	*Causal organism*
I. Wilt of chillies	(A) *Fusarium oxysporum* f. sp. *cepae*
II. Blight of chillies	(B) *Alternaria solani*
III. Black mold in onion	(C) *Aspergillus niger*
IV. Basal rot of onion	(D)*Fusarium annuum*

Answer

I	*II*	*III*	*IV*
D	*B*	*C*	*A*

65. Match the following crop disease and their casual organism.

Disease, crop	*Causal organism*
I. White blister/white rust of crucifers	(A) *Albugo candida*
II. Club root (Finger and toe disease) of crucifers	(B) *Plasmodiophora brassicae*
III. Blue mold rot of onion	(C) *Penicillium* spp
IV. Downey mildew of cucurbits	(D) *Pseudopernospora cubensis*

Answer

I	*II*	*III*	*IV*
A	*B*	*C*	*D*

66. Match the following crop disease and their casual organism.

Disease, crop	*Causal organism*
I. Rust of onion	(A) *Macrophomina phaseoli*
II. Downy mildew of onion	(B) *Peronospora destructor*
III. Black scurf of potato	(C) *Rhizoctonia solani*
IV. Charcoal rot of potato	(D) *Puccinia porri*

Answer

I	*II*	*III*	*IV*
D	*B*	*C*	*A*

67. Match the following crop disease and their casual organism.

Disease, crop	*Causal organism*
I. Wart of potato	(A) *Fusarium sambucinum* and *F. solani* var. *coeruleum*
II. Dry rot of potato	(B) *Synchytrium endobioticum*

III. Powdery scab of potato (C) *Spongospora subterranea*
IV. Leaf blotch of onion (D) *Cladosporium allii*

Answer

I	II	III	IV
B	A	C	D

68. Match the following crop disease and their casual organism.

Disease, crop	*Causal organism*
I. Black leg and end soft rot of potato	(A) *Erwinia carotovora*
II. Neck rot of onion	(B) *Botryotinia allii*
III. Charcoal rot of sweet potato	(C) *Macrophomina phaseoli*
IV. Soft rot of sweet potato	(D) *Rhizopus nigricans*

Answer

I	II	III	IV
A	B	C	D

69. Match the following crop disease and their casual organism.

Disease, crop	*Causal organism*
I. Black rot of sweet potato	(A) *Ceratostomella fimbriata*
II. Fruit rot of tomato	(B) *Phytophthora* spp
III. Foot rot and wilt of tomato	(C) *Sclerotium rolfsii*
IV. Powdery mildew of tomato	(D) *Leveillula taurica*

Answer

I	II	III	IV
A	B	C	D

70. Match the following crop disease and their casual organism.

Disease, crop	*Causal organism*
I. Net blotch of barley	(A) *Pyrenophora teres*
II. Spot blotch of barley	(B) *Ustilago hordei*
III. Root rot and foot rot of barley	(C) *Cochliobous sativus*
IV. Covered smut of oat	(D) *Bipolaris sorokiniana*

Answer

I	II	III	IV
A	C	D	B

71. Match the following crop disease and their casual organism.

Disease, crop	*Causal organism*
I. Blast of rice	(A) *Piricularia oryzae*
II. Brown spot of rice	(B) *Helminthosporium sigmoidium*
III. Irregular stem rot of rice	(C) *Cochliobolus miyabeanus*
IV. Stem rot of rice	(D) *Leptosphaeria salvinii*

Answer

I	*II*	*III*	*IV*
A	*C*	*B*	*D*

72. Match the following crop disease and their casual organism.

Disease, crop	*Causal organism*
I. Sheath blight rot of rice	(A) *Rhizoctonia solani*
II. Foot rot or bakanae disease	(B) *Gibberella fujikuroi*
III. Bunt of rice	(C) *Neovossia horrida*
IV. Stack burn disease of rice	(D) *Trichoconis padwickii*

Answer

I	*II*	*III*	*IV*
A	*B*	*C*	*D*

73. Match the following crop disease and their casual organism.

Disease, crop	*Causal organism*
I. Udbatta disese of rice	(A) *Ephelis oryzae*
II. Leaf blotch of wheat	(B) *Septoria tritici*
III. Black mold of wheat	(C) *Cladosporium herbarum*
IV. Glume blotch of wheat	(D) *Septoria nodorum*

Answer

I	*II*	*III*	*IV*
A	*B*	*C*	*D*

74. Match the following crop disease and their casual organism.

Disease, crop	*Causal organism*
I. Foot rot of wheat	(A) *Helminthosporium sativum* and *Fusarium culmorum*
II. Brown spot of maize	(B) *Physoderma maydis*
III. Head smut of maize	(C) *Sphacelotheca reiliana*
IV. Charcoal rot of maize	(D) *Macrophomina phaseolina*

Answer

I	*II*	*III*	*IV*
A	*B*	*C*	*D*

75. Match the following crop disease and their casual organism.

Disease, crop	*Causal organism*
I. Seedling blight of maize	(A) *Pythium aphanidermatum, Fusarium moniliforme*
II. Green ear disease of bajra	(B) *Sclerospora graminicola*
III. Ergot of bajra	(C) *Claviceps microcephala*
IV. Smut of bajra	(D) *Tolyposporium penicillariae* and *T. senegalense*

Answer

I	*II*	*III*	*IV*
A	*B*	*C*	*D*

76. Match the following crop disease and their casual organism.

Disease, crop	*Causal organism*
I. Zonate leaf spot of bajra	(A) *Gloeocercospora sorghi*
II. Grain smut of jowar	(B) *Sphacelotheca sorghi*
III. Downy mildew of jowar	(C) *Sclerospora sorghi*
IV. Shooty strip of jowar	(D) *Ramularie sorghi*

Answer

I	*II*	*III*	*IV*
A	*B*	*C*	*D*

77. Match the following crop disease and their casual organism.

Disease, crop	*Causal organism*
I. Latter leaf spot of sorghum	(A) *Cercospora fusimaculans*
II. Ergot of sorghum	(B) *Claviceps sorghi*
III. Anthracnose of cowpea	(C) *Glomerella lindemuthianum*
IV. Die black of cowpea	(D) *Colletotrichum capsici*

Answer

I	*II*	*III*	*IV*
A	*B*	*C*	*D*

78. Match the following crop disease and their casual organism.

Disease, crop	*Causal organism*
I. Dry root rot of cowpea	(A) *Pelliculari filamentosa*
II. Root rot of cowpea	(B) *Macrophomina phaseoli*
III. Rust rot of cowpea	(C) *Ascochyta rabiei*
IV. Blight of gram	(D) *Uromyces appendiculatus*

Answer

I	*II*	*III*	*IV*
B	*A*	*D*	*C*

79. Match the following crop disease and their casual organism.

Disease, crop	*Causal organism*
I. Root rot of gram	(A) *Rhizoctonia bataticola*
II. Wilt of gram	(B) *Fusarium oxysporum* var. *ciceri*
III. Collar rot of groundnut	(C) *Pellicularia (Sclerotium) rolfsii*
IV. Leaf spot of sunflower	(D) *Alternaria tenuis*

Answer

I	*II*	*III*	*IV*
A	*B*	*C*	*D*

80. Match the following crop disease and their casual organism.

Disease, crop	*Causal organism*
I. Grey mildew of cotton	(A) *Ramularia areola*
II. Powdery mildew of sunn hemp	(B) *Golovinomyces cichoracearum*
III. Black band disease of jute	(C) *Diplodia corchori*
IV. Stem rot of mesta	(D) *Sclerotinia sclerotiorum*

Answer

I	*II*	*III*	*IV*
A	*B*	*C*	*D*

81. Match the following crop disease and their casual organism.

Disease, crop	*Causal organism*
I. Crazy top downy mildew of sorghum	(A) *Sclerophthora macrospora*
II. Leaf spot of methi	(B) *Cercospora traversiana*
III. Rust of sorghum	(C) *Puccinia purpurea*
IV. Pink root of onion	(D) *Phoma terrestris*

Answer

I	*II*	*III*	*IV*
A	*B*	*C*	*D*

82. Match the following entry of important disease into India.

Disease	*Introduced from/Year*
I. Leaf rust of coffee	(A) Sri Lanka (1879)
II. Late blight of potato	(B) Europe (1883)
III. Flag smut of wheat	(C) Australia (1906)
IV. Downey mildew of grape	(D) Europe (1910)

Answer

I	*II*	*III*	*IV*
A	*B*	*C*	*D*

83. Match the following entry of important disease into India.

Disease	*Introduced from/Year*
I. Downey mildew of cucurbits	(A) Dutch East (1938)
II. Downey mildew of maize	(B) Java (1912)
III. Powdery mildew of rubber	(C) Malaya (1938)
IV. Black shank of tobacco	(D) Sri Lanka (1910)

Answer

I	*II*	*III*	*IV*
D	*B*	*C*	*A*

84. Match the following entry of important disease into India.

Disease	*Introduced from/Year*
I. Onion smut	(A) Europe (1958)
II. Crown gall of apple and pear	(B) The Netherlands (1953)
III. Wart disease of potato	(C) England (1940)
IV. Bunchy top of banana	(D) Sri Lanka (1940)

Answer

I	*II*	*III*	*IV*
A	*C*	*B*	*D*

85. Match the important antagonistic microorganisms that reduce the amount of pathogenic inoculums:

Antagonist	*Pathogen*
I. *Laetisaria arvalis* (*Corticium* sp.)	(A) *Rhizoctonia* and *Pythium* sp.
II. *Sporidesmium sclerotivorum*	(B) *Sclerotinia sclerotiorum*
III. *Coniothyrium minitants*	(C) *Sclerotinia sclerotiorum*
IV. *Chaetomium* sp.	(D) *Venturia inaequalis*

Answer

I	*II*	*III*	*IV*
A	*B*	*C*	*D*

86. Match the important antagonistic microorganisms that reduce the amount of pathogenic inoculums:

Antagonist	*Pathogen*
I. *Verticillium lecanii*	(A) Powdery mildew fungi
II. *Ampelomyces quisqualis*	(B) Rust fungi
III. *Nectrai inventa*	(C) *Alternaria* sp.
IV. *Streptomyces* sp.	(D) *Pythium* sp.

Answer

I	*II*	*III*	*IV*
B	*A*	*C*	*D*

87. Match the important antagonistic microorganisms that reduce the amount of pathogenic inoculums:

Antagonist	*Pathogen*
I. *Tuberculina maxima*	(A) *Cronartium ribicola*
II. *Tilletiopsis* sp.	(B) *Sphaerotheca fuligena*
III. *Bacillus ceres*	(C) *Gaeumannomyces graminis*
IV. *Scytalidium uridinicola*	(D) *Poria carbonica* (decay of Douglas fir)

Answer

I	*II*	*III*	*IV*
A	*B*	*C*	*D*

88. Match the following phenolic compounds formed in response to injury or infection into the host.

Phenol	*Host*
I. Caffeic acid	(A) Sweet potato
II. Umbliferon	(B) Carrot
III. Phloretin	(C) Apple
IV. Hydroquinone	(D) Pear

Answer

I	*II*	*III*	*IV*
A	*B*	*D*	*C*

89. Correctly pair the synonyms of some common plant pathogens in section 'A' to those given under section 'B'

Section 'A'	*Section 'B'*
I. *Alternaria padwickii*	(A) *Macrosporium solani*
II. *Alternaria solani*	(B) *Tricoconis padwickii*

III. *Ascochyta pisi* (C) *Ascochyta pisicola*

IV. *Colletotrichum lindemuthianum* (D) *Gloeosporium lindemuthianum*

Answer

I	*II*	*III*	*IV*
B	*A*	*C*	*D*

90. Correctly pair the synonyms of some common plant pathogens in section 'A' to those given under section 'B'

Section 'A'	*Section 'B'*
I. *Diplodia maydis*	(A) *Cladosporium fulvum*
II. *Drechslera poae*	(B) *Diplodia zeae*
III. *Fulvia fulva*	(C) *Helminthosporium vagans*
IV. *Gaeumannomyces graminis*	(D) *Ophiobolus graminis*

Answer

I	*II*	*III*	*IV*
B	*C*	*A*	*D*

91. Correctly pair the synonyms of some common plant pathogens in section 'A' to those given under section 'B'

Section 'A'	*Section 'B'*
I. *Gloeocercospora annosum*	(A) *Ramulispora andropogonis*
II. *Heterobasidion annosum*	(B) *Fomes annosus*
III. *Koleroga noxia*	(C) *Corticium koleroga*
IV. *Macrophomina phaseolina*	(D) *Rhizoctonia bataticola*

Answer

I	*II*	*III*	*IV*
A	*B*	*C*	*D*

92. Correctly pair the synonyms of some common plant pathogens in section 'A' to those given under section 'B'

Section 'A'	*Section 'B'*
I. *Magnaporthe salvinii*	(A) *Leptosphaeria salvinii*
II. *Monilinia fructicola*	(B) *Sclerotinia fructicola*
III. *Mycosphaerell pinodes*	(C) *Didymella pinodes*
IV. *Phlebia gigantea*	(D) *Peniophora gigantea*

Answer

I	*II*	*III*	*IV*
A	*B*	*C*	*D*

93. Correctly pair the synonyms of some common plant pathogens in section 'A' to those given under section 'B'

	Section 'A'		*Section 'B'*
I.	*Phoma medicaginis*	(A)	*Ascochyta imperfecta*
II.	*Phytophthora nicotianae*	(B)	*Phytophthora parasitica*
III.	*Pithomyces chartarum*	(C)	*Sporidesmium bakeri*
IV.	*Puccinia recondita*	(D)	*Puccinia dispersa*

Answer

I	*II*	*III*	*IV*
A	*B*	*C*	*D*

94. Correctly pair the synonyms of some common plant pathogens in section 'A' to those given under section 'B'

	Section 'A'		*Section 'B'*
I.	*Rhizopus stolonifer*	(A)	*Rhizopus nigricans*
II.	*Puccinia striiformis*	(B)	*Puccinia glumarum*
III.	*Pyricularia oryzae*	(C)	*Dactylaria oryzae*
IV.	*Tilletia barclayana*	(D)	*Neovosia horrida*

Answer

I	*II*	*III*	*IV*
A	*B*	*C*	*D*

95. Match the following phytopathagenic toxin/host plant with target site.

	Phytopathogenic toxin/host plant		*Action or target site*
I.	AK toxin/Strawberry	(A)	Membrane
II.	AT toxin/Tobacco	(B)	Mitochondria
III.	Ten toxin/Potato	(C)	Chloroplast
IV.	Fusaric acid/Vegetables	(D)	Water permeability

Answer

I	*II*	*III*	*IV*
A	*B*	*C*	*D*

96. Match the following phytopathagenic toxin/host plant with target site.

	Phytopathogenic toxin/host plant		*Action or target site*
I.	Tab toxin/Tobacco	(A)	Glutamine
II.	Phaseolo toxin/Bean	(B)	Arginine synthesis
III.	Gibberellin/Rice	(C)	Plant hormone
IV.	Helminthosporol/Barley	(D)	Respiration

Answer

I	*II*	*III*	*IV*
A	*B*	*C*	*D*

97. Match the information of group 'A' with that of group 'B'

Group 'A'	*Group 'B'*
I. Vander Plank	(A) Gene for gene hypothesis
II. Flor	(B) Vertical and horizontal resistance
III. Painter	(C) Mechanism of insect resistance
IV. Striga	(D) Sorghum

Answer

I	*II*	*III*	*IV*
B	*A*	*C*	*D*

98. Match the information of group 'A' with that of group 'B'

Group 'A'	*Group 'B'*
I. Vertical resistance	(A) Major gene resistance
II. Horizontal resistance	(B) Minor gene resistance
III. Antixenosis	(C) Non preference
IV. Physiological race	(D) Pathotype

Answer

I	*II*	*III*	*IV*
A	*B*	*C*	*D*

99. Match the following crop disease and their casual organism.

Disease, crops	*Causal organism*
I. Soft rot (Rhizome rot) of turmeric	(A) *Pythium* sp.
II. Bligh of fennel	(B) *Ramularia foeniculi*
III. Brown root disease of coffee	(C) *Phellinus noxius*
IV. Powdery mildew of grape	(D) *Uncinula necator*

Answer

I	*II*	*III*	*IV*
A	*B*	*C*	*D*

100. Match the following crop disease and their casual organism.

Disease, crops	*Causal organism*
I. Rust of grape	(A) *Phakopsora euvitis*
II. Panama wilt of banana	(B) *Fusarium oxysporum* var. *cubense*
III. Sigatoka of banana	(C) *Mycosphaerella fijiensis*
IV. Bed leaf spot of jowar	(D) *Colletotrichum graminicola*

Answer

I	*II*	*III*	*IV*
A	*B*	*C*	*D*

Answers (Multiple choice questions)

1.	(A)	33.	(A)	65.	(A)	97.	(A)
2.	(A)	34.	(A)	66.	(D)	98.	(B)
3.	(B)	35.	(C)	67.	(A)	99.	(D)
4.	(B)	36.	(B)	68.	(A)	100.	(A)
5.	(C)	37.	(D)	69.	(D)	101.	(B)
6.	(A)	38.	(C)	70.	(C)	102.	(B)
7.	(D)	39.	(A)	71.	(C)	103.	(C)
8.	(B)	40.	(B)	72.	(A)	104.	(A)
9.	(A)	41.	(C)	73.	(A)	105.	(B)
10.	(B)	42.	(A)	74.	(A)	106.	(B)
11.	(A)	43.	(C)	75.	(B)	107.	(A)
12.	(B)	44.	(D)	76.	(C)	108.	(C)
13.	(A)	45.	(B)	77.	(A)	109.	(B)
14.	(B)	46.	(B)	78.	(B)	110.	(A)
15.	(C)	47.	(A)	79.	(A)	111.	(D)
16.	(A)	48.	(A)	80.	(B)	112.	(A)
17.	(D)	49.	(A)	81.	(C)	113.	(C)
18.	(D)	50.	(A)	82.	(A)	114.	(A)
19.	(C)	51.	(A)	83.	(C)	115.	(C)
20.	(C)	52.	(C)	84.	(B)	116.	(C)
21.	(D)	53.	(B)	85.	(C)	117.	(A)
22.	(D)	54.	(C)	86.	(B)	118.	(B)
23.	(B)	55.	(C)	87.	(D)	119.	(A)
24.	(B)	56.	(A)	88.	(A)	120.	(B)
25.	(C)	57.	(B)	89.	(C)	121.	(A)
26.	(D)	58.	(A)	90.	(A)	122.	(A)
27.	(B)	59.	(A)	91.	(A)	123.	(D)
28.	(B)	60.	(B)	92.	(D)	124.	(A)
29.	(C)	61.	(C)	93.	(B)	125.	(B)
30.	(A)	62.	(A)	94.	(A)	126.	(A)
31.	(C)	63.	(C)	95.	(B)	127.	(A)
32.	(B)	64.	(A)	96.	(A)	128.	(B)

129.	(C)	163.	(D)	197.	(A)	231.	(B)
130.	(A)	164.	(A)	198.	(C)	232.	(A)
131.	(C)	165.	(A)	199.	(B)	233.	(A)
132.	(B)	166.	(B)	200.	(D)	234.	(A)
133.	(B)	167.	(C)	201.	(D)	235.	(B)
134.	(D)	168.	(B)	202.	(D)	236.	(B)
135.	(A)	169.	(A)	203.	(C)	237.	(A)
136.	(D)	170.	(B)	204.	(B)	238.	(A)
137.	(A)	171.	(D)	205.	(D)	239.	(A)
138.	(A)	172.	(A)	206.	(C)	240.	(A)
139.	(A)	173.	(B)	207.	(A)	241.	(A)
140.	(B)	174.	(D)	208.	(C)	242.	(B)
141.	(A)	175.	(A)	209.	(A)	243.	(A)
142.	(A)	176.	(C)	210.	(B)	244.	(B)
143.	(B)	177.	(D)	211.	(C)	245.	(A)
144.	(A)	178.	(C)	212.	(A)	246.	(B)
145.	(A)	179.	(B)	213.	(D)	247.	(A)
146.	(A)	180.	(B)	214.	(D)	248.	(A)
147.	(B)	181.	(C)	215.	(B)	249.	(B)
148.	(B)	182.	(A)	216.	(C)	250.	(B)
149.	(B)	183.	(C)	217.	(A)	251.	(A)
150.	(C)	184.	(C)	218.	(D)	252.	(D)
151.	(A)	185.	(C)	219.	(B)	253.	(C)
152.	(C)	186.	(C)	220.	(A)	254.	(A)
153.	(A)	187.	(A)	221.	(A)	255.	(C)
154.	(C)	188.	(C)	222.	(B)	256.	(B)
155.	(A)	189.	(A)	223.	(A)	257.	(A)
156.	(D)	190.	(B)	224.	(C)	258.	(B)
157.	(B)	191.	(B)	225.	(C)	259.	(A)
158.	(A)	192.	(A)	226.	(B)	260.	(B)
159.	(D)	193.	(B)	227.	(B)	261.	(C)
160.	(A)	194.	(A)	228.	(B)	262.	(C)
161.	(B)	195.	(A)	229.	(C)	263.	(A)
162.	(D)	196.	(B)	230.	(A)	264.	(A)

265.	(B)	299.	(A)	333.	(B)	367.	(A)
266.	(A)	300.	(A)	334.	(A)	368.	(C)
267.	(A)	301.	(A)	335.	(B)	369.	(B)
268.	(A)	302.	(D)	336.	(C)	370.	(D)
269.	(B)	303.	(A)	337.	(C)	371.	(C)
270.	(C)	304.	(D)	338.	(D)	372.	(C)
271.	(A)	305.	(D)	339.	(A)	373.	(D)
272.	(B)	306.	(A)	340.	(A)	374.	(A)
273.	(A)	307.	(C)	341.	(B)	375.	(C)
274.	(A)	308.	(B)	342.	(B)	376.	(C)
275.	(A)	309.	(A)	343.	(A)	377.	(D)
276.	(C)	310.	(A)	344.	(A)	378.	(A)
277.	(B)	311.	(C)	345.	(A)	379.	(B)
278.	(D)	312.	(C)	346.	(B)	380.	(C)
279.	(B)	313.	(A)	347.	(B)	381.	(C)
280.	(B)	314.	(B)	348.	(A)	382.	(A)
281.	(C)	315.	(D)	349.	(A)	383.	(A)
282.	(B)	316.	(A)	350.	(C)	384.	(B)
283.	(A)	317.	(A)	351.	(C)	385.	(D)
284.	(A)	318.	(C)	352.	(B)	386.	(B)
285.	(B)	319.	(A)	353.	(D)	387.	(C)
286.	(B)	320.	(B)	354.	(B)	388.	(A)
287.	(B)	321.	(B)	355.	(A)	389.	(D)
288.	(A)	322.	(A)	356.	(B)	390.	(A)
289.	(A)	323.	(B)	357.	(A)	391.	(A)
290.	(C)	324.	(A)	358.	(C)	392.	(D)
191.	(B)	325.	(A)	359.	(A)	393.	(A)
292.	(A)	326.	(D)	360.	(C)	394.	(B)
293.	(C)	327.	(B)	361.	(A)	395	(B)
294.	(A)	328.	(B)	362.	(B)	396.	(C)
295.	(B)	329.	(B)	363.	(C)	397.	(A)
296.	(C)	330.	(A)	364.	(D)	398.	(A)
297.	(B)	331.	(C)	365.	(B)	399.	(A)
298.	(A)	332.	(A)	366.	(A)	400.	(B)

401.	(B)	435.	(B)	469.	(B)	503.	(A)
402.	(C)	436.	(D)	470.	(D)	504.	(A)
403.	(A)	437.	(A)	471.	(D)	505.	(B)
404.	(A)	438.	(A)	472.	(A)	506.	(C)
405.	(A)	439.	(A)	473.	(B)	507.	(D)
406.	(D)	440.	(B)	474.	(A)	508.	(A)
407.	(D)	441.	(A)	475.	(A)	509.	(C)
408.	(C)	442.	(D)	476.	(C)	510.	(D)
409.	(A)	443.	(B)	477.	(A)	511.	(D)
410.	(B)	444.	(C)	478.	(C)	512.	(A)
411.	(A)	445.	(A)	479.	(B)	513.	(C)
412.	(B)	446.	(C)	480.	(B)	514.	(B)
413.	(A)	447.	(A)	481.	(A)	515.	(A)
414.	(B)	448.	(C)	482.	(D)	516.	(C)
415.	(C)	449.	(B)	483.	(A)	517.	(A)
416.	(B)	450.	(A)	484.	(A)	518.	(C)
417.	(D)	451.	(A)	485.	(B)	519.	(B)
418.	(D)	452.	(B)	486.	(A)	520.	(A)
419.	(C)	453.	(A)	487.	(D)	521.	(B)
420.	(A)	454.	(A)	488.	(B)	522.	(D)
421.	(B)	455.	(B)	489.	(A)	523.	(A)
422.	(A)	456.	(A)	490.	(C)	524.	(A)
423.	(B)	457.	(D)	491.	(A)	525.	(D)
424.	(A)	458.	(C)	492.	(A)	526.	(C)
425.	(C)	459.	(A)	493.	(D)	527.	(A)
426.	(A)	460.	(B)	494.	(A)	528.	(B)
427.	(B)	461.	(A)	495.	(C)	529.	(B)
428.	(C)	462.	(B)	496.	(A)	530.	(A)
429.	(D)	463.	(D)	497.	(A)	531.	(C)
430.	(A)	464.	(C)	498.	(A)	532.	(D)
431.	(D)	465.	(D)	499.	(D)	533.	(A)
432.	(C)	466.	(A)	500.	(B)	534.	(D)
433.	(D)	467.	(C)	501.	(C)	535.	(A)
434.	(A)	468.	(C)	502.	(A)	536.	(C)

537.	(B)	571.	(D)	605.	(B)	639.	(B)
538.	(A)	572.	(A)	606.	(C)	640.	(A)
539.	(A)	573.	(D)	607.	(D)	641.	(A)
540.	(B)	574.	(C)	608.	(A)	642.	(A)
541.	(A)	575.	(C)	609.	(C)	643.	(B)
542.	(A)	576.	(A)	610.	(A)	644.	(A)
543.	(D)	577.	(A)	611.	(A)	645.	(D)
544.	(A)	578.	(C)	612.	(D)	646.	(A)
545.	(A)	579.	(B)	613.	(A)	647.	(B)
546.	(C)	580.	(A)	614.	(A)	648.	(A)
547.	(C)	581.	(B)	615.	(C)	649.	(B)
548.	(A)	582.	(B)	616.	(B)	650.	(A)
549.	(C)	583.	(A)	617.	(B)	651.	(D)
550.	(B)	584.	(B)	618.	(A)	652.	(A)
551.	(B)	585.	(A)	619.	(A)	653.	(B)
552.	(A)	586.	(D)	620.	(A)	654.	(C)
553.	(D)	587.	(A)	621.	(B)	655.	(B)
554.	(B)	588.	(A)	622.	(C)	656.	(B)
555.	(A)	589.	(D)	623.	(C)	657.	(B)
556.	(C)	590.	(C)	624.	(A)	658.	(A)
557.	(A)	591.	(B)	625.	(A)	659.	(A)
558.	(D)	592.	(A)	626.	(B)	660.	(B)
559.	(A)	593.	(C)	627.	(C)	661.	(A)
560.	(C)	594.	(A)	628.	(A)	662.	(C)
561.	(B)	595.	(A)	629.	(A)	663.	(A)
562.	(D)	596.	(B)	630.	(D)	664.	(A)
563.	(A)	597.	(A)	631.	(A)	665.	(D)
564.	(D)	598.	(B)	632.	(C)	666.	(A)
565.	(A)	599.	(B)	633.	(A)	667.	(C)
566.	(A)	600.	(A)	634.	(A)	668.	(A)
567.	(B)	601.	(C)	635.	(B)	669.	(C)
568.	(B)	602.	(A)	636.	(D)	670.	(A)
569.	(B)	603.	(B)	637.	(C)	671.	(A)
570.	(B)	604.	(C)	638.	(A)	672.	(B)

673.	(C)	685.	(A)	697.	(A)	709.	(A)
674.	(A)	686.	(C)	698.	(B)	710.	(A)
675.	(B)	687.	(C)	699.	(A)	711.	(A)
676.	(A)	688.	(D)	700.	(A)	712.	(A)
677.	(C)	689.	(D)	701.	(B)	713.	(C)
678.	(B)	690.	(A)	702.	(B)	714.	(D)
679.	(B)	691.	(A)	703.	(A)	715.	(A)
680.	(A)	692.	(C)	704.	(A)	716.	(C)
681.	(A)	693.	(A)	705.	(A)	717.	(A)
682.	(A)	694.	(A)	706.	(A)		
683.	(C)	695.	(B)	707.	(B)		
684.	(B)	696.	(C)	708.	(A)		

Chapter 2
Phytobacteriology

Queries with Timeline of Phytobacteriology

1. During the 13th century the first simple lenses was invented by Roger Bacon. It is not known if bacteria were unseen by these inventors. History of bacteriology started when Antonie Philips van Leeuwenhoek, Dutch worker (1632-1723) observed bacteria for the first time in 1675 with microscope. He was the first person to see the bacteria and protozoa under his lenses and invented first simple microscope (Holland). He saw the bacteria in body fluids, in water and called these discoveries of 'animalcules' or 'little animal'.

2. In 1857, Louis Pasteur demonstrated that microorganisms are capable of causing fermentation. Pasteur introduced the terms 'aerobic' and 'anaerobic' *i.e.* life in presence or absence of oxygen. He also demonstrated that in presence of oxygen, sugar was converted to CO_2 by yeasts and alcohol + CO_2 were formed only in the absence of oxygen.

3. The history of phytobacteriology can be traced to late 19th century. Thomas J. Burrill, the American plant pathologist of the University of Illinois, was the first to describe bacterial diseases of plants in 1878 and proved by an inoculation test that fire blight of pear was caused by a bacterium, which he named as *Micrococcus amylovorus* (today known as *Erwinia amylovora*) in 1882. Jan Wakker (1883) in Holland, provided strong evidence of the involvement of Bacterium *Hyacinthi* (today *Xanthomonas hyacinthi*) as the causal agent of yellow disease of hyacinth.

4. However, the idea that bacteria are able to cause diseases in plants was still not accepted by many phytopathologists. This occurred despite the fact that in 1876 Robert Koch had already demonstrated that the bacterium *Bacillus*

anthracis is responsible for anthrax disease. In 1885, Joseph C. Arthur in the U.S., confirmed Burrill's results with the fire blight.

5. Robert Koch (1875) established pathogenicity rules (proved that a particular organism is the cause of a particular disease) known as 'Koch's Postulates', known as the father of 'Microbial Techniques' for his outstanding contributions in microbiology. Robert Koch (1876) proved bacterial etiology of anthrax disease. He also demonstrated the biological specificity of the disease agent.
6. An American, Erwin F. Smith (1895) finally proved the fact that *Erwinia amylovora* was indeed the causal agent of fire blight disease. He was first to noticed the study of crown gall disease (*Agrobacterium tumefaciens*) in 1893-94 and also worked on the bacterial wilt of cucurbits. He also investigated important plant pathogenic bacteria *i.e. Xanthomonas, Pseudomonas.* In view of these significant contributions, Smith is considered the 'Father of Modern Phytobacteriology'. He wrote a book '**Bacteria in Relation to Plant Diseases**'. He developed various bacteriological techniques/methods.
7. Alfred Fisher (1897, 1901) a professor of botany at Leipzing in Germany had the opinion that bacteria could not enter uninjured plant body and hence, could not cause disease. He challenged the work of Burrill and Smith. However, E.F. Smith (1899, 1901) meticulously planned and executed experiments and demolished the prejudiced notions of Fisher.
8. C.O. Jensen (1910) demonstrated evidences ascertaining similarity between crown gall disease of plants and cancer of animals. G.H. Koons and J.E. Kotila (1925) isolated bacteriophages of *Bacillus carotovorus.* R.M. Klein (1954) discovered basic process of transformation in crown gall. A.C. Braun (1955) reported about wild fire toxin and its action. A.K. Chatterjee and M.P. Starr (1972) discovered conjugative gene transfer in bacteria. P.B. New and A. Kerr (1972) reported bio-control of crown gall (*Agrobacterium radiobacter* Strain 84).
9. I.M.Windsor and I.M. Black in 1972 observed Rickettsia like bacteria (RLB or RLO) in the phloem of clover plants infected with the club leaf disease. In the following year similar organisms were observed in the xylem vessels of a grape plant infected with pierces, in peach infected with phony peach disease, alfalfa infected with alfalfa dwarf disease etc.
10. Clayton in 1934 confirmed an earlier suggestion made by Johnson and Marwin (1925) that the bacterium, *Pseudomonas tabaci* which causes the wildfire disease of tobacco produces a toxin which is responsible for the bacteria free chlorotic zone (halo) surrounding the necrotic lesion that contain the bacteria.
11. D.W. Dye *et al.* (1980) introduced pathovar (pv.) system in taxonomy of plant pathogenic bacteria, a specific epithet of classification is the pathovar (pv.) designation, which is defined as 'an infrasubspecific term referring to a group of phytopathogenic bacteria differentiated principally on the basis of their host range'. The pathovar classification is very useful to define host specificity of a particular bacterium.

12. In 1952, Hershey and Chase showed that in bacterial viruses (phages) only the DNA of the phages entered the host bacteria and carried information into the bacteria.

13. Klambt *et al.* and Helgson and Leonard in 1966, showed that bacterium *Corynebacterium fascians* which causes fasciations disease of pea, produces a cytokinin and IAA. Although IAA (auxin) and cytokinin were shown to accumulate in plants affected by disease that showed symptoms of overgrowth, wilts and various malformations and stunted plants, they also causes tumor formation in crown gall disease of grapes (*Agrobacterium tumefaciens*) due to high concentration of IAA and cytokinin.

14. The first case of altered permeability was reported by Hutchinson (1913) for Rangpur tobacco wilt caused by *Pseudomonas solanacearum* (now *Ralstonia solanacearum*). Permeability phenomena in plant diseases have also been reviewed by Wheeler (1976).

15. The first antibiotic, penicillin was discovered in 1929 by Alexander Fleming. H.W. Florey and E. Chain, developed the technique of industrial production of penicillin during world war-II. Streptomycin antibiotic was discovered in 1943 by A. Schatz and S.A. Waksman. Streptomycin is effective against both gram negative and positive bacteria.

16. Bacterial blight of cotton was first reported in Alabama, USA in 1891 and in India, it was firstly reported in 1918 from Rajapalayam, Tamil Nadu., Bacterial blight disease of rice was first observed in 1956, in India. The first bacterial pathogen to be recorded as seed borne was *Xanthomonas stewarti* from corn by Smith in 1909. Seed transmission of *X. phaseoli* causing common bean blight was reported by Orton in 1931.

17. M.D. Chilton *et al.* (1977) first to introduced plasmid DNA of *Agrobacterium tumefaciens* into the cell of higher plants and proved that part of this DNA inserts into DNA of plant cell.

18. In 1977, Sanger *et al.*, determined the complete sequence of the 5386 base pair of the DNA phage f x 174.

19. *Pseudomonas tabaci* on tobacco, *Erwinia amylovora* infecting apples and pear, *Xanthomonas campestris* pv. *phaseoli* infecting *Phaseolus* spp, *Xanthomonas campestris* pv. *malvacearum* infecting cotton and *Pseudomonas phaseolicola* infecting *Phaseolus* spp enter the host plants through film of water in stomata.

20. *Streptomyces scabies* (common scab of potato) enters in young developing tubers through stomata or lenticels. *Pectobacterium carotovorum* (it causes beet vascular necrosis and blackleg of potato) enters through wound or lenticels. *Erwinia amylovora* enters through nectaries of apple and pear flowers. *Corynebacterium michiganense* (canker of tomato) enters through trichomes.

21. *Xanthomonas campestris* pv. *oryzae* enters through hydathodes. Vascular wilt of cucurbits is caused by *Erwinia tracheiphila* which overwinters in hibernating

adults of cucumber beetles. *Erwinia amylovora* causes fire blight of apple and pear leaves which overwinter in cankers and carried to the blossoms by insects.

22. Pectic enzymes are most important in soft rot diseases in which the parenchymatous tissue of the host plants are rapidly invaded by *Pectobacterium carotovorum* and *Pseudomonas marginalis* responsible for soft rot. Cellulolytic enzymes have been implicated in wilt disease caused by *Pseudomonas solanacearum*.

23. Roux and Yersin in 1888, who first visualized the presence of a toxin in a disease caused by *Corynebacterium diptheriae*. Wild fire disease of tobacco is caused by *Pseudomonas syringae* pv. *tabaci*, produces a pathotoxin called tabtoxin. Woolly *et al.* (1952) the wild fire toxin was the first toxin to be isolated in pure form.

24. Tabtoxin inactivates the enzymes glutamine synthetase, which leads to depleted levels of glutamine and as a consequence, causes accumulation of ammonia to toxic levels. *Pseudomonas solanacearum* the cause of wilt of solanaceous plants induces a 100 fold increase in the IAA level of disease plants compared with that of healthy plants. It is also the causal agent of moko disease of banana.

25. The knot disease of olive (Hyperplastic disease) caused by *Pseudomonas savastanoi*, the pathogen produces IAA which induces infected plants to produce galls. A cytokinin is partly responsible for several bacterial galls of plants, such as 'leafy' gall disease of sweet pea caused by *Rhodococcus* (*Corynebacterium fascians*).

26. In the fruit of banana infected with *Pseudomonas solanacearum*, the ethylene content increases proportionately with (premature) yellowing of the fruit, whereas no ethylene can be detected in healthy fruits.

27. Transgenic potato plants expressing the gene for antibacterial enzyme T4 lysozyme exhibited resistance to soft rot and blackleg caused by the *Erwinia carotovora* pv. *atroseptica*. Wilting in moko disease of banana caused by *Pseudomonas solanacearum*, it is one of the two causes (i) vascular occlusion and (ii) a toxic effect which would influence water retention by host cells.

28. Pea shoots resistance to *Erwinia amylovora* is due to presence of arbutin phenols. Tumor cells formed by *Agrobacterium tumefaciens* contain higher amount of IAA than normal ones. *Erwinia amylovora* does not produce ethylene in culture but when cauliflower florets are inoculated with this bacterium, it stimulates ethylene production in them.

29. Formation of an abscission layer (upon infections) in peach leaves by the bacterium, *Xanthomonas arboricola* pv. *pruni* leads to prevention and further spread of the pathogen resulting greater resistance. In case of resistance of pears to fire blight caused by *Erwinia amylovora*, it has been found that resistance in certain tissues is related to the presence of phenolic glucoside arbutin.

30. Potato tubers contain higher amount of chlorogenic acid which inhibits the growth of *Streptomyces scabies*. Pathogenic races of *Xanthomonas campestris* pv.

malvacearum (angular leaf spot of cotton) contain more antigens in cotton leaves than non-pathogenic races of bacteria.

31. In 1928, Frederick Griffith was the first experiment suggested that bacteria are capable of transferring genetic information through a process known as transformation. He worked with the bacterium *Pneumococcus pneumonia* (then called *Diplococcus*), the causal organism of *Pneumonia*.

32. Avery Macleod and Mc Carty (1944) the American workers, chemically identified the transformation principle, that was DNA. He proved first time that DNA was the genetic material.

33. Transduction was discovered in 1953 by Joshua Lederberg and Norton Zinder during their search for evidence of conjugation in *Salmonella typhimurium*. Transduction was also demonstrated in 1955 by Okabe and Goto in a plant pathogenic bacterium (*Pseudomonas solanacearum*), transduction has a role in producing new biotypes of plant pathogenic bacteria in nature. Harris-Warrich and Lederberg in 1978, reported inter specific transduction in *Bacillus*.

34. Conjugation process was first postulated by Joshua Lederberg and Edward Tatum in a series of experiments in 1946, who studied two strains of *Escherichia coli* with different nutritional requirement. Strain A would grow on a minimal medium only if the medium were supplemented with methionine and biotin; strain B would grow on a minimal medium only if it were supplemented with threonine, leucine, and thiamine. Cells of type A or type B cannot grow on an unsupplemented (minimal) medium (MM), because A and B each carry mutations that cause the inability to synthesize constituents needed for cell growth.

35 Tatum and Lederberg in 1946, reported genetic recombination in *Escherichia coli*, during conjugation recombination process. The frequency of recombination was up to 10,000 times in *E. coli*, greater than observed in F^+ strains. These were called Hfr (High fertility rate) strains. Recombination of genetic factor occurred during the meiotic division of the zygotes.

36. Lelliott and Stead (1987) wrote a book entitled 'Methods for the Diagnosis of Bacterial Diseases of Plants' and 'Guide to Plant Pathogenic Bacteria' written by Bradbury in 1986. G. Rangaswami worked on bacterial disease and authored the book entitled '**Disease of Crop Plants in India**' and '**Bacterial Plant Disease in India**'.

37. Discovery of mutation in bacteria by Delbruck and Luria (USA) in 1943. The 'hrp' genes found only in gram negative bacteria so far and several 'hrp' gene coded proteins called harpins. The 'avr Bs2' gene from the bacterium *Xanthomonas campestris* pv. *vesicatoria*, encoded proteins that are also necessary for pathogenicity. The best known 'avr' gene group is the *Xanthomonas* 'avr' gene family called 'pth' (for pathogenicity) genes. The avr/pth genes cause the hypersensitivity response and they are also required for induction of angular leaf spot symptoms of cotton and for citrus canker disease.

38. Gene 'pth' is required for pathogenicity of *Xanthomonas citri* on citrus to cause the typical citrus canker disease and as avirulence genes for example by causing pth A transformed strains of *X. phaseoli* and *X. campestris* pv. *malvacearum* that respectively, infect bean or cotton but not citrus, to cause the hypersensitivity response on their respective hosts bean and cotton while remaining non-pathogenic on citrus. Pathogenicity factors encoded by 'pathogenicity genes' (pat) and 'disease specific gene (dsp) are those involved in steps crucial for the establishment of disease.

39. Wilt of Solanaceous plants caused by *Pseudomonas solanacearum* are favored by high temperature. *Erwinia* and *Pseudomonas* are favored by wet conditions but not flood. *Streptomyces scabies* (common scab of potato) is favored by dry soil or water deficient soils, it may become severe at pH 5.2 to 8.2 and it fails to develop when the soil pH falls below 5.2.

40. High amount of nitrogen increases the susceptibility of pear to fire blight (*Erwinia amylovora*). Reduced availability of nitrogen increases the susceptibility of wilt of Solanaceoons plants, caused by *Pseudomonas solanacearum*. *Streptomyces scabies* (scab of potato) is more serious when nitrate nitrogen is applied in the soil and severity is reduced by the application of phosphorus. Calcium increases the severity of this disease.

41. Copper applications to the soil reduced the stem melanosis caused by *Pseudomonas chicorrii*. Ring rot of potato caused by *Corynebacterium sepidonicum* and brown rot of potato (*Pseudomonas solanacearum*) is carried through infected potato tubers.

42. *Xanthomonas campestris* (black rot of cabbage), *Corynebacterium michiganense* (blight of tomato) and *Erwinia amylovora* (fire blight of pear) disseminate through splashing during rains. *Xanthomonas campestris* pv. *malvacearum* (angular leaf spot of cotton) is spread by windblown rains.

43. *Erwinia amylovora* (fire blight of apple and pear) also disseminated through insects (flies and ants). Cucurbit wilt is caused by *Erwinia tracheiphila* disseminated by cucumber beetle while *Erwinia stewartii* (corn wilt) overwinters inside the flea beetle (*Chaetocnema pulicaria*) is spread by this insect and also transmitted on seed.

44. Blackleg or soft rot of potato caused by *Erwinia carotovora* which is spread by insect (seed corn maggot). Citrus canker is caused by *Xanthomonas campestris* pv. *citri* which is carried by the leaf minor from diseased to healthy plants. Gummosis of sugarcane is caused by *Xanthomonas campestris* pv. *vasculorum* and is transmitted through insects and cane sets also.

45. Leaf gall of herbaceous plants caused by *Corynebacterium fascians* is carried by nematode (*Aphelenchoides*). *Clavibacter* (*Corynebacterium*) *tritici* is the causal agent of yellow ear rot of wheat and is spread by the cockle nematode (*Anguina tritici*).

46. *Pseudomonas lachrymans* (angular leaf spot of cucumber) and *Clavibacter michiganense* subsp. *michiganense* (tomato bacterial canker) is spread by man.

47. Streptomycin binds to the bacterial ribosomes and restricts protein synthesis. Polyene antibiotics (nystatin, filipin, pimaricin etc) are bound to sterol in cell membrane of sensitive organisms resulting increased permeability and loss of metabolites. The polyene antibiotics have no effect on true bacteria.

48. Peritrichous flagella are seen in the genus *Erwinia, Agrobacterium, Bacillus* and *Clostridium. Bacillus, Enterobacter* and *Pseudomonas* parasitizes or inhibits the pathogenic fungi *viz. Pythium* sp., *Sclerotium cepivorum, Phytophthora* sp. and *Gaeumannomyces graminis.*

49. Root knot nematode, *Meloidogyne javanica* is parasitized by the bacterium *Pasteuria* (*Bacillus*) *penetrans.* Dry heat treatment of barely seed at 72°C for 7-10 days eliminates the leaf streak and black chaff causing bacterium *Xanthomonas campestris* pv. *translucens* from the seed.

50. *Agrobacterium radiobacter* K-84 sold as Gallex or Galltrol, is used against crown gall (*Agrobacterium tumefaciens*). Control is based on production by strain K84 of a bacteriocin called agricin 84. *Pseudomonas fluorescens* sold as Dagger G, which is used against *Rhizoctonia* and *Pythium* (damping off of cotton seedlings).

51. *Bacillus subtilis* (strain A 13) sold as Kodiak, which is used as a seed treatment for the control of damping off and root of several crops including potato, sugarbeet and wheat. *Streptomyces* sp. protected the plants against root pathogens. *Streptomyces* and *Pseudomonas* parasitize and/or inhibit the pathogenic fungi *Pythium* sp. and *Gaeumannomyces tritici. Bacillus cereus* strain UM 85 provides effective control of damping off diseases of legumes.

52. Fire blight of apple (*Erwinia amylovora*) is partially controlled with spray of *Erwinia herbicola. Xanthomonas traslucens* pv. *oryzicola* is reduced with sprays of isolates of *Erwinia* and *Pseudomonas.* Spraying with *Bacillus subtilis* reduced infection of apple leaf scars by *Nectria galligena* and of grape by *Eutypa lata. Pseudomonas fluorescens* is used for control of *Drechslera dictyoides,* the causative agent of net blotch of ryegrass.

53. Spraying of peanut or tobacco plants with *Pseudomonas cepacia* or *Bacillus* sp. reduced *Cercospora* and *Alternaria* leaf spot on these hosts, respectively. Take all disease of wheat (*Ophiobolus graminis*) is controlled with spray of *Bacillus cereus.* Fungi, *Sclerotium cepivorum, Fusarium roseum, Pythium ultimum, Rhizoctonia solani, Phytophthora cinamomi,* and *Nectria galligena* is controlled by *Bacillus subtilis.*

54. *Pseudomonas syringae, P. fluorescens* and *Erwinia herbicola* act as ice–nucleation active (INA) catalysts which cause ice formation inside the cells of the host plants at as high temperature as -1°C, causing frost injury. However, non-ice-nucleation active (non INA) bacteria are able to antagonize INA bacteria. If these non-INA bacteria are applied to plant surface frost injury can be protected.

55. Antibiotic streptomycin is produced by the actinomycete, *Streptomyces griseus*. Streptomycin or streptomycin sulphate is sold as agrimycin, phytomycin ortho-streptomycin and it is also effective against fungal pathogens like *Phytophthora parasitica* var. *piperina* (foot rot and leaf rot of piper betle), *Pseudoperonospora humuli* (hop downy mildew) and *Phytophthora infestans* (late blight of tomato).

56. Cyclohexamide is sold as 'Actidione', it is a byproduct of streptomycin and is produced by different species of *Streptomyces griseus* and *S. nouresi*. It is very effective against a wide range of fungi and yeast but is inactive against bacteria. Griseofulvin is produced by *Penicillium griseofulvum, P. patulum, P. nigricans* and *P. janczewski* and it is effective against plant pathogenic fungi.

57. Blasticidin S is produced by *Streptomyces griseochromogenes* and effective against fungi like *Pyricularia grisea*. Antibiotic aureofungin is produced in submerged culture by *Streptoverticillium cinnamomeum* var. *terricolum* and it is active against wide range of fungi.

58. Tetracyclines are produced by a number of species of *Streptomyces*. Derivates of tetracycline are (i) Terramycin (oxytetracycline) is produced by *S. rimosus* (ii) Aureomycin (chlorotetracyclin) is produced by *S. aureofaciens* and (iii) Actinomycin (tetracycline) produced by *Streptomyces* var. *terricola*.

59. Fifteen per cent streptomycin sulphate and 1-5 per cent terramycin mixture is Agrimycin-100 and agrimycin-500 is 1.755 per cent streptomycin sulphate, 9.176 per cent terramycin and 42.4 per cent metallic copper, and it is very effective against *Erwinia amylovora, Pseudomonas syringae* pv. *phaseolicola, Xanthomonas campestris* pv. *citri. X. campestris* pv. *malvacearum* and *Erwinia carotovora*.

60. Nystatin is a polyene (tetraene) antifungal antibiotic produced by *Streptomyces noursei*. Bulbiformin antibiotic produced by *Bacillus subtitlis*, it is an antifungal polypeptide antibiotic and was reported by Vasudeva *et al.* in 1958.

61. Walled prokaryotes are true bacteria which have unbounded nucleus. Size of a bacteria is about 0.1-0.3 to 15-16-0.5-2.0 µm. Bacteria has three distinct layers (i) The cytoplasmic membrane responsible for stain reaction (ii) Cell wall responsible for proper shape and (iii) Slime layer (Capsule)-outermost layer of bacterial cell.

62. Flagella can be easily detached from the cell, they are directed backward to the direction of motion at an angle of 45° and reversal of direction occurs by swinging the flagella through an angle of 90°. Reproduction in bacteria by binary fission, the process may be repeated in every 20 or 30 minutes.

63. The F factor is now known to be a circular DNA molecule. In conjugation process, F^+ and F^- factor involved. The bacterial chromosomes are not involved, only physical contact was involved and the DNA from one passed into other cell through conjugation tube. Only duplicate F factor passes into the female cell and female is not converted into a male as the F^+ factor is not passed on. The F factor is an infective element, now called as episome. The F factors (episome) always found at the near end never goes to the recipient cell.

64. When a F male conjugates with F^- (recipient) cell, the F-factor is transferred from donor to the recipient cell, and such a recipient bacterial cell becomes heterozygous (merozygous) for that part of the bacterial chromosome, which the F-factor had obtained during its anomalous dissociation. Genetic recombination of this type, mediated by F-factor is called sexdution or F-duction (Jacob and Alderberg, 1959).

65. Term bacteriophage coined by de Herelle in 1917. Viruses that attack bacteria are called bacteriophages (or only phages), which means bacteria eater. Bacteriophage head contain ds DNA and six tail (helical) fibers are associated with adsorption of the phage on the bacterial wall. Phages have two types of cycles *i.e.* lytic (T series of phages infecting E. *coil*) and lysogenic cycle (λ phages).

66. Lambda (λ) phages contain single tail (helical) fiber; it shows dual or binal symmetry of the protein coat. Lambda phage consists of a DNA fitted head; it is hollow tube without a contractile sheath.

67. Bacteriophages attacking *E. coil* were termed as T-type by Max Delbruck in 1938. While Andre Lwoff coined the name Lambda (λ) phages. Gas vacuoles produced by cynophages bacteria. Viruses that attack blue green algae are called cynophages. In 1963, the first cynophages was discovered by Safferman and Morris. These are DNA viruses, consisting of a head, icosahedrons in shape, long helical tail and show dual or binal symmetry.

68. Growth cycle of bacteriophages can be divided into two groups. The virulent and temperate phages. The virulent phages (T-series) show lytic cycle and in lysogenic cycle (temperate phage), does not multiply and there is no death of the host cells. Lysogenic cycles are called temperate phages and the process is called lysogey. The cells do not lyse and no virus particles are formed. A kind of 'symbiotic' association develops which is called lysogenic state.

69. Endospore forming bacteria are gram positive *e.g. Bacillus, Clostridium. Xanthomonas* and *Pseudomonas* are rod shaped and monotrichous (polar), but only the former produces a water soluble yellow pigment. *Erwinia* is peritrichous (flagella scattered all over the outer surface). It is gram negative and belongs to order Eucbaterials and Enterobacteriaceae family.

70. *Clavibacter* (*Corynebacterium*) is club shaped, rod shaped, non motile and gram positive; *Streptomyces* (with rudimentary hyphae) is rod shaped gram positive. Majority of phytothogenic bacteria (*Pseudomonas, Xanthomonas, Erwinia* and *Agrobacterium*) are gram negative, non-endospore forming, rod shaped cells. However, some gram positive bacteria (*Streptomyces, Corynebacterium*) are phytopathogenic. In 1886, E. Haeckel proposed third kingdom Protist.

71. Corynebacterium are V-form, gram positive, aerobic, non acid fast, non spore forming bacteria, constitute a major group collectively known as 'coryneform bacteria'. This group includes or has now been transferred to other genera such as *Arthrobacter, Rhodococcus, Phodococcus, Nocardia* and *Streptomyces.*

72. Endospore is dormant cell. Exospores are formed by some bacteria outside the cell by budding at one end of the cell. Cysts are formed by *Azotobacter*. Conidiophores and sporangiospores are reproductive spore formed by only *Streptomyces*. Twisting movement found in *Pseudomonas*.

73. Cytoplasmic membrane is typically double tracks (unit) membrane made up of phospholipids and protein. Plasmids are extra chromosomal genetic material found in bacteria. They regulate their own replication, within bacterial cell. The plasmids are circular ds DNA molecule ranging in size from 1113 x 106 Daltons and weight approximately 10^6-10^8.

74. Term plasmid coined by Lederberg in 1952. These types of plasmid are found (i) F factor. It is responsible for genetic transformation (ii) R factor. It is responsible medicine resistant and (iii) Colocin factor. It is responsible for production of bacteriocins.

75. In 1972, American scientist Paul Berg and coworkers synthesized the first recombinant DNA (rDNA) and in 1980, he got Nobel Prize in Chemistry for his protein synthesis and rDNA work. He transferred the viral DNAs (SV-40 virus) into *Escherichia coli* bacteria.

76. Most walled bacterial cells are rigid. The rigidity is imported by high peptidoglycan content of the cell wall, and flexibility is attributed to low amount of peptidoglycan in the cell wall. All prokaryotes (except mycoplasma and L-form bacteria) are a layer of a complex polymer known as peptidoglycan or murein.

77. Gram's staining techniques was developed by Christian Gram in 1884 (Danish physician) into two major groups, the gram positive and gram negative. All bacteria take up certain stains such as crystal violet or gentian violet. Those that retain the Gram's stain after washing with alcohol/or alcohol treatment are called Gram positive (GP) while those that lose the stain after treatment of alcohol are designated as Gram negative (GN).

78. Gram positive bacteria contain more muramic acid and teichoic acid or greater amount of mucopeptide and do not contain lipids. Gram negative bacteria has less amount of muramic acid and no teichoic acid, it contains more lipids. Gram negative bacteria become less rigid due to plastic nature of lipid protein, polysaccharide complex.

79. Gram positive bacteria retain the stain while gram negative bacteria get decolorized. When counter stained with safranin or carbol-fuschin, the gram negative becomes red while the gram positive bacteria remain deep purple.

80. The high lipid content of cell wall of gram negative bacteria allows the alcohol to pass through the cell wall and reach the cytoplasm to remove the stain. In gram positive bacteria, due to the absence of lipids in the cell wall, the alcohol fails to reach the stained cytoplasm.

81. Gliding motility is a surface translocation in non-flagellated bacteria. Capsule and slime formation is widespread in *Bacillus, Erwinia, Psuedomonas, Xanthomonas,* and *Agrobacterium.*

82. Prokaryotes have no membrane bound means in bacteria, the material (DNA) is not bound with proteins to form chromosomes. The bacterial genomes consist of a single closed ring about 1000μ long.

83. The word nucleoid or genome is used for nuclear apparatus of prokaryotes. The bacterial cell in which the endospore is formed is termed a sporangium; usually one endospore is formed in each cell.

84. The concentrated structure is called nucleoid. There is no nuclear membrane and spindles are not formed during cell division. Lamellae or vesicles (300 A° dia.) found in photosynthetic bacteria, it is also called chromatophores instead of chloroplast.

85. The term genophore is used to represent the bacterial nucleus or chromosomes. In the absence of nuclear membrane, the genophore is the morphologically distinct nucleoplasm. It consists of double stranded DNA fibrils. Genophore is a closed loop, *i.e.* circular chromosome and has one linkage group.

86. *Rhodomicrobium* is a budding bacterium and *Actinomycetes* is a mycelial eubacteria. *Nocardia* is an aerobic bacterium. *Streptomyces* reproduces by conidia, *Actinoplanes* by sporangiospores and *Mycobacterium* reproduces by binary fission.

87. The bacteria which are prokaryotic, in cellular organization are placed in a single class. Schizomycetes and classified into ten orders according to the Bergey's Manual of Determinative Biology.

88. Nitrifying bacteria oxidize the ammonia into nitrate. Denitrifying bacteria reduce the nitrates to molecular N_2. Free living nitrogen fixing bacteria reduce the nitrogen of the atmosphere to amino compounds.

89. Bacterial cells are surrounding by a wall made up of mucopetide. The cell wall is surrounded in some bacteria by a polysaccharide in the form of definite layer called capsule.

90. The gram positive bacterial cell wall component contains 85 percent mucopeptides, rest being simple polysaccharides like teichoic acid. However, gram negative bacteria contain small quantity, the major portion being formed by lipoprotein and lipopolysaccharide.

91. Pilli (fimbriae) which are minute hair like superficial appendages, covering the surface of most of the gram negative bacteria is helpful to holding the cells to water surface and during conjugation.

92. Circular chromosome, long DNA molecule present in bacteria, which are attached to the cell membrane. In bacteria, lacks an organized nucleus and other organelles such as chloroplast, mitochondria, Golgi body, endoplasmic reticulum and lysosomes.

93. The photosynthesis bacteria have lamellae (thalakoids) and vesicle, which constitute the photosynthesis apparatus.

94. Mesosomes are particles concerned with bacterial genome replication and septum formation during the cell division and ribosome's having the sedimentation coefficient of 70 S, lie free in the cytoplasm. The cell sap contains high molecular weight (*e.g.* various soluble enzymes and t-RNA and low molecular weight (amino acid, nucleotide) substances.

95. Bacterial cell (structurally) divided into five regions, (i) Surface appendages *i.e.* flagella, pill (ii) Surface adherent *i.e.* capsule and slime layers (iii) Cell wall (iv) Cytoplasm and organelles and (v) Special structure *i.e.* endospores, stalks.

96. Bacterial flagella 4-5μ long, flagella and cilia morphologically and physiologically similar, cilia are shorter, flagella diameter 120-150 A°, cylindrical, hallow strand, made up of protein molecule called flagellin, each flagellin molecule is 40 A° diameters. Several (usually 3-8) longitudinal chains of flagellin molecules are found in bacteria. A cross section of the flagellum reveals 8 flagellin molecules around a central space.

97. Bacterial flagellum is single stranded (made up of several parallel protein fibrils), hallow, helical and no sheath. The antigenic property of some bacteria *e.g. Salmonella* resides in the flagellin molecules.

98. Mostly bacteria have narrow genome size; range around 1-3 x 10^9 Daltons. The prokaryotes have two types of nucleic acid DNA (deoxyribose sugar in DNA) and RNA (ribose sugar in RNA). In addition to sugar, the nucleic acid contains nitrogen bases: A (adenine), G (guanine), T (thymine), C (cytosine) and U (uracil) as well as phosphate.

99. Sugar molecules are linked by phosphate. Each sugar molecules attached to one of the four bases A, G, C and T in DNA and A, G, C and U in RNA. One set of sugar phosphate, and based attached to the sugar is called nucleotide. A=T and G=C.

100. Ribosome's are organelles present in all types of cells. They consist of RNA and protein. They are the work bench for protein synthesis through ribosome's RNA (r RNA) by translating the message encoded in mRNA from the DNA in the nucleoid.

101. Streptomycin, tetracycline and chloramphenicol inhibit activity of 70S ribosomes, hence are effective against prokaryotes, but not all eukaryotes. However, in eukaryotes the organelles like chloroplast and mitochondria have70 S ribosomes and are affected by these antibiotics.

102. In binary fission, the DNA replication is followed by equal division of the cytoplasm and synthesis of the cell wall while in budding, the DNA replication is followed by unequal division of cytoplasm. In mycoplasma and spiroplasma both processes were described by Masover and Hyflick, in 1981.

103. A bacterial cell may divide in 5 hours and its ribosomal content may be 200 per cell but when it divides in 25 minutes, the ribosomal content may be about 70,000 per cell. This is because more protein synthesis is required in rapidly multiplying cells.

104. Polyene antibiotics (nystatin, filipin, pimaricin etc.) are bound to sterols in the cell membrane of sensitive organism, resulting in increased permeability and loss of metabolites.

105. Protoplast is a completely wall free cell which is bound only by the plasmalemma or plasma membrane. It obtained from gram positive bacteria. Sphaeroplast is a bacterial cell with damaged cell wall; the wall is not completely removed. It obtained from gram negative bacteria.

106. Gram positive bacteria have only mucopeptide in their wall and therefore, enzymes (*e.g.* lysozyme) and antibiotics (*e.g.* penicillin and cephalosporin) which attack the mucopeptide, can remove the wall completely and form protoplast.

107. In gram negative bacteria, the wall consists of lipoprotein, lipopolysacchride layers which are not attacked by these agents. Thus, the wall of gram negative bacteria is damaged but not easily removed completely.

108. L from-Klienberger Nobel in 1935 discovered the L-forms (L-for lister institute). It is protoplast or sphaeroplast like body capable of growth and multiplication. L-forms can be produced in the laboratory from both gram positive and gram negative bacteria by the action of antibiotics. Penicillin is most widely used for this purpose. The L-forms do not multiply by binary fission it multiply by budding. Plant pathogenic bacteria which reported to produce L-forms are *Agrobacterium tumefaciens* (crowns gall disease) and *Erwinia carotovora* pv. *atroseptica* (blackleg of potato).

109. Vitamin A, C and D are not required by fungi and bacteria while vitamin B group are required by some bacteria.

110. Cysts are less resistant than endospores. The whole cell is transformed into a cyst, *i.e. Azotobacter*, Maxobacteriales *e.g. Maxococcus* and *Chondromyces*.

111. Bergey's Manual of Determinative Bacteriology started in 1923. The first volume of Bergey's Mannual of Sytematic Bacteriology edited by N.R. Krieg, published in 1984. The 8th edition of Bergey's Manual of Determinative Bacteriology was edited by R.E. Buchanan and Gibbons in 1974.

112. In the first volume of Bergey's Mannual of Systematic Bacteriology Murry (1984) has proposed the higher taxa for prokaryotes.

113. *Pseudomonas* produce green blue pigment, *Xanthomonas* produce yellow pigment, *Erwinia* produce white or yellow, blue and pink and *Corynebacterium* produce white or yellow, orange and blue pigment.

114. *Pseudomonas* is straight to curved rods 0.5-1.0 by 1.5-4.0 μm. Some Plant pathogenic *Pseudomonas* species (*e.g. Ps. syringae*) are called *Pseudomonads*

because on a medium of low iron content, they produce yellow-green, diffusible fluorescent pigments. Others *i.e. Ps. solanacearum* do not produce fluorescent pigment.

115. Family Pseudomonadaceae was established by Winslow *et al.* in 1917. The type genus of family is *Pseudomonas* Migula 1894. *Pseudomonas* causes wilt disease in solanaceous crop; *e.g.* wilt of potato caused by *Pseudomonas soanacearum.*

116. Other fluorescent species are *Pseudomonas aeruginosa, Ps. fluorescens* (with 3 biotypes) –*Ps. chlororaphis, Ps. putida, Ps. aereofaciens* (all non-phytopathogenic) and phytopathogenic fluorescent species are *Ps. syrinagae* and *Ps. cichorii.*

117. International committee on Systematic Bacteriology published its report in the form of International standard for Naming of pathovars of phytopathogenic bacteria and a list of pathovar names and pathotype listed by D.W. Dye *et al.* in 1980.

118. *Pseudomonas solanacearum* is a wilt causing pathogen. It has 3 Races, Race 1 infects soloanaceous plants, Race 2 banana and *Heliconia* sp. and Race 3 infect potato.

119. Bacteriocin and phages may be useful in the identification at subspecies level, but not for species delimitation. Serology has used primarily as a tool to confirm the identity of strain.

120. *Pseudomonas syringae* Von Hall 1902 constitutes the largest group of fluorescent phytopathogenic *Pseudomonas*. It produces yellow green pigment that fluoresces under ultra violet light.

121. *Pseudomonas solanacearum* species was first described by E.F. Smith in 1896 as *Bacillus solanacearum*. In 1914, he confirmed as *Pseudomonas solanacearum* species *Pseudomonas cepacia* was described by Burkholder in 1950.

122. *Xanthomanas* species belongs to the family *Pseudomonadaceae.* The first observation of phytopathogenic bacterium *Xanthomonas* was made by J.H. Wakker in 1881 (Germany) for the organism association with yellow disease of hyacinth. It produces acid in a lactose medium.

123. In 1939, W.J. Dowson erected the genus *Xanthomonas* and *Agrobacterium tumefaciens* erected by Smith and Townsend Conn in 1942, it induced plant tumor (crown gall). The bacterium was discovered by E.F. Smith in 1907 and the causal bacterium was firstly isolated by Smith and Townsend, while *A. radiobacter* erected by Beijerinck and Van Dladen, Conn in 1942 and *A. rhizogenes* by Ricker *et al.,* Conn in 1942.

124. *Pseudomonas solanacearum* produces IAA, ethylene and pectic and cellulotic enzymes during pathogenesis. However, brown rot of potato is not a typical soft rot.

125. The bacteria *Erwinia* are straight rods of 0.5-1.0 x 1.0-3.0 µm, motile, peritrichous flagella and belong to family Enterobacteriaceae. Some *Erwinia* do not produce enzymes yellow pigment but do cause necrosis or wilt, whereas other *Erwinia*

have strong pectolytic capacity they cause soft rot are called the 'Carotovora group' or 'soft rot group' or the pectolytic group. *Erwinia carotovora* and *E. dissolvens* produces large amount of H_2S gas from carbohydrate.

126. The genus *Corynebacterium* (now *Clavibacter*) was proposed by Lehmann and Neumann in 1896 with the *Diphtheria* bacterium, *Corynebacterium diphtheriae* the type species. These bacteria were subsequently transferred to other genera such as *Arthrobacter* (Collins *et al.*, 1981), *Aureobacter* (Collins *et al.*, 1983), *Curtobacterium* (Collins *et al.*, 1983), *Clavibacter* and *Rhodococcus*. These species are gram positive and *Agrobacterium, Erwinia, Pseudomonas, Xanthomonas* and *Xyllela* are gram negative.

127. K. Maramorosch (1973) first time reported that ratoon stunting disease (RSD) was caused by a rickettsia like organism/rickettsia like bacteria (RLO/RLB). In 1974, He coined the term 'Mollicute like organism'. In 1984, Davis *et al.*, characterized the bacterium causing RSD of sugarcane and also the bacterium causing a similar stunting disease of bermuda grass (*Cynodan dactylon*) are created the species *Clavibacter xyli* (both xylem inhabiting).

128. Gelatin was the first substance to be used for the solidification of culture media. Chemically gelatin is a protein, Robert Koch's introduced the use of agar as a solidifying agent in the year 1881. Agar is a dried mucilaginous substance extracted from several species of geladium (red algae). Chemically agar is a galactan *i.e.* kind of polysaccharide, insoluble in cold water, it dissolves at 98°C and does not solidify until the temperature drops about 45°C.

129. Pink eye disease of potato and soft rots of other fleshy vegetables are caused by *Pseudomonas marginalis*, slipper skin disease of onion is caused by *Ps. gladioli* pv. *allicola* and sour skin of onion is caused by *Ps. cepacea*. Gumming disease of sugarcane is cuased by *Xanthomonas vascularum*.

130. *Pseudomonas syringae* whose pathovars causes wild fire of tobacco (*Ps. syringae* pv. *tabaci*), angular leaf spot of cucumber (*Ps. syringae* pv. *lacrymans*), halo blight of bean (*Ps. syringae* pv. *phaseolicola*), bacterial blight of oats (*Ps. syringae* pv. *coronafaciens*), bacterial blight of pea (*Ps. syringae* pv. *pisi*), bacterial blight of soybean (*Ps. syringae* pv. glycinea), fruit spot of apple (*Ps. syringae* pv. *populans*) and bacterial speck of tomato (*Ps. syringae* pv. *tomato*).

131. *Xanthomonas campestris* whose pathovars causes common blight of beans (*Xanthomonas campestris* pv. *phaseoli*), angular leaf spot of cotton (*Xanthomonas campestris* pv. *malvacearum*), bacterial leaf blight of rice (*Xanthomonas campestris* pv. *oryzae*), bacterial blight or stripe of cereals (*Xanthomonas campestris* pv. *translucens*), bacterial leaf streak of rice (*Xanthomonas campestris* pv. *oryzicola*), bacterial spot of stone fruits (*Xanthomonas campestris* pv. *gummisudans*) and walnut blight (*Xanthomonas campestris* pv. *juglandis*).

132. *Agrobacterium* causing crown gall of many woody plants primarily stone fruits, pome fruits and grape (*Agrobacterium tumefaciens*), gall of raspberries and black berries (*Agrobacterium rubi*), hairy root apple (*Agrobacterium rhizogenes*). All three species produces ti-plasmid (tumor-inducing plasmid), induce crown

gall but so far only two strains of *Agrobacterium tumefaciens* and *Agrobacterium rhizogenes* have found to contain Ri-plasmid (root inducing plasmid), induce hair root symptoms.

133. *Erwinia tracheiphila* is the causal agent of wilt cucurbits. It survives for only a few weeks in infected plant debris and it also overwinters in the intestines of striped cucumber beetles (*Acalymma vittata*) and spotted cucumber beetles (*Diabrotica undecimpunctata*), in which it hibernates.

134. *Erwinia amylovora* is the causal agent of fire blight of apple and pear. Infected flowers become water soaked, twigs and suckers are also affected. The bark of cankers appears water soaked at first later darker sunken and dry. The bacteria overwinter at the margins of cankers and possibly in buds and apparently healthy wood tissue.

135. Bacterial canker and wilt of tomato is caused by *Clavibacter michiganensis* subsp. *michiganensis*. The bacteria overwinter in or on seeds and in some areas, in plant refuse in the soil. *C. michiganensis* subsp. *sepedonicus* (ring rot of potato) overwinter mostly in infected tubers as dried slime on machinery, crates and sacks. In India, the first bacterial disease brown rot of potato (*Pseudomonas solanacearum*) was recorded by Cappel in 1892.

136. *Pseudomonas solanacearum* in the causal agent of southern wilt of solanaceous plants and moko disease of banana, it overwinters in diseased plant debris, in vegetative propagative organs such as potato tubers and banana rhizomes, on the seeds of some crops, in wild host plants and probably in the soil.

137. Bacterial canker and gummosis of stone fruit trees is caused by *Pseudomonas syringae* pv. *syringae*, it produce the phytotoxins syringomycin and many strains of *Ps. syringae* are ice-nucleation active that is, they serve as nuclei for ice formation and therefore, cause frost injury to plants. It overwinters in active cankers, in infected buds and leaves.

138. Citrus canker is caused by the pathogen *Xanthomonas axonopodis* (formerly *X. campestris* pv. *citri*), it over season in leaves, twig and fruit canker lesions. The bacteria enter through stomata or wound. The most recent classification of the bacterium was proposed by Harting *et al.* (1996).

139. The severity of common scab of potato (*Streptomyces scabies*) increases at the soil pH from 5.2 to 8.0 and decreases beyond these limits. Control of common scab of potato is through the use of certified scab free seed potatoes or through seed treatment PCNB (banned in India since 1998) or with maneb –zinc dust. In the plains of India it was first seen at Patna in 1958.

140 The root nodule bacteria are *Rhizobium, Bradyrhizobium, Azorhizobium* and *Frankia*, are irregular, club shaped, no flagella, most are gram negative while *Frankia* are gram positive, filamentous and they produce spores.

141. *Rhodospirillum, Rhodomicrobium, Chromatium, Chlorobium* are phototrophic bacteria. *Sphaerotillus* and *Leptothrix* are sheathed bacteria. *Bdellovibrio, Spirillum* are spiral and curved bacteria. *Bdellovibrio* are parasitic on a variety of bacteria.

142. Citrus canker symptoms appear on leaves, twigs, thorns, older branches and fruits. Leaf lesions first appear as small, round, watery, translucent spots. It first develop on the lower surface of the leaf and then on both the surfaces as yellowish brown.

143. Bacterial blight, angular leaf spot or black arm of cotton (*Xanthomonas campestris* pv. *malvacearum*) was first reported from Alabama State of USA in 1891. In India, it was first observed in Tamil Nadu in 1918. The pathogen is both externally and internally seed borne. A few phylloplane bacteria of cotton (namely species of *Aeromonas, Flavobacterium* and *Pseudomonas*) on pre inoculation protected the leaves from infection. *Thurbaria thespesoides, Eriodendron anfractuosum, Jatropha curcas* and *Lochnera pusilla* are reported to be collateral hosts of the bacterium but their role in disease cycle is not yet established.

144. Bacterial leaf blight (BLB) of Paddy (*Xanthomonas oryzae* pv. *oryzae* Ishiyama, Dye) was first recorded on farmer's field of Japan in 1884 and in India, it was first recorded in 1951 in the Khopali area of Mumbai. Optimum temperature for growth is 25-30°C.

145. Bacterial leaf blight is a seed borne. Infected seeds and *Leersia sayanuka* (weed, alternate host) is one of the major sources of primary inoculum. The secondary spread is brought about through wounds and stomata by bacterial cells disseminated by wind, rain drop splashes, by irrigation water. The leafhopper (*Nephotetix virescence*) and the grasshopper (*Hieroglyphus banian*) can transmit the bacterium.

146. Bacterial leaf streak disease of rice (*X. campestris* pv. *oryzicola*) was first described by Reinking in 1918 from the Philippines and in India, it was first reported by Srivastava *et al.* (1967). Identification and description of the causal organism was given by Fang *et al.* (1957) from China. The first sign of the disease is the appearance of fine, water soaked to translucent interveinal streaks which may be as long as 1-10 cm. It grows favorably at 28°C.

147. Bacterial leaf spot, blight and canker of mango (*X. campestris* pv. *mangiferae indicae*) was first reported in 1948 from Maharashtra. Optimum temperature for growth is 27°C. Wilt and brown rot of potato or tomato is caused by a bacterium *Pseudomonas solanacearum* (now known as *Ralstonia solanacearum*). In India, the disease was recorded on potato by Cappel in 1892 and was the first bacterial disease from India. Optimum temperature for growth is 35-37°C. Red stripe of sugarcane (*Pseudomonas rubrilineans*) was first reported in India in 1933. It also infects sorghum, pearl millet, maize etc.

148. Yellow ear rot or tundu disease of wheat (*Clavibacter tritici*) was first described by Hutchinson (1917) from Punjab in India. It is also known as bacterial spike blight, gumming disease and yellow slime disease. The disease occurs due to the combined action of wheat nematode (ear cockle disease) *Anguina tritici* and *Clavibacter tritici*. In India, the disease does not occur on wheat unless the nematode is also present.

149. Blackleg wilt and soft rot of potato caused by *Erwinia carotovora* subsp. *carotovora* and *E. carotovora* subsp. *atroseptica.* Optimum temperature for growth is 27 to 30°C. The bacteria survive in soil, as saprophytes for 20 years or more. *E. carotovora* subsp. *atroseptica* can live in the body of (during all growth stages of) seed corn maggot, *Hylema platura.*

150. Stalk rot of maize (*Erwinia chrysanthemi* pv. *zeae*) was first reported in the USA by Rosen in 1922 and in India, it was reported in Pusa, Bihar by H.H. Prasad in 1980. Streptocycline and Agrimycin 100 as well as bleaching powder were found effective in controlling the disease.

151. The fastidious vascular bacteria (previously known as RLO/RLB) that cause plant diseases and cannot be grown on simple culture media in the absence of host cells. Fastidious phloem limited bacteria was first reported in 1972, in the phloem of clover and periwinkle plants affected with the clover club leaf disease and later, in citrus plants affected with citrus greening disease.

152. Fastidious xylem limited bacteria were first observed in 1973 in the xylem vessels of grape plants affected with pierce's disease and alfalfa affected with alfalfa dwarf.

153. The fastidious vascular bacteria are rod shaped cells of 0.2 to 0.5 µm in diameter by 1 to 4 µm in length. They have no flagella. All fastidious vascular bacteria are gram negative. Xylem limited bacteria have placed in the recently created genus *Xylella.*

154. Only the xylem inhabiting bacteria causing ratoon stunting disease (RSD) of sugarcane and bermuda grass stunting are gram positive and they are classified as members of the genus *Clavibacter xyli* pv. *xyli.*

155. Gram negative, phloem inhabiting fastidious bacteria cause clover club leaf and citrus greening diseases have not grown in culture media so far, but all the G.N. xylem inhabiting bacteria could be grown in culture on complex nutrient media.

156. A gram negative xylem inhabiting fastidious bacterium, *Xylella fastidiosa* causes pierce's of grape, almond leaf scorch, plum leaf scald, citrus variegation chlorosis, phony peach disease and also causes leaf scorch diseases on elm, sycamore, oak and mulberry. These diseases are transmitted by xylem feeding insects such as leafhoppers, sharp hoppers and spittle bugs.

157. No insect vector is known for the gram positive xylem inhabiting fastidious bacteria so far, but for at least one of them which cause RSD can be transmitted mechanically by cutting implements during harvest. Leafhoppers and *Psyllid* insects are the vectors of clover club leaf and citrus greening, respectively.

158. In India, ratoon stunting disease was first observed in Cos 510 at Golagokharannath (U.P.) by S.J.P. Chilton, in 1956. Man is the primary agent in spreading the bacterium (*Clavibacter xyli*). Citrus greening is caused by a Gram negative phloem inhabiting fastidious bacterium *Xylella fastidiosa.* This

bacterium is transmitted through vegetative propagation and by two species of citrus psylla, *Diaphorina citri* and *Trioza erytreae*. The bacterium transmitted from sweet orange to periwinkle (*Vinca rosea*) through dodder (cuscuta).

159. *Agrobacterium radiobacter* K 84 applied as a root dip treatment to transplants of apple, peach, rose and other woody plants to protect against crown gall caused by *A. tumefaciens*. *Bacillus subtillis* GBO3 and MBI 600, *Pseudomonas cepacia* type Wisconsin and *Trichoderma harzianum* RL-AG-2 are all applied as seed treatments. *Streptomyces griseoviridis* K 61 and *Gliocladium virens* GL 21 also used as seed treatment.

160. Fluorescent *Pseudomonas* strains have shown potential to colonize roots 10-15 cm below the seed when introduced as an inoculum source on the seed. *Bacillus* sp. L 324-92, with broad spectrum activity against wheat root pathogens, has shown potential to colonize roots 5-10 cm below the seed when introduced as an inoculum source on the seed.

161. Bacterial blight of cotton (BBC) caused by *Xanthomonas axonopodis* pv. *malvacearum* (Xam). Bacterial blight resistant gene (B. genes) minimized the genotype environment interaction. Xam R-32 is the most virulent race of cotton which is neutralize by five B-genes (i.e. B7, B4, Bz, Bi, BN), contains five plasmids (size; 60, 40, 10, 5.5 and 2.2 kb each). Moderately virulent race-26 neutralize three B. genes (B4, B2, B1,) contains three plasmids (60, 40, 10 kb) while neutralized only one B-gene (Bin) contains one plasmid of 10 kb.

162. *Escherichia coli* strain containing such a chimeric intermediate vector, *Agrobacterium tumefaciens* (LAB 4404) harboring a disarmed Ti plasmid and an *Escherichia coli* strain such as pRK-2013 were subjected for triple mating procedure.

163. Polysaccharides are known to be produced by *Xanthomonas phaseoli*, *Pseudomonas solanacearum* and species of *Corynebacterium*, they induce wilting. The soft rot bacteria are *Erwinia carotovora* pv. *carotovora*, *E. carotovora* pv. *atroseptica*, *E. aroidae*, *E. chrysanthemi* and other also cause soft rot at high temperatures and high humidity *e.g. Bacillus polymyxa* reduces potatoes to a yellow sticky mass with a characteristic fishy odour.

164. Phospholipase. C enzyme appears to have direct effect on intact membranes and this enzyme from *Erwinia carotovora* lyses intact cucumber protoplasts.

165. Growth inhibitors abscisic acid (ABA) was observed in fungal, viral and bacterial diseases that there is an increase in inhibitor content in infected tissue. Tobacco plants infected with *Pseudomonas solanacearum* showed that stunting was correlated with enhance levels of ABA originating in the host. Since ABA has a prime involvement in abscission there in an obvious role for it in cotton defoliation, plants under water stress undergo stomatal closure and ABA is responsible for such stomatal closure. Thus, the enhanced level of ABA in diseased plants may be in response to stress as a result of infection.

Multiple Choice Questions (Choose the correct answer)

1. **Sedimentation coefficient of bacterial ribosomes is:**
   (A) 80S (B) 50 S
   (C) 70 S (D) 30S

2. **Transformation in bacteria was discovered by Griffith in:**
   (A) *E. coli* (B) *Staphylocccus avreus*
   (C) *Streptococcus pneumoniae* (D) *Salmonella*

3. **Who established for the first time the bacteria can cause plant disease:**
   (A) T.J. Burill (B) Anton de Bary
   (C) Leeuwenhock (D) Woronin

4. **Which of the following genera of bacteria is gram positive:**
   (A) *Erwinia* (B) *Xanthomonas*
   (C) *Corynebacterium* (D) *Agrobacterium*

5. **The flagellation in *Erwinia* is:**
   (A) Peritrichous (B) Monotrichous
   (C) Lophotrichous (D) Atrichous

6. **The genus *Erwinia* belongs to the family:**
   (A) Enterobacteriaceae (B) Pseudomonadaceae
   (C) Corynebacteriaceae (D) Rhizobiaceae

7. **Optimum temperature for the growth of plant pathogenic bacteria is:**
   (A) 10°C (B) 35°C
   (C) 15°C (D) 25°C

8. **Green fluorescent pigment in the culture medium is produced by:**
   (A) *Erwinia* (B) *Pseudomonas*
   (C) *Agrobacterium* (D) *Corynebacterium*

9. **Sex factor in bacteria is known as:**
   (A) Episome (B) Mesosome
   (C) Chromosome (D) Nucleosome

10. **Cell wall of bacteria are characterized :**
    (A) Glutamic acid (B) Nicotinic acid
    (C) Glycin (D) Muramic acid

11. Which of the following pathogen of potato causes 100 fold increase in auxin content as compared to healthy plants:

(A) Potato virus
(B) *Rhizoctonia solani*
(C) *Pseudomonas solanacearum*
(D) *Phytophthora infestans*

12. Flagella of gram positive bacteria contains:

(A) One ring
(B) One pair of rings
(C) Three rings
(D) Four rings

13. The flagellum of gram negative bacterium contains:

(A) One ring
(B) Two pair of rings
(C) One pair of rings
(D) No ring

14. Gas vacuoles/gas vesicles are found in:

(A) *Xanthomonas*
(B) *Clostridium*
(C) *Pseudomonas*
(D) *Cynobacteria*

15. λ (Lambda) phage contains which type of nucleic acid:

(A) ssDNA
(B) ds DNA
(C) ds RNA
(D) ss RNA

16. Bacterial rot of wheat is caused by:

(A) *Xanthomonas*
(B) *Pseudomonas*
(C) *Corynebacterium triitici*
(D) *Erwinia aroidae*

17. *Erwinia carotovora* causes:

(A) Black leg of potato
(B) Wilt of potato
(C) Stalk rot of maize
(D) Scab of potato

18. The prokaryotic organisms are characterized by:

(A) 80 S ribosomes
(B) Well developed nucleus
(C) Nucleus without nuclear membrane
(D) More than one chromosome in the nucleus

19. The characteristic of constituent of bacterial cell wall is:

(A) Peptidoglycan
(B) Chitin
(C) Cellulose
(D) Pectin

20. In antigen-antibiotic reaction precipitin bands are formed in:

(A) Tube precipitin test
(B) Ouchterlony gel diffusion test
(C) Slide agglutination test
(D) ELISA

21. Ice-nucleation active bacteria are:

(A) *Pseudomonas syringae* (B) *Xanthomonas campestris*

(C) *Streptomyces scabies* (D) None of these

22. Transformation experiments were discovered with the work on:

(A) Actinomycetes (B) *E. coli*

(C) Diplococcus (D) *Pseudomonas*

23. Bdellovibrio multiplies in the:

(A) Cytoplasm (B) Intraperiplasm

(C) Nucleus (D) Protoplasm

24. Bdellovibrio discovered by:

(A) H. Stolp (1952) (B) E.F. Smith (1895)

(C) T.J. Burril (1882) (D) Robert Koch (1876)

25. Bdellovibrio parasitizes and kills:

(A) Gram positive bacteria (B) Gram negative bacteria

(C) Virus (D) Viroid

26. After passing through gram staining procedure *Clavibacter* stains:

(A) Red (B) Violet

(C) Yellow (D) Green

27. Bacterial flagellum posses:

(A) Antigenic properties (B) No antigenic properties

(C) Antibody properties (D) None of these

28. Bacterial flagellum consists almost:

(A) Lipid (B) Polysaccharide

(C) Vitamin (D) Entirely of protein

29. Mucopeptide is a polymer made up of alternating units on N-acetyl glucosamine and N-acetyl muramic acid joined by:

(A) β, 1-4 linkages (B) λ, 1-3 linkages

(C) α, 1-2 linkages (D) β, 1-6 linkages

30. The cell wall of gram negative bacteria contains:

(A) One layer (B) Two layer

(C) Three layer (D) Four Layer

31. The Tn-3 (in *E. coli*) is:

(A) 5000 bp long (B) 10000 bp long

(C) 15000 bp long (D) 20000 bp long

32. The soft rot bacteria can grow and are active over a range of temperature from:

(A) 5-35°C
(B) 36-40°C
(C) 40-415°C
(D) None of these

33. The cell wall of gram positive bacteria contains:

(A) One layer
(B) Two layer
(C) Three layer
(D) Four layer

34. The cell wall of gram positive bacteria contains:

(A) One Layer
(B) Two Layer
(C) Three Layer
(D) Four Layer

35. Citrus canker is a serious disease caused by:

(A) Mycoplasma
(B) Fungus
(C) Bacterium
(D) Virus

36. Most of the bacteria contain ________nuclei per cell during exponential phase:

(A) 2-4
(B) 5-6
(C) 6-8
(D) 8-10

37. Transduction was discovered in 1952 in:

(A) *Escherichia coli*
(B) *Salmonella typhimurium*
(C) *Enterococcus faecalis*
(D) None of these

38. Wilt disease of jute is caused by:

(A) *Verticillium alboatrum*
(B) *Pseudomonas solanacearum*
(C) *Fusarium oxysporum*
(D) *Hemeleia vastatrix*

39. Which of the following bacteria possesses common virulence factors for pathogenic in plants and animals:

(A) *Bacillus subtilis*
(B) *Agrobactorium tumefaciens*
(C) *Erwinia carotovora*
(D) *Pseudoonas aeroginosa* strain 4CBPPPA14

40. Bacterial streak of rice is transmitted through:

(A) Soil
(B) Seed
(C) Air
(D) Rain

41. Bacterial blight of rice is transmitted through:

(A) Soil
(B) Seed
(C) Air
(D) Rain

42. **Black rot of cauliflower is caused by:**
(A) Virus (B) Fungi
(C) Bacteria (D) Actinomycetes

43. **Pathological wilting caused by *Ralstonia solanacearum* is mainly due to:**
(A) Extra cellular slime (B) Pectic enzyme
(C) Cellulose (D) Proteases

44. **Brown rot disease of potato is caused by:**
(A) Fungi (B) Bacteria
(C) Virus (D) Nematode

45. **Bacteria having a viral genome integrated in its DNA is termed as:**
(A) Lysogenic (B) Lytic
(C) Transduction (D) Transformation

46. **Match Archaebacteria:**
(A) Mendosicutes (B) Tenericutes
(C) Gracillicutes (D) Firmicutes

47. **Match Gracillicutes:**
(A) Scotobacteria (B) Anoxyphotobacteria
(C) Oxyphotobacteria (D) All of the above

48. **Match Firmicutes:**
(A) Gram positive (B) Gram negative
(C) Both (A) and (B) (D) None of these

49. **Match oxyphotobacteria:**
(A) Actinomycetes (B) Cynobacteria
(C) Archeobacteria (D) None of these

50. **Match Firmicutes:**
(A) Thallobacteria (B) Cynobacteria
(C) Archaebacteria (D) None of these

51. **Match Thallobacteria:**
(A) *Actinomycetes* (B) *Rhizobium*
(C) *Pseudomonas* (D) None of these

52. **The antibiotic produced by *Streptomyces griseus* is:**
(A) Tetracyclines (B) Polypeptide
(C) Polyenes (D) Aminoglycoside

53. Match Corynebacterium:

(A) *Rhodococus*
(B) *Pseudomonas*
(C) *Erwinia*
(D) None of these

54. Match pyrothin type antibiotic:

(A) Nystatin
(B) Thiolutin
(C) Bulbiformin
(D) None of these

55. Antibiotic blasticidin is produced by:

(A) *Penicillium griseofulvum*
(B) *Streptomyces griseochromogenes*
(C) *Streptoverticillium cinnamomeum*
(D) None of these

56. Antibiotic griseofulvin is produced by:

(A) *Penicillium griseofulvum*
(B) *Streptomyces griseochromogenes*
(C) *Steptoverticillium cinnamomeum*
(D) None of these

57. Match glutarimide antibiotic:

(A) Auriofungin
(B) Blasticidin
(C) Cyclohexamide
(D) None of these

58. Extra chromosomal DNA in bacteria may code for:

(A) Virulence
(B) Cell wall protein
(C) Drug resistance
(D) All of these

59. *Erwinia amylovora* enters through:

(A) Hydethodes
(B) Lenticels
(C) Nectaries
(D) Directly

60. Black rot of crucifers' bacterial pathogen enters host through:

(A) Lenticels
(B) Nectaries
(C) Directly
(D) Hydethodes

61. Soft rot bacteria enter through:

(A) Wounds
(B) Nectaries
(C) Directly
(D) None of these

62. Who showed that the *Corynebacterium fasciens* which causes fasciation disease of pea produce a cytokinin and IAA:

(A) Yabuta and Hayashi (1939) (B) Kurusawa (1926)
(C) Kalmbt *et al.* (1966) (D) None of these

63. Black rot of crucifers caused by *Xanthomonas campestris* was introduced in India from Jawa during:

(A) 1912 (B) 1929
(C) 1932 (D) 1935

64. Pasteur is related with the discovery of:

(A) Yeast (B) Budding
(C) Fermentation (D) Aerobic respiration

65. Streptomycin antibiotic is obtained from:

(A) *Streptomyces griseus* (B) *S. erythreus*
(C) *S. helsidii* (D) *S. fradiae*

66. Endospore forming bacteria are:

(A) *Corynebacterium* (B) *Clostridium* and *Bacillus*
(C) *Pseudomonas* (D) *Erwinia*

67. Polyene antibiotic is:

(A) Pimaricin (B) Streptomycin
(C) Penicillin (D) None of these

68. Absence of flagella is called:

(A) Atrichous (B) Amphitrichous
(C) Lophotrichous (D) Peritrichous

69. If one flagellum at each end of the cell called:

(A) Atrichous (B) Amphitrichous
(C) Lophotrichous (D) Peritrichous

70. If two or more flagella at one or both ends of the cell is called:

(A) Atrichous (B) Amphitrichous
(C) Lophotrichous (D) Pertrichous

71. A large number of flagella over the cell is called:

(A) Monotrichous (B) Peritrichous
(C) Lophotrichous (D) Amphitrichus

72. Match a single flagellum at one end of the cell:

(A) Monotrichous (B) Lophotrichous

(C) Atrichous (D) Peritrichous

73. Defined nucleus are termed:

(A) Prokaryote (B) Protozoa

(C) Eukaryote (D) Fungi

74. Black arm of cotton is caused by:

(A) *Xanthomonas campestris* pv. *oryzae* (B) *X. campretis* pv. *oryzaecola*

(C) *X. campestris* pv. *malvacearum* (D) *Pseudomonas solanacearum*

75. Crown gall bacterium overwinters in:

(A) Infected leaves (B) Infested soil

(C) Infected buds (D) Infected fruits

76. The only genera belonging to gram positive group which causes plant disease is:

(A) *Xanthomonas* (B) *Erwinia*

(C) *Pseudomonas* (D) *Clavibacter*

77. Waksman was awarded Nobel prize for the discovery of:

(A) Chloromycetin (B) Streptomycin

(C) Aureofungin (D) Penicillin

78. Transformation was discovered with the experiment on the:

(A) *Streptococcus* (B) *Pseudomonas*

(C) *E. coli* (D) *Xanthomonas*

79. Crown gall bacterium is:

(A) Air-borne (B) Seed-borne

(C) Soil-borne (D) Fruit-borne

80. Stalk rot of maize is caused by:

(A) *Agrobacterium tumefaciens* (B) *Erwinia carotovora*

(C) *Fusarium oxysporum* (D) None of these

81. Routine method used for measuring bacterial growth is:

(A) Total count (B) Dry weight

(C) Viable count (D) Tubidimetry

82. Who gave the term pasteurization?

(A) Louis Pasteur (B) K.C. Mehta

(C) De Barry (D) Leeuwenhoek

83. Latest edition (8th) of Bergey's manual of Deteminative Bacteriology was edited by:

(A) J.M. Young and A.M. Pathon
(B) R.E. Buchanan and N.E. Gibbons
(C) D.c. Sands and M.P. Star
(D) None of these

84. Which of the following is the most important character of prokaryotes?

(A) Absence of cell wall
(B) Absence of chloroplast
(C) Pleomorphic shape
(D) Absence of nuclear membrane

85. The genus *Xanthomonas* was edited by:

(A) Erwin/F. Smith
(B) W.J. Dowson
(C) R.A. Lelliott
(D) M.P. Star

86. Antibacterial antibiotic oxytetracycline is produced by:

(A) *Streptomyces griseus*
(B) *Steptomyces rimosus*
(C) *Streptomyces aureofungin*
(D) *Pencillium notatum*

87. Match terramycin:

(A) Oxytetracyclin
(B) Chlorotetracyclin
(C) Tetracyclin
(D) None of these

88. Match aureomycin:

(A) Oxytetracyclin
(B) Chlorotetracyclin
(C) Tetracyclin
(D) None of these

89. Match actinomycin:

(A) Oxytetracyclin
(B) Chlorotetracyclin
(C) Tetracyclin
(D) None of these

90. Photo reactivation phenomenon is related to:

(A) Nucleic acid
(B) Protein
(C) Lipid
(D) Carbohydrate

91. Bacteriocin-colicin is generally produced by the members of family:

(A) Enterobacteriaceae
(B) Pseudomonadaceas
(C) Corynebacteriaceae
(D) Mycoplasmataceae

92. Which of the following bacterial genus is gram positive:

(A) *Agrobacterium*
(B) *Xanthomonas*
(C) *Pseudomonas*
(D) *Corynebacterium*

93. What is the mode of action of wild fire toxin produced by *Pseudomonas tabaci*:

(A) Inhibits respiratory enzyme

(B) Chelates iron respiratory enzyme

(C) Interferes with methionine metabolism

(D) Affects catalase activity

94. Black rot of cauliflower caused by:

(A) *Erwinia carotovora* (B) *Xanthomonas phaseoli*

(C) *Xanthomonas campestris* (D) *Pseudomonas solanacearum*

95. Tundu disease of wheat is caused by:

(A) *Anguina tritici* (B) *Corynebacterium tritici*

(C) Both (A) and (B) (D) None of these

96. Bacterial blight of rice epidemic occurred in Bihar in the year:

(A) 1960 (B) 1963

(C) 1965 (D) 1970

97. For the control of bacterial blight of cotton seeds should be given hot water treatment at 56°C for:

(A) 5 minutes (B) 10 minutes

(C) 20 minutes (D) 30 minutes

98. A bacterium having a cluster of flagella at one end is called:

(A) Amphitrichous (B) Polytrichus

(C) Ditrichus (D) Lophotrichous

99. Siderophores are produced by:

(A) *Trichoderma* (B) *Gliocladium*

(C) *Chaetomium* (D) *Pseudomonas fluorescens*

100. Bacilli are:

(A) Spherical (B) Rod

(C) Spiral (D) None of these

101. Cocci are:

(A) Rod (B) Spiral

(C) Spherical (D) None of these

102. An example of plant disease that is caused by gram positive bacterium is:

(A) Ear cockle of wheat
(B) Black rot of crucifers
(C) Fruit canker of tomato
(D) Ring rot of potato

103. Acid and gas are produced from lactose source by:

(A) *Erwinia carotovora*
(B) *Escherichia coli*
(C) *Xanthomonas campestris* pv. *campestris*
(D) *Pseudomonas solanacearum*

104. Moko disease of banana is caused by:

(A) *Agrobacterium tumefaciens*
(B) *Pseudomonas solanacearum*
(C) *Erwinia carotovora*
(D) None of these

105. Bacterial blight of rice is primarily:

(A) Seed borne
(B) Soil borne
(C) Air borne
(D) None of these

106. *Xanthomonas* species produce:

(A) Yellow diffusible pigment
(B) Yellow non diffusible pigment
(C) Pink non diffusible pigment
(D) Pink diffusible pigment

107. Cells of *Erwinia* have:

(A) Monotrichous flagella
(B) Peritichous flagella
(C) Lophotrichous flagella
(D) Amphitrichous flagella

108. The most convenient method for testing pathogenicity of *Xanthomonas camprestris* pv. *oryzae* is:

(A) Leaf cut method
(B) Stab inoculation method
(C) Pin prick method
(D) Spray inoculation

109. *Agrobacterium* belongs to the family:

(A) Bacillacae
(B) Pseudomonadaceae
(C) Corynebacteriaceae
(D) Rhizaobiaceae

110. Which of the following is a typical wound parasite?

(A) *Xanthomonas campestris* pv. *oryzae*
(B) *X. campestris* pv. *oryzicola*
(C) *X. camkprestis* pv. *camprestris*
(D) *Agrobacterium tumefaciens*

111. Fertility factor is transmitted only by:

(A) Conjugation
(B) Transformation
(C) Transduction
(D) All of these

112. The F-factor (fertility factor) is transmitted only by directed:

(A) Cell to cell contract (B) Plasmodesmata

(C) Periplasm (D) None of these

113. The intercellular passage of DNA is highly specific and consists of only:

(A) Other cellular material (B) DNA

(C) RNA (D) None of these

114. Prior to conjugation the plasmid DNA is:

(A) ssDNA (B) dsDNA

(C) Broken DNA (D) None of these

115. Who coined the term plasmid?

(A) Laderberg (B) Tatum

(C) Smith (D) None of these

116. The integration of F (fertility) and chromosome (by crossing over) occurs in about once per 10^5 cells in each generation. These are known as:

(A) High frequency of recombination (B) Complementary strands

(C) Transformation (D) Transduction

117. Who discovered the regulation of genes:

(A) J. Lederbeg (1952) (B) S.A. Waksman (1952)

(C) J. Watson (1962) (D) F. Jacob (1965)

118. Phenol production by wounded plant cells is essential for the infection of:

(A) *Clavibacter* (B) *Erwinia*

(C) *Xanthomonas* (D) *Agrobacterium*

119. *Corynebacterium michiganense*, which causes the bacterial canker of tomato plants, often enters through:

(A) Nectarthodes (B) Hydathodes

(C) Trichomes (D) Lenticels

120. The shikimic acid pathway was elucidated first in:

(A) Fungi (B) Bacteria

(C) Virus (D) MLO

121. Bacterial canker of tomato is caused by *Corynebacterium michiganense* and spread by:

(A) Insect (B) Mite

(C) Man (D) Wind

122. Antibiotic blasticidins is produced by:

(A) *Penicillium nigricans*

(B) *Streptomyces griseochromogenes*

(C) *Penicillium patulum*

(D) None of these

123. Foot rot of wheat is caused by:

(A) *Pseudocercosporella herpotrichoides* (B) *Aphanomyces euteiches*

(C) *Sphaerotheca fuligena* (D) None of these

124. Black arm of cotton is caused by:

(A) *Xanthomonas campestris* pv. *malvacearum*

(B) *X. campestris* pv. *phaseoli*

(C) *X. campestris* pv. *translucens*

(D) None of these

125. Bacterial stripe of sorghum and corn is caused by:

(A) *Pseudomonas andropogonis* (B) *Ps. avenae*

(C) *Ps. atrofaciens* (D) None of these

126. Basal glumes rot of cereals is caused by:

(A) *Pseudomonas rubrilinens* (B) *Pseudomonas atrofaciens*

(C) *Xanthomonas albilineans* (D) None of these

127. Which of the following bacteria is generally used genetic engineering for transfer of characters from one organism to another:

(A) *Agrobacterium tumefaciens* (B) *Bacillus thuringiensis*

(C) *Pseudomonas syringae* (D) None of these

128. Fire blight in pear is caused by:

(A) *Xanthomonas amylovora* (B) *Erwinia carotovora*

(C) *Erwinia amylovora* (D) *Pseudomonas syringae*

129. Ice-nucleation-active bacteria are:

(A) *Pseudomonas syringae* (B) *Xanthomonas sp.*

(C) *Erwinia amylovora* (D) None of these

130. Gummosis of stone fruit is caused by:

(A) *Pseudomonas syringae* pv. *morsprunorum*

(B) *Xanthomonas syringae*

(C) *Erwinia carotovora*

(D) None of these

131. Ratio of RNA to DNA in gram positive bacteria is:

(A) 8:1
(B) 10:2
(C) 12:5
(D) 10:10

132. *Agrobacterium tumefaciens* and *A. radiobacter* are identical except for the plant pathogenic nature that contains:

(A) Single stranded RNA
(B) Double stranded RNA
(C) Circular double stranded DNA
(D) Circular single stranded DNA

Fill in the Blanks with Correct Answer

1. In citrus, canker is caused by________ and the bacterium survives in canker on the standing trees.

   Xanthomonas citri

2. The primary inoculums of cucurbits wilt (*Erwinia tracheiphila*) by striped and spotted cucumber beetles in their________

   Body

3. Tab toxin produced by the________

   Pseudomonas syringae* pv. *tabaci

4. *Pseudomonas solanacearum* (bactrerial wilt) induces a 100 fold, increase in the________levels.

   IAA

5. Phaseolo toxin produced by________causes halo blight of bean.

   Pseudomonas syringae* pv. *phaseolicola

6. Moko diseases of banana is caused by________

   Pseudomonas solanacearum

7. Halo ring in citrus is due to the attack of________

   Xanthomonas citri

8. Crown gall of apple is caused by________

   Agrobacterium tumefaciens

9. Citrus canker is the most serious problem of________

   Acid lime

10. The crown gall caused by *Agrobacterium tumefaciens* has served as a classic model to study the role of auxins, IAA and related compounds in plants.

    Pathogenesis

11. *Escherichia coli* is the most common ________attacked by bactreriophages.

    Bacteria

12. Flagella are commonly found in ________shaped bacteria.

    Rod or spiral

13. *Xanthomonas malvacearum* is________more similar to its host cotton than to other species of *Xanthomonas*.

    Antigenically

14. Dahlbeck asdn Stahl (1979) reported mutation for changes of race in culture of________

    Xnathomonas vesicatoria

15. In the fruit of banana infected with *Pseudomonas solanacearum*, the ethylene content________with the yellowing (pre mature) of the fruit.

    Increases

16. The bacteria that retain the primary stain (appear dark blue or violet) are called________

    Gram positive

17. Agrosin 84 chemical produced by ________of strain K84.

    Agrobacterium radiobacter

18. Bacteriocin is a part of ________

    Antibiotics

19. Quantum 4000 is a trade name of commercial bio-control agent of________ which control wilt of cereals and vegetables.

    Bacillus subtillis

20. Barkholdaria cepacea is synonyms of ________

    Pseudomonas cepacea

21. A non-bacteriocin producing strain of________gave control of fire blight of apple.

    Erwinia herbicola

22. Sex factor in bacteria is________

    Episome

23. Gas vacuole found in which bacteria________

    Cynobacteria

24. Ti and Ri plasmids as vectors for________

    Higher plants

25. Ti (tumor inducing) plasmid found in virulent strains of________

    Agrobacterium tumefaciens

26. Ri (root inducing) plasmid found in virulent strains of________

    Agrobacterium rhizogenes

27. Siderophore produced by________in bio-control of *Gaeumannomyces graminis* var. *tritici* causes take all disease of wheat and *Fusarium oxyeporum* f. sp. *lini* causes flax wilt.

    Fluorescant pseudomonas

28. The F factor is an infective element now called as________

    Episome

29. Bdellovibrio is a genus of________chaemoorganotrophic bacteria of uncertain taxonomic affinity.

    Gram negative

30. ________was the first phytopathogenic bacterium in which the plasmid was detected and correlated with pathogenicity.

    Agrobacterium tumefaciens

31. Aureofungin produced in submerged cultures of________

    Steptomyces cinnamomeus* var. *terricola

32. Pyrothin type antibiotic produced by ________

    Streptomyces albus

33. Blight of leek in onion is caused by________

    Pseudomonas syringae* pv. *porri

34. Pyrroles metabolites produced by fluorescent *Pseudomonas* spp and active against________

    Pythium ultimum/cucumber

35. Commercial name of Streptocyclin is________

    Tetracyclin

36. Aurecomycin antibiotic is produced by________

    Streptomyces aureofaciens

37. Terramycin antibiotic is produced by________

    Streptomyces romosus

38. Pyceyanine metabolites produced by fluorescent *Pseudomonas* spp and active against________

    Septoria tritici/wheat

39. Many prokaryotes and some of the lower eukaryotes also carry small circular molecules of DNA, called ________

    Plasmids

40. Leaf scald of sugarcane is caused by________

    Xanthomonas albilineans

41. Bacterial canker and gummosis is one of the most important diseases of

    Stone fruit trees

42. Split genes was discovered in________

    1977

43. In., foreign genes were introduced with the introduction of the gene gun in 1987.

    Chloroplast

44. Herbert Boyer and Stanley Cohen________inserted foreign gene (DNA) into plasmid.

    1973

45. Founder of modern bacteriology is________

    Louis Pasteur

46. In________an antibiotic resistant gene was inserted into tobacco, leading to the first genetically engineered plant.

    1983

47. .________is considered the father of bacteriological techniques

    Robert Koch (1843-1910)

48. Split gene is discovered by ________

    Richard J. Roberts and Phillip A. Sharp in 1977

49. Spiral or helix form bacteria are________

    Spirillum, vibrio* and *spirochete

50. Spiral bacteria have________

    One or more twists

51. Spiral bacteria is in________

    Spirillum shaped

52. Spirochete have a helical shape and flexible bodies, it is a ________

    Spiral bacteria

53. Chitin is present in the________of many bacterial species.

    Cell wall

54. ________variability is common among races of bacteria.

    Phenotypic

55. The first cynophages was discovered in 1963 by Safferman and Morris, which parasitized three blue green algae. *Lyngbya*, *Phomidium* and *Plectonema* henced named

    LPP-1

56. For most bacteria, the suitable pH ranges from________

    5.0 to 9.0

57. Bacteria can survive at________temperature.

    0 to 85°C

58. Surface appendages in bacteria are________and________

    Flagella and Pilli

59. Surface adherent in bacteria are________and________

    Capsule and slime layer

60. The word nucleoid is used for the nuclear apparatus of________

    Prokaryotes

61. All cells of bacterial species are killed at given temperature called________

    Thermal death point

62. The best known pressure to which bacteria respond is________

    Osmotic pressure

63. Genus *Xanthomonas* survive in the soil for________ years, in the absence of host.

    4-5

64. Bacteriochlorophyllus absorb light in the near infra-red region.

    660 to 870 nm

65. Cilia associated with________

    Bacillus

66. A formulation of streptomycin mixed with oxytetracyclin (Terramycin) is marked under the name________

    Agrimycin

67. Streptomycin is effective against________

    both gram negative and gram positive bacterial plant pathogens

68. Bdellovibrio has a________polar flagellum.

    Single

69. The________developed by Luria and Delbruck (1943), established that the bacterial mutations are rare chance events and not directed by environment.

    Fluctuation test

70. Bacteria do not store the recessive genes because they are________

    Haploid

True or False

1. Prokaryotic cells have one type of ribosomes. (**True**)
2. Bacteria are the smallest organism. (**False)**
3. Bean bacterial blight (*Xanthomonas phaseoli*) spread through irrigation and drainage water. (**True**)
4. Bacteria are most common in boiled water. (**False**)
5. Bacteria commonly occur in air and soil. (**True**)
6. Chromosomes replicate by mitosis in prokaryotic cell. (**False**)
7. All the bacteria are unicellular. (**False**)
8. Bacterial cytoplasm is vacuolated. (**False**)
9. Flagella of bacteria and fungi are similar. (**False**)
10. Bacteria cannot be grown artificially. (**False)**
11. Bacterial flagella are not similar to those found in other plants. (**True)**

12. Gene encoding for lytic peptide is transferred for bacterial resistance. (**True**)
13. Angular leaf spot of cotton (*Xanthomonas malvacearum*) and black shank of tobacco (*Phytophthora nicotianae* var. *nicotianae*) do not spread through irrigation and drainage water. (**False**)
14. *Botrytis* strains are not resistant to benomyl, dichloran, iprodion. (**False**)
15. *Bacillus subtilis* in soybeen, Corynebacterium michiganense pv. *sepedonicum* in tomato, *Pseudomonas syringae* pv. *pisi* in pea, *Xanthomonas campestris* pv. *malvacearum* in cotton are found in seed coats. (**True**)
16. Natural opening such as the hilum is not highly absorptive and provides entry of bacteria. (**False**)
17. Bacteria, viruses and fungal pathogens do not obstruct photosynthesis by adversely affecting chloroplasts. (**False**)
18. Firmicutes bacteria are mostly double called, and gram negative. (**False**)
19. *Xylophilus* causes the bacterial necrosis and canker of grapevines. (**True**)
20. Gracillicutes bacteria have triple cell membrane and gram positive. (**False**)
21. Rod shaped phytopathogenic bacteria reproduce by the asexual process known as binary fission. (**True**)
22. Restriction fragment length polymorphism (RFLP), profiles may be characteristics of the bacterium and therefore, can be used to identify the bacterium. (**True**)
23. Bacterial stripe of sorghum and corn is caused by *Pseudomonas avenae*. (**False**)
24. Halo blight of oats and other cereals is caused by *Pseudomonas syringae* pv. *coronafaciens*. (**True**)
25. Leaf blight of cereals of caused by *Pseudomonas andropogonis*. (**False**)
26. The bacteria (*Xanthomonas campestris* pv. *arboricola*) overwinter in twig lesions and in the buds. (**True**)
27. Fire blight of pome fruits is caused by *Erwinia amylovora*. (**True**)
28. Ring rot of potato is caused by *Clavibacter michiganense* subsp. *sepedonicum*. (**True**)
29. Common scab of potato (*Streptomyces scabies*) incidence is favored by water deficiency or dry soil. (**True**)
30. Potato ring rot is avoided through the use of healthy seed tubers. (**True**)
31. The black rot bacteria (*Xanthomonas campestris* pv. *campestris*) overwinter in infected plant debris and on or in the seed. (**True**)
32. Bacterial canker and gummosis is one of the most important diseases of potato. (**False**)

33. Mollicutes are sensitive to antibiotics, particularly those of the tetracycline group. (**True**)
34. Acidic soil favors the incidence of common scab (*Streptomyces scabies*) of potato. (**False**)
35. Bacteria seem to be less effective than fungi as plant pathogens. (**True**)
36. In the bacterial wilt and brown rot of potato caused by *Pseudomonas solanacearum* the bacterium grows in vascular bundles of the tuber and contain 100 times more IAA than the healthy plants. At the same time phenolic compounds (scopoletin) also increase 10 times. **(True)**
37. In bacterial brown rot and wilt of potato (*Pseudomonas solanacearum*) the amount of polysaccharide produced by the pathogen is proportional to the severity of the symptoms. **(True)**
38. Aureofungin is a heptaene antibiotic produced in the submerged cultured of *Streptomyces cinnamomens* var. *terricola*. **(True)**
39. The cause of bacterial wilt is *Pseudomans solanacearum*. (**True**)
40. *Bacillus polymyxa* causes soft rot in potato at high temperature and high humidity and produce yellow sticky mass with a characteristics fishy odour. (**True**)
41. Teichoic acid is absent in gram positive bacteria. (**False**)
42. More muramic acid is found in gram negative bacteria. (**False**)
43. Gram positive bacteria retain the stain violet whereas the gram negative bacteria are decolorized. (**True**)
44. *Erwinia amylovora* grows well in yeast extract medium. (**True**)
45. *Azospirillus* bacteria, produces siderophores which are of low molecular weight iron binding compounds under iron deficient conditions. (**True**)
46. Bacteria *Azospirillus*, produces growth promoting substances *i.e.* IAA, GA and vitamins. (**True**)
47. Endotoxin is produced by gram positive bacteria. (**False**)
48. Exotoxin is produced by gram positive bacteria. (**False**)
49. Kingdom monera is true prokaryotes. (**True**)
50. All gram positive bacteria are strictly coryneform. (**False**)
51. Archaebacteria is not prokaryote. (**True**)
52. Bacillus is the rod shaped bacterial cell. (**True**)
53. Shape of prokaryotic cells (rigid or flexible) is governed by the presence and chemical composition of the cell wall. (**True**)
54. V-form bacteria are not corynebacteria (irregular shape). (**False**)

55. The L-form or L-shape (*Agrobacterium* and *Erwinia*) of these bacteria lacks rigid cell wall. (**True**)
56. The cell wall is necessary for the flageller movement. If the wall is removed the flageller moment ceases. (**True**)
57. Eukaryotic flagellum made up of axoneme and sheath (9+2). (**True**)
58. Eukaryotic flagellum posses' antigenic properties. (**False**)
59. Bacterial flagellum consists almost entirely of protein while eukaryotic flagellum contains about 70 per cent protein, 20 per cent lipid and 10 per cent polysaccharides. (**True**)
60. Bacterial flagellum does not show the presence of certain amino acids (*e.g.* cysteine) while eukaryotic flagellus, the protein contains all the usual amino acids. (**True**)
61. Isolated flagella posses ATPase activity while eukaryotes flagellum do not show ATPase activity. (**False**)
62. The antibiotic cyclohexamide inhibits the function of 80S ribosomes, so it is effective against eukaryotes but not against prokaryotes. **(True)**
63. During nuclear division just before cell division two double helices of the RNA are formed. (**False**)
64. Strongly bactericidal rays are those between 2500-2800A°. (**True**)
65. *Pseudomonas* and *Xanthomonas* is pigment producer. (**True**)
66. Antibiotics that affect bacteria often inhabit mitochondria or chloroplast but not interfere with the other functions of eukaryotes plant cells. (**True**)
67. Pink eye disease of potato and soft rot of other fleshy vegetable *Pseudomonas gladioli*. (**False**)
68. *Pseudomonas* causing the southern bacterial wilt of solanaceous crop and moko disease of banana. (**True**)
69. Cancer type symptoms caused by *Agrobacterium tumefaciens*. (**True**)
70. *Corynebacterium* causing fasciations or leaf gall on many annual or perennial herbaceous ornamentals. (**True**)
71. Black rot or black vein of crucifers is caused by *Pseudomonas syringae*. (**False**)
72. *Pseudomonas* causing the olive knot disease and other the bacterial gall or canker of oleander (*Pseudomonas syringae* pv. *savastanoi*). (**True**)
73. The plant pathogenic species of *Streptomyces* cause only scabs or lesions of below ground crops. (**True**)
74. Bacterial leaf blight is characterized by the drying of the leaf tip and inward rolling and twisting of the leaf blades. (**True**)
75. *Pseudomonas savastanoi* (olive knot) spread by only olive fly. (**True**)

76. Syringomycine phytopathogenic toxin produced in peach plant which is associated with target site of membrane. (**True**)

77. Coronatine toxin is produced by *Pseudomonas atropurpurea.* (**True**)

78. Bacterial chromosome is composed of dsDNA. (**True**)

79. Bacteria multiply by binary fission. (**True**)

Match the Following

1. **Correctly pair the terms in section 'A' to those given under section 'B':**

Section 'A'	*Section 'B'*
I. Citrus	(A) Sexual reproduction
II. Conjugation	(B) Canker
III. Plasma membrane	(C) Symbiotic bacteria
IV. Root nodules	(D) Respiration

Answer

I	*II*	*III*	*IV*
B	*A*	*D*	*C*

2. **Match disease and host with the parasite:**

Disease Host	*Parasite*
I. Canker, citrus	(A) *Xanthomonas campestris* pv. *malvacearum*
II. Angular leaf spot, cotton	(B) *Xanthomonas campestris* pv. *citri*
III. Bacterial leaf blight, rice	(C) *Xanthomonas campestris* pv. *oryzicola*
IV. Bacterial leaf streak	(D) *Xanthomonas oryzae* pv. *oryzae*

Answer

I	*II*	*III*	*IV*
B	*A*	*D*	*C*

3. **Match disease and host with the parasite**

Disease Host	*Parasite*
I. Bacterial leaf spot or blight, mango	(A) *Clavibacter xyli* sub. sp. *xyli*
II. Red stripe, sugarcane	(B) *Pseudomonas solanacearum*
III. Wilt of potato and brown rot	(C) *Pseudomonas rubrilineans*
IV. Ratoon stunning disease sugarcane	(D) *Xanthomonas campestris* pv. *mangifera-indicae*

Answer

I	*II*	*III*	*IV*
D	*C*	*B*	*A*

4. Match disease and host with the parasite

Disease Host	*Parasite*
I. Wild fire of tobacco	(A) *Erwinia amylovora*
II. Halo blight, bean	(B) *Pseudomonas syringae* pv. *tabaci*
III. Spot of tomato	(C) *Pseudomonas syringae* pv. *phaseolicola*
IV. Fire blight, pome fruit	(D) *Xanthomonas camprestris* pv. *vesicatoria*

Answer

I	*II*	*III*	*IV*
B	*C*	*D*	*A*

5. Match the antibiotic with antagonistic pathogen producing by it:

Antibiotics	*Producing antagonistic pathogen*
I. Bulbiformin	(A) *Bacillus subtilis*
II. Phenazine	(B) *Pseudomonas fluorescens*
III. Gliovirin	(C) *Gliocldium virens*
IV. Agrocin 84	(D) *Agrobacterium radiobacter*

Answer

I	*II*	*III*	*IV*
A	*B*	*C*	*D*

6. Match the following bacterial group with their genus:

Bacterial group	*Genus*
I. Firmibacteria	(A) *Clavibacter*
II. Rhizobiaceae	(B) *Xanthomonas*
III. Pseudomonadaceae	(C) *Agrobacterium*
IV. Thallobacteria	(D) *Bacillus*

Answer

I	*II*	*III*	*IV*
D	*C*	*B*	*A*

7. Match the following bacterial group with their genus:

Bacterial group	*Genus*
I. Firmibacteria	(A) *Curtobacterium*
II. Enterobacteriaceae	(B) *Xylophilus*
III. Pseudomonadaceae	(C) *Erwinia*
IV. Thallobacteria	(D) *Clostridium*

Answer

I	*II*	*III*	*IV*
D	*C*	*B*	*A*

8. Match the following bacterial group with their genus:

Bacterial group	*Genus*
I. Proteobacteria	(A) *Escherichia*
II. Enterobacteriaceae	(B) *Sphingomonas*
III. Pseudomonadaceae	(C) *Rhizobacter*
IV. Thallobacteria	(D) *Streptomyces*

Answer

I	*II*	*III*	*IV*
A	*B*	*C*	*D*

9 Match the following bacterial genus with their characteristics features:

Bacterial genus	*Characteristics*
I. *Streptomyces*	(A) Filamentous and gram positive
II. *Clavibacter*	(B) Non motile and gram positive
III. *Xylella*	(C) Non motile and gram negative
IV. *Agrobacterium*	(D) Motile and one to four peritrichous flagella

Answer

I	*II*	*III*	*IV*
A	*B*	*C*	*D*

10. Match the following bacterial genus with their characteristics features:

Bacterial genus	*Characteristics*
I. *Xanthomonas*	(A) Utilizes starch
II. *Erwinia*	(B) Firmentive and peritrichous
III. *Pseudomonas*	(C) Polar flagella
IV. *Arthrobacter*	(D) Branching and gram positive

Answer

I	*II*	*III*	*IV*
A	*B*	*C*	*D*

11. Match the following which are related to their features:

I. *Pathovar*	(A) Based on biochemical differentiation
II. *Race*	(B) Pathogenicity on different host cultivar
III. *Serovar*	(C) Based on serological differentiation
IV. *Biovar*	(D) Pathogenicity on different host species of genera

Answer

I	*II*	*III*	*IV*
D	*B*	*C*	*A*

12. Match the following inhibitors of protein synthesis in prokaryotes with their mode of action:

I. Streptomycin — (A) Inhibits aminoacyl-tRNA binding at A site

II. Chloramphenicol — (B) Inhibits peptidyl transferase activity

III. Tetracycline — (C) Premature peptidyl-tRNA dissociation

IV. Erythromycin — (D) Prevents formation of initiation complexes

Answer

I	*II*	*III*	*IV*
D	*B*	*A*	*C*

13. Match the following prokaryotes inhibitors with their mode of action:

I. Fusidic acid — (A) Inhibits EF-G: GDP dissociation from ribosome

II. Bacitran — (B) Inhibit cell wall synthesis

III. Sulfonamides — (C) Inhibit folic acid synthesis

IV. Streptomycin — (D) Codon misreading

Answer

I	*II*	*III*	*IV*
A	*B*	*C*	*D*

14. Match the following group of bacteria with their environment:

I. Neutrophiles — (A) Grow at around at 7.0 pH

II. Halophiles — (B) Adapted to high salt concentration

III. Psychrophiles — (C) Adapted to low temperature

IV. Microaerophiles — (D) Adapted to low oxygen

Answer

I	*II*	*III*	*IV*
A	*B*	*C*	*D*

15. Match the source of energy to the group of bacteria:

I. CO_2 as sole carbon source — (A) Autotrophs

II. Facultative autotrophs — (B) Mixiotrophs

III. Light — (C) Organotrophs

IV. Organic compound as electron donor — (D) Phototrophs

Answer

I	*II*	*III*	*IV*
A	*B*	*D*	*C*

16. **Match the group of bacteria with their source of energy:**

I. Biotrophs	(A) Require metabolically active host cells
II. Auxotrophs	(B) Require amino acid
III. Necrotrophs	(C) Obtain energy from moribund cells
IV. Paratrophs	(D) Obtain energy from biosynthetic reactions

Answer

I	*II*	*III*	*IV*
A	*B*	*C*	*D*

17. **Match disease and host with the parasite**

Disease Host	*Parasite*
I. Fruit blotch in watermelon	(A) *Acidovorax avenae* subsp. *citrulli*
II. Seed rot in chickpea	(B) *Bacillus subtilis*
III. White blotch in wheat	(C) *Bacillus megaterium* pv. *cerealis*
IV. Canker in tomato	(D) *Clavibacter michiganensis* subsp. *michiganensis*

Answer

I	*II*	*III*	*IV*
A	*B*	*C*	*D*

18. **Match disease and host with the parasite**

Disease Host	*Parasite*
I. Bacterial stripe of sorghum	(A) *Pseudomonas andropogonis*
II. Top rot of sugarcane	(B) *Pseudomonas rubrilineans*
III. Bacterial wilt in cucumber	(C) *Erwinia tracheiphila*
IV. Bacterial disease of maize	(D) *Pantoea stewartii*

Answer

I	*II*	*III*	*IV*
A	*B*	*C*	*D*

19. Match disease and host with the parasite

Disease Host	*Parasite*
I. Bacterial blight of walnut	(A) *Pseudomonas syringae* pv. *coronafaciens*
II. Halo blight of oat	(B) *Pseudomonas rubrilineans*
III. Black rot of crucifers	(C) *Erwinia tracheiphila*
IV. Bacterial gall in carrot	(D) *Rhizobacter daucus*

Answer

I	*II*	*III*	*IV*
D	*A*	*D*	*C*

Answers (Multiple choice questions)

(1)	(C)	(34)	(A)	(67)	(A)	(100)	(B)
(2)	(C)	(35)	(C)	(68)	(A)	(101)	(C)
(3)	(A)	(36)	(A)	(69)	(B)	(102)	(A)
(4)	(C)	(37)	(A)	(70)	(C)	(103)	(A)
(5)	(A)	(38)	(B)	(71)	(B)	(104)	(B)
(6)	(A)	(39)	(D)	(72)	(A)	(105)	(A)
(7)	(B)	(40)	(B)	(73)	(A)	(106)	(A)
(8)	(B)	(41)	(B)	(74)	(C)	(107)	(B)
(9)	(A)	(42)	(C)	(75)	(B)	(108)	(C)
(10)	(D)	(43)	(B)	(76)	(B)	(109)	(D)
(11)	(C)	(44)	(B)	(77)	(B)	(110)	(D)
(12)	(A)	(45)	(A)	(78)	(A)	(111)	(A)
(13)	(B)	(46)	(A)	(79)	(C)	(112)	(A)
(14)	(D)	(47)	(D)	(80)	(B)	(113)	(C)
(15)	(B)	(48)	(A)	(81)	(C)	(114)	(B)
(16)	(C)	(49)	(B)	(82)	(A)	(115)	(A)
(17)	(A)	(50)	(A)	(83)	(B)	(116)	(A)
(18)	(C)	(51)	(A)	(84)	(D)	(117)	(D)
(19)	(A)	(52)	(A)	(85)	(B)	(118)	(B)
(20)	(B)	(53)	(A)	(86)	(B)	(119)	(C)
(21)	(A)	(54)	(B)	(87)	(A)	(120)	(B)
(22)	(C)	(55)	(B)	(88)	(B)	(121)	(C)
(23)	(B)	(56)	(A)	(89)	(C)	(122)	(B)
(24)	(A)	(57)	(C)	(90)	(A)	(123)	(A)
(25)	(B)	(58)	(A)	(91)	(A)	(124)	(A)
(26)	(B)	(59)	(C)	(92)	(D)	(125)	(A)
(27)	(A)	(60)	(D)	(93)	(A)	(126)	(B)
(28)	(D)	(61)	(A)	(94)	(C)	(127)	(A)
(29)	(A)	(62)	(C)	(95)	(C)	(128)	(C)
(30)	(C)	(63)	(B)	(96)	(B)	(129)	(B)
(31)	(A)	(64)	(C)	(97)	(B)	(130)	(A)
(32)	(A)	(65)	(A)	(98)	(D)	(131)	(A)
(33)	(A)	(66)	(B)	(99)	(D)	(132)	(C)

Chapter 3
Phytovirology

Queries with Timeline of Phytovirology

1. In 1886, Adolf Edward Mayer, the German born scientist worked in Holland on tobacco mosaic. He reported that the tobacco mosaic disease was neither due to microorganism nor due to nutritional imbalance. He demonstrated the contagious nature of the causal agent by artificial inoculation and finally concluded that tobacco mosaic was probably caused by a bacterium. He called the name of disease as 'Mosaikkrankheit'.
2. In 1891, E.F. Smith for the first time demonstrated that budding or grafting could be another method of transmission of contagious disease (peach yellows virus).
3. In 1892, Dmitri Ivanovsky, a Russian botanist, showed that the causal agent of tobacco mosaic could pass through a filter that retains bacteria. He believed that the disease was caused by a toxin secreted bacteria or by small bacteria that passed through the pores of the filter.
4. Martinus Willem Beijerinck, Dutch scientist, confirmed the findings of Mayer and Ivanovsky in 1898, and finally concluded that tobacco mosaic was not caused by a microorganism but by a 'Contagium vivum fluidum' *i.e.* infectious living fluid which he called a 'virus'. He is regarded as the Father of Plant Virology.
5. In 1929, H.O. Holmes provided a tool by which the virus could be measured by showing that the amount of virus present in the plant sap preparation is proportional to the number of lesions produced on appropriate host plant leaves rubbed with the sap.

6. The first major contribution to the nature of viruses was made by W.M. Stanley (1935). He treated the sap from diseased leaves of tobacco with ammonium sulphate and obtained a crystalline protein which he thought to be gamma glubuline and when placed on healthy tobacco leaves could reproduce the disease. This discovery finally proved that viruses are not living micro-organism (neither have nucleus nor cytoplasm but either DNA or RNA) because no living can be chemically treated and crystallized and still remain viable, for this discovery he received the Nobel Prize in 1946.
7. In 1936, F.E. Bawden and N.W. Pirie discovered the real nature of Tobacco Mosaic Virus. He demonstrated that the crystalline preparation of the virus actually consisted of protein and RNA.
8. Watson and Crick (1953) first to showed that DNA exists in a double helix, which boosted the molecular level studies. Rosalind Franklin (1955) discovered the full DNA structure of the virus. Heinz Fraenkel-Conrat and Robley William (1955) showed that purified TMV RNA and its coat protein can assemble by themselves to form functional viruses.
9. In 1956, Gierrer and Schramm, showed that the protein could be removed from the virus and that the nucleic acid carried the genetic information, so that inoculation with nucleic acid alone could cause infection and reproduce complete virus.
10. In1939, Kausche and his colleagues saw virus particles for the first time with electron microscope.
11. In 1962, Agar double diffusion serological test was observed. In 1977, Enzyme linked Immunosorbent Assay (ELISA) developed by A. Voller *et al.* and Clark and Adams. In 1963, some nucleic acid of viruses has been found as double stranded RNA (dsRNA). Charles Otley and Gary Stewart (1965) produced the first marketable scanning electron microscope. R.J.F. Shepherd *et al.* (1968) reported that viruses also contain double stranded DNA (dsDNA) in cauliflower. H.C. Harrison *et al.* (1977), first to discovered single stranded DNA (ssDNA) virus. In 1980, cauliflower mosaic virus, whose genomes is circular dsDNA chromosomes was the first plant virus for which the exact sequence of all its 8000 base pairs was determined. In 1982, the complete sequence of the bases in the ss TNV RNA was determined.
12. Many viral diseases are more virulent when the temperature is between 20°C and 25°C. Reduced light intensity generally increases the susceptibility of plants to virus infections. Phosphorus increases the severity of Cucumber Mosaic Virus on spinach.
13. Katte or marble disease of cardamom, mosaic and grassy shoot of sugarcane, mosaic, streak of wheat, maize mosaic, barley mosaic, bean mosaic, cowpea mosaic, bunchy top of banana, papaya mosaic, potato necrosis, potato leaf roll, tomato mosaic and chilli mosaic are disseminated by aphids.

14. Tobacco leaf curl, bean yellow mosaic, moong yellow mosaic, bottle gourd mosaic and yellow mosaic of *Dolichos lablab* are transmitted by white fly. Bottle gourd mosaic is transmitted by red pumpkin beetle while tungro of rice is disseminated by leafhopper.

15. Pigeon pea sterility virus (PSV), wheat streak mosaic virus, peach mosaic virus, fig mosaic virus are disseminated by mite (Eriophyidae). Cauliflower mosaic virus transmitted by aphid in a non-persistent manner.

16. Elongated plant viruses are rigid rods about 15 by 300 nm, but most appears as long, thin, flexible threads, 10 to 13 nm wide and range in length from 480-2000 nm, *e.g.* Rhabdoviruses are short, bacillus like, cylindrical rods, about 52-75 by 300-380 nm and rough surface.

17. Spherical viruses are polyhedral particles, ranging in diameter from about 17 nm (Tobacco Necrosis Satellite Virus) to 60 nm (Wound Tumor Virus) and Tomato Spotted Wilt Virus is surrounded by a membrane, flexible, spherical, about 100 nm in diameter.

18. Many plant viruses have split genomes, consisting of two or more distinct nucleic acid strands. For eg. Tobacco Rattle Virus consists of two rods, a long one (195 by 25 nm) and a shorter one (43 by 25 nm) while, Alfalfa Mosaic Virus consists of four components of different sizes.

19. Plant viruses consist of nucleic acid (5 to 40 per cent) and protein (60 to 95 per cent). Elongated plant viruses contain less nucleic acid and more protein, while spherical viruses contain more nucleic acid and less protein.

20. The rhabdoviruses and a few spherical viruses are provided with an outer lipoprotein envelope or membrane. Inside membrane is the nucleocapsid, consisting of nucleic acid and protein subunit.

21. The viral protein consists of amino acid. Virus protective coat or shell called capsid. The total mass of the nucleoprotein of different virus particles varies from 4.6 to 73 million Daltons (Da).

22. The weight of nucleic acid alone ranges between 1 and 3 million (1-3 x 10^6) Da. per virus particle for most viruses, some have up to 6 x 10^6 Da. and the 12-component wound tumor virus nucleic acid approximately 16 x 10^6 Da.

23. The protein subunit of Tobacco Mosaic Virus (TMV) consists of 158 amino acids in a constant sequence and it has mass of 17600 Da. Turnip Yellow Mosaic Virus (TYMV) consists of 189 amino acids.

24. TMV protein subunits are arranged in a helix containing $16^1/_3$ subunits per turn (or 49 subunits per three turns). Each TMV particle consists of 130 helix turns of protein subunits.

25. The nucleic acid is packed lightly between the helices of protein subunits. In the Rhabadoviruses the helical nucleoproteins are enveloped in a membrane. In the Polyhedral plant viruses, protein subunits are tightly packed.

26. The nucleic acid of most plant viruses consist of either RNA or DNA. Both RNA and DNA are long chain-like molecules consisting of hundreds or more often thousands of units called nucleotide. Each nucleotide consists of a ring compound called the base attached to a five carbon sugar [ribose (I) in RNA, deoxyribose (II) in DNA], which in turn is attached to phosphoric acid.

27. In viral RNA only one of four bases, adenine, guanine, cytosine and uracil, can be attached to each ribose molecule. The first two, adenine and guanine are purines and other two uracil and cytosine are pyrimidies. In RNA, thymine is absent.

28. Viruses are not organism. All organisms are cellular while viruses are not cellular. Viruses are macromolecules. Mature particles of plant viruses are generally called virion, whole infective particle is nucleotide. Viruses are not cellular organisms and have neither nucleus nor the cytoplasm, but either of DNA or RNA (never both) is always present.

29. During replication virus do not use the internal environment created immediate function of gene. This environment termed a 'Genosphere'.

30. The term 'satellite virus' was first used by Kassanis in 1961, to describe the small virus that was dependent on Tobacco Necrosis Virus (TNV). The term is now usually used to refer to a virus or nucleic acid that is unable to multiply in the cells without the assistance of a specific helper virus, it is not necessary for the multiplication of the helper virus.

31. The satellite RNAs are small, linear or circular RNAs found inside virions of certain multi-component viruses. Satellite RNAs are not related or are only partially related to the RNA of the virus, satellite RNAs may increase or decrease the severity of viral infections.

32. Satellite virus is a polyhedral single stranded, linear of circular RNA virus, approximately 17 nm in diameter, isometric particles, having 60 protein subunits which are pentamer clusters giving pointed capsomere appearance. Satellite virus is unable to replicate in a plant host, only when that host is infected by TNV. A satellite virus of mamavirus that inhibits the replication of its host has been termed a virophage.

33. The protein coat of a virus plays a role in determining vector transmissibility of a virus and the kinds of symptoms it causes. Protein itself has no infectivity, but its presence generally increases the infectivity of the nucleic acid. Protein also provides a protective sheathing for the nucleic acid of the virus.

34. The important plant viruses such as Potato Virus X, Tobacco Mosaic Virus and Cucumber Mosaic Virus are transmitted through sap in the field and may cause severe losses.

35. Nematodes such as *Longidorus*, *Paralongidorus* and *Xiphinema* transmit several polyhedral shaped viruses known as nepoviruses such as grapes fan leaf, tobacco and tomato ring spot and others, whereas *Trichodorus* and

Paratrichodorus transmit at least two rod shaped Tobraviruses, Tobacco Rattle Virus and Pea Early Browning Virus.

36. Tobamovirus genome consists of one positive single stranded RNA (+ssRNA); of approximately 6400 nucleotides (6.4 kb). Tobacco Mosaic Virus, Pepper Green Mottle Virus and Odontoglossum Ring Spot Virus of orchids are tobamoviruses and are easily transmitted mechanically.

37. Tobacco Mosaic Virus is stable and over seasons in infected tobacco stalks, leaves in the soil, on the contaminated seeds and for many years in cigars, cigarettes etc, it can be controlled by sanitation.

38. Tobraviruses consist of two rod-shaped particles, each particle contains positive-single stranded RNA of the long particles (6.8 kb) contains four genes. Tobacco Rattle Virus and Pea Early Browning Virus are tobraviruses and are transmitted in nature by nematodes of the genera *Trichodorus* and *Paratrichodorus*.

39. Furoviruses such as soil borne Wheat Mosaic Virus and Peanut Clump Virus both transmitted by fungus *Polymyxa graminis*, Potato Mop Top Virus transmitted by *Spongospora subterranea*, Beet Necrotic Yellow Vein Virus transmitted by *Ploymyxa betae*, Tobacco Necrosis Virus (TNV), TNV satellite and Cucumber Necrosis Virus transmitted by *Olpidium brassicae*. Fungus *Polymyxa* and Spongospora are plasmodiophoromycetes which are now classified as protozoa rather than fungi. The term furoviruses, therefore, is basically incorrect.

40. Furoviruses consist of two, three or four rod shaped positive ssRNA (+ssRNA) and measures from 65 to 390 nm long by 18 to 24 nm in diameter. Fields sanitation and virus free seed is essential for control of furoviruses.

41. Hordeiviruses *e.g.* Barley Stripe Mosaic Virus (BSMV) consist of three rigid rod shaped particles about 100 to150 nm long and 20 nm in diameter and over season in infected seeds for several years and in perennial hosts.

42. Potato Virus X (PVX) is a Potexviruses consist of a single flexuous rod and genome is a positive single stranded RNA (5.8-7.0kb), and over season in perennial hosts, vegetative propgative organs of their hosts and transmitted by seed.

43. Carnation Latent Virus, Pea Streak Virus and Popular Mosaic Virus are Carlaviruses group and are transmitted by vegetative propagative organ and by aphids in the non-persistent manner and sometimes by seed also.

44. Capilloviruses (thin or hair like viruses) such as apple stem grooving virus and citrus tatter leaf virus, consist of ssRNA and no vectors are known. Apple Chlorotic Leaf Spot Virus is a Trichoviruses (hair like) and transmitted through aphid or mealy bugs.

45. Plant disease caused by Potyviruses including Potato Virus Y, Bean Common Mosaic Virus, Bean Yellow Mosaic Virus, Beet Mosaic Virus, Celery Mosaic

Virus, Lettuce Mosaic Virus, Papaya Ring Spot Virus, Plum Pox Virus, Sugarcane Mosaic Virus, Tobacco Etch Virus, Turnip Mosaic Virus, Watermelon Mosaic Virus, Zucchini Yellow Mosaic Virus, Dasheen Mosaic Virus and Tulip Breaking Virus. Potyvirus consists a single flexous, rod- shaped +ssRNA and transmitted by aphids in the non-persistent manner.

46. Bean Common Mosaic Virus is transmitted through bean seeds and can also be transmitted by pollen while Bean Yellow Mosaic Virus is transmitted by aphid. Lettuce Mosaic Virus is transmitted by several species of aphids and by 1-8 per cent by the seed.

47. Rymoviruses include Rye Grass Mosaic Virus, Wheat Streak Mosaic Virus, Agropyron Mosaic Virus and Oat Necrotic Mottle Virus is transmitted by mite (*Aceria* and *Abacus*). Bymoviruses include Oat Mosaic Virus, Rice Necrosis Mosaic Virus, Wheat Spindle Streak Mosaic Virus and Barley Yellow Mosaic Virus are soil borne, transmitted by the fungus *Polymyxa graminis*.

48. Closteroviruses have long, thin, very flexuous threadlike; included Beet Yellows, Citrus Tristeza and Lettuce Infections Yellow Virus. Closteroviruses are transmitted by aphid in the semi-persistent manner while Lettuce Infectious Yellow Virus transmitted by whitefly (*Bemisia tabaci*) and Grapevine Leaf Roll Virus spread by mealy bug. Citrus Tristeza Virus also transmitted by budding or grafting.

49. Rice Tungro Spherical Virus and Maize Chlorotic Dwarf Virus are Waikaviruses, isometric, ssRNA genome about 30 nm in diameter. Rice Tungro is the result of infection by two viruses, ssRNA virus rice tungro spherical waikavirus (RTSV) and dsDNA virus rice tungro bacilliform Badnavirus (RTBV). Both viruses are transmitted by leafhoppers, particularly *Nephotettix virescens* in the semi-persistent manner.

50. Maize Chlorotic Dwarf Virus (MCDV) is transmitted by the leafhopper (*Graminella nigrifrons*). Luteoviruses are isometric ssRNA viruses, included Potato Leaf Roll, Barley Yellow Dwarf Virus, Beet Western Yellows and are transmitted by aphid in the persistent manner.

51. Comoviruses such as Cowpea Mosaic Virus (affects bean, pea, cowpea, soybean, and clover), Squash Mosaic Virus, Radish Mosaic Virus, Bean Pod Mottle Virus are transmitted by beetles and cause mosaic, stunting and malformation of varying severity whereas fabaviruses are transmitted by aphids.

52. The important Nepoviruses are Tomato Ring Spot, Cherry Leaf Roll, Grapevine Fan Leaf, Tobacco Ring Spot Virus, Arabis Mosaic Virus, Raspberry Ring Spot Virus, Raspberry Yellow Dwarf and Raspberry Leaf Curl.

53. Tomato Ring Spot Virus is transmitted by nematode *Xiphinema*, Cherry Leaf Role Virus is transmitted by budding, grafting, seed, pollen and also by nematode *Xiphinema*. Grapevine Leaf Roll Virus is transmitted by cuttings and also by nematodes *Xiphinema*.

54. Arabis Mosaic Virus, Raspberry Ring Spot Virus, Tomato Black Ring Virus are Nepoviruses. Arabis Mosaic Virus is transmitted by *Xiphinema* nematode and tomato Black Ring Virus is transmitted by *Longidorus* nematodes.

55. Bromoviruses infect gramineous and legume plants and Alfamoviruses infect mainly legumes. Alfamoviruses (sometimes ilaviruses) have four particles each that are 18 nm in diameter and are mostly bacilliform ranging in length from 30 to 57 nm. Some bromoviruses is transmitted by beetles.

56. Cucumoviruses such as Cucumber Mosaic Virus, Peanut Stunt Virus and Tomato Aspermy Virus are transmitted by aphids in the non-persistent manner.

57. ILaviruses such as Tobacco Streak Virus, Prunus Necrotic Ring Spot Virus, Apple Mosaic Virus, Rose Mosaic Virus, Prune Dwarf Virus, Plum Line Pattern Virus, Citrus Leaf Rugose Virus and Citrus Variegation Virus have no known vectors and transmitted through seed and vegetative propagation.

58. Family Reoviridae viruses infect humans, animals, insects and plants. The family Reoviridae contains three genera of reoviruses: Phytoreovirus, Fijivirus and Oryzavirus. All Reoviridae are transmitted form plant to plant by leaf hoppers in the persistent, propagative manner but phytoreviruses are transmitted to new generations through the egg. All Reoviridae viruses contain dsRNA.

59. The plant Reoviridae (Rice dwarf virus) causes galls or tumors on their hosts. Phytoreovirus include wound tumor rice gall dwarf virus and rice dwarf virus.

60. Fijiviruses such as Rice Black Streaked Dwarf Virus, Maize Rough Dwarf Virus and Oat Sterile Dwarf Virus and Oryzavirus, for example Rice Ragged Stunt Virus also cause stunting, leaf distortion and galls on the leaf veins.

61. White Clover Cryptic Virus I is an Alfacryptovirus and White Clover Cryptic Virus II is example of Betacryptovirus. Partitivirus and Crysovirus infect fungi, including *Penicillium*, *Rhizoctonia*, *Gaeumannomyces* and *Helminthosporium*.

62. Reoviruses are isometric, 65 to 70 nm in diameter and genome consists of 12 dsRNAs in Phytoreovirus and 10 dsRNAs in Fijivirus and the Oryzaviruses. The size of the RNAs from 800 to 2000 base pairs (bp).

63. The important Rhadboviruses are Lettuce Necrotic Yellows Virus, Potato Yellow Dwarf Virus, Rice Transitory Yellowing Virus and Wheat Striate Mosaic Virus and are transmitted by the circulative, propagative manner by leafhopper or plant hopper or by aphids and few transmitted by lace bugs or mite. Rhabdoviruses are bacilliform and are largest among the plant viruses size varies from about 50 to 95 nm in diameter and 200 to 500 nm long.

64. Single stranded negative RNA viruses (–ssRNA) are Rhabdovirus, Tospovirus and Tenuivirus.

65. Tomato Spotted Wilt Virus, Watermelon Silver Mottle Virus and Impatiens Necrotic Spot Virus are Tospoviruses and have three linear ssRNAs and are transmitted by several species of thrips.

66. Rice Stripe Virus, Rice Hoja Blanca (White Leaf) Virus, Rice Grassy Stunt Virus and Maize Stripe Virus are Tenuiviruses and are transmitted by plant hopper in the circulative-propagative manner.

67. Cauliflower Mosaic Virus, Carnation Etched Ring Virus, and Dahlia Mosaic Virus are caulimoviruses having circular dsDNA particles and are transmitted by aphid in the non-persistent manner.

68. Sugarcane Bacilliform Virus, Rice Tungro Bacilliform Virus, Cacao Swollen Shoot Virus and Banana Streak Virus are Badnaviruses, size 30 to 300 nm and contain dsDNA.

69. Rice Tungro Bacilliform Virus is transmitted by leafhoppers in the semi-persistent manner. Banana Streak Virus is transmitted by mealy bugs. While Cacao Swollen Shoot Virus is a Bacilliform virus, size 142 by 27 nm and is transmitted by mealy bugs in the semi-persistent manner.

70. Geminiviruses are single stranded DNA viruses (ssDNA), which include Bean Golden Mosaic Virus, Squash Leaf Curl Virus, African Cassava Mosaic Virus, Tobacco Leaf Curl Virus, Tomato Mottle Virus, Tomato Yellow Leaf Curl Virus, Beet Curly Top Virus, Tobacco Yellow Dwarf Virus, Maize Streak Virus And Wheat Dwarf Virus and contain geminate (twin) particles have 200 nucleotide sequences.

71. Beet Curly Top and Maize Streak Geminivirus are transmitted by leafhopper in the persistent manner. African Cassava Mosaic Virus, Bean Golden Mosaic Virus, Squash Leaf Curl Virus, Tomato Mottle Virus, Tomato Yellow Leaf Curl Virus, Tobacco Leaf Curl Virus are transmitted through whitefly (*Bemisia tabaci*) in the persistent manner.

72. Banana Bunchy Top is a single stranded DNA virus and other ssDNA viruses are Subterranean Clover Stunt Virus, Coconut Foliar Decay Virus and Faba Bean Necrotic Virus. These viruses are isometric, small (18-20 nm in diameter). Banana Bunchy Top Virus is transmitted over long distances by propagative materials such as rhizomes, suckers and over short distances by the aphid (*Pentalonia nigronervosa*).

73. Barley Yellow Dwarf, Arabis Mosaic, Tobacco Ring Spot, Velvet Tobacco Mottle, Subterranean Clover Mottle, Lucerne Transient Streak are satellite RNAs.

74. Vein Banding Mosaic of potato and Rugose Mosaic Virus are transmitted by aphid. Yellow Vein Mosaics of Okra, Leaf Curl of Papaya, Chilli Leaf Curl Virus are transmitted by whitefly (*Bemisia tabaci*).

75. Protoplasts are useful for studies on viral pathogenesis. Citrus Psorosis Virus is a multicomponent ssRNA virus with probably three genomic RNAs, ranging in size from about 10 to 1.5 kb, 3 nm diameters. Indian Citrus Ring Spot Virus (ICRSV) contains a single species of ssRNA, 7-8 kb in size and single coat protein (cp) with an apparent size of about 34 kda.

76. Papaya Leaf Curl Virus (PLCV) is transmitted by whitefly (*Bemicia tabaci* Genn). It is a bipartite geminivirus. PLCV is a distinct geminivirus.

77. Tomato Leaf Curl Virus (TLCV) is transmitted by whitefly, it is bipartite geminivirus. This geminivirus genome is composed of DNA-A, DNA-B nucleotide sequence and pattern of amino acid sequence. The stretch in both DNA-A and DNA-B includes a stable hairpin structure and the loop of the hairpin includes the conserved non-nucleotide sequence TAATATTAC present in all geminivirus so far reported.

78. Three different cloned geminiviral DNAs *viz.* Indian Tomato Leaf Curl Virus, Tomato Leaf Curl Virus, Tomato Golden Leaf Curl Virus were used as probes (with radioactive labeling).

79. Cucumber Mosaic Virus (CMV) is a multicomponent virus belongs to Cucumovirus group, it has single stranded plus sense RNA (+ssRNA) genome, consisting of 3 RNA species (1-3). CMV has complete nucleotide sequences (tripartite ssRNA genome).

80. The coat protein (CP) gene of Tomato Leaf Curl Virus prepare a construct in a binary vector pBin 121 under the influence of a eukaryotic promoter (CaMV 35 S) and NOS poly (A) terminator.

81. Rice Ragged Stunt Virus (RRSV) is transmitted exclusively by the brown plant hopper (*Nilaparvata lugens*). RRSV dsRNA genome profile (10 segments).

82. Pumpkin Yellow Vein Mosaic Virus is transmitted by whitefly (*Bemisia tabaci*). Watermelon Mosaic Virus-1 and 2 is a Potyvirus and transmitted by aphid (*Myzus persicae*). Cucumber Vein Yellowing Virus is a Geminivirus and Bottle Gourd Mosaic Virus is a Potyvirus group.

83. Tetrazolium test is used for the diagnosis of presence of Banana Bunchy Top Virus. Banana Streak Virus is a serious problem and is transmitted through mealy bug (*Planococcus citri*).

84. Dicson (1925) showed first time that winter streak disease of tomato plants was due to the combined effect of the Tobacco Mosaic Virus and a Potato Mosaic Virus and this is known as double virus disease. For the first time, the disease Leaf Curl of Papaya (ssDNA) was observed in Tamil Nadu by Krishna Swamy in 1939 and Papaya Mosaic Ring Spot in India was first reported by Kapoor and Verma in 1948 from Maharashtra.

85. Tobacco Mosaic the first recognized viral disease of plants is worldwide in distribution. The virus is rod shaped, measuring about 280 x 15-18 nm with a central hollow tube of about 20 A° in diameter, 40 millions of molecular weight and consists of 6400 nucleotide.

86. Yellow Vein Mosaic of Okra (Bhindi) is a serious disease in India first reported in 1924. The virus is transmitted in nature by whitefly (*Bemisia tabaci*) and virus particles persist in the female whitefly throughout its life, if the insect is allowed to feed on diseased plants for 4-6 hours.

87. Leaf Roll of Potato is responsible for heavy losses in the yield of potato in India and this was first observed in India in 1943 (Pal). The virus is transmitted in nature by infected seed tubers and through the agency of insects (Aphid, *Myzus persicae*). Leaf Roll Virus is a member of Luteovirus group.

88. Mild or Latent Mosaic of Potato Virus is transmitted by sap (contact) by core grafting and by dodder. No insect vector is known.

89. Rugoe Mosaic of Potato is caused by a combination of two viruses' PVX and PVY. Size of PVX is 470 to 580 nm long and by 11-12 nm in diameter and PVY 680-90 nm long and 12 nm in diameter. The virus consists of a single positive RNA (+ssRNA) species and easily transmitted by aphid.

90. Bunchy Top Disease of Banana (Banana Virus and Musa Virus 1) was first reported from Australia. In India, the disease was reported for the first time in 1940 from Kerala. In India, it introduced through infected suckers brought from Sri Lanka into the state of Kerala.

91. Sugarcane Mosaic Virus was first noted in India at Pusa (Bihar) in 1921 and virus is transmitted by aphid in non- persistent manner.

92. Tobacco Necrosis Virus is transmitted by *Olpidium brassicae*. Sugarbeet Curly Top Virus transmitted by *Cuscuta subinclusa* and squash plants with stone fruit ring spot virus through pollen.

93. Cardamom Katte disease, Cardamom Chirke Disease, Chilli, Pea, Barley Mosaic, Dwarf or Foorkey disease of cardamom transmitted by aphid in non-persistent manner.

94. Squash Mosaic, Turnip Yellow Mosaic, Turnip Crinckle, Turnip Rosette and Potato Viruses X are transmitted by beetles.

95. Decrease in the activity of glycolic acid oxidase occurs at later stages of disease with the decreasing photosynthetic activity in Cucumber Mosaic Virus, rust and wilt diseases.

96. In RNA viruses, the purines (adenine, guanine) are liked with pyrimidines (cytosine, uracil), while in DNA viruses' uracil is replaced by thymine. The pentose sugar (either ribose or deoxyribose) are bonded to the phosphate group of atom by esterification involving hydroxyl group of its 32 and 52 position of carbon atoms.

97. Four different bases (adenine guanine, cytosine, and uracil) and 64 triplets are possible for just 20 amino- acids.

98. The genome of Tobacco Mosaic Virus (TMV) consists of a single molecule of positive sense (+ssRNA) and +ssRNA having 6395 nucleotide. The Ilavirus, Tobacco Streak Virus (TSV) contains four molecules of positive sense ssRNA. The potyvirus genome consists of single stranded positive sense RNA. Gemini viruses consist of circular ssDNA molecules and multipartite genomic organization.

99. The satellite RNAs act as molecular parasites that appear to compete efficiently for viral replication with adverse effect on both virus (Cucumber Mosaic Virus and Peanut Stunt Virus) in terms of reduction in virus genome concentration and attenuation of symptom expression.

100. Wound Tumor Virus (WTV) replicates in both insects and plants. On infection it produces the typical tumors in plants along with other symptoms and its (WTV) losses vector transmissibility.

101. Genome is single stranded positive (+) RNA and monopartite viruses are Tobamoviruses (rigid), Tymoviruses, Tombusviruses, Sobemoviruses, Tobacco Necrosis Viruse, Luteoviruses and Machloviruses (isometric) and Potexviruses, Potyviruses, Carlaviruses, Closteroviruses all are flexuous and monopartite.

102. Genome is single stranded positive (+) RNA and multipartite viruses are Hordeiviruses, Tobraviruses, Soilborne Wheat Mosaic Virus, Peanut Clum Virus, Comoviruses, Nepoviruses, Pea Enation Mosaic Virus, Dianthoviruses, Cucumoviruses, Bromoviruses and Ilaviruses.

103. Genome is single stranded negative (-) RNA is Cytorhabdovirus (Lettuce Necrotic Yellows Virus), Nucleorhabdovirus (Potato Yellow Dwarf Virus), Tospovirus (Tomato Spotted Wilt Virus) and Tenuivirus (Rice Stripe Virus).

104. Double stranded RNA (ds RNA) isometric viruses are Alphacryptovirus (White Clover Cryptic Virus I), Betacryptovirus (White Clover Cryptic Virus II), Fiji Virus (Rice Fiji Disease Virus), Oryzavirus (Rice Ragged Stunt Virus) and Phytoreovirus (Wound Tumor Virus).

105. Single stranded DNA (ssDNA), geminate (geminivirus) twin particles viruses are Maize Streak Virus, Beet Curly Top Virus, Bean Golden Mosaic Virus and single stranded isometric particles virus is Banana Bunchy Top Virus.

106. Double stranded DNA (dsDNA), isometric, circular virus's is Caulimoviruses (Cauliflower Mosaic Virus) and non enveloped bacilliform particles virus is Badnavirus (Rice Tungro Bacilliform Virus, Banana Streak Virus and Cacao Swollen Shoot Virus).

107. Bipartite (Virus with multiple genomes) viruses are Comoviruses (Cowpea Mosaic Virus), Nepoviruses (Tobacco Ring Spot Virus), Pea Enation Mosaic Virus, Dianthovirus (Carnation Ring Spot Virus) and tripartite viruses are Cucumovirus group (Cucumber Mosaic Virus), Bromovirus Group (Brome Mosaic Virus), Ilavirus group (Tobacco Streak Virus) and Alfalfa Mosaic Virus (Bacilliform).

108. The first authentic record of an aphid as a vector of plant viruses was that of *Aphis gossypii* from the U.S.A. in 1916, the virus being Cucumber Mosaic.

109. Brome Mosaic, Cowpea Chlorotic Mottle, Broadbean Mottle Virus, Turnip Yellow Mosaic and Eggplant Mosaic are transmitted by beetles.

110. Tomato Spotted Wilt Virus (TSWV) transmitted by thrips (*Thrips tabaci, Frankliniella schultzei, F. fusa, F. occidentalis* and *Scirtothrips dorsalis*). Both *T.*

tabaci and *F. schultzei* are highly polyphagous and are known to transmit every strain of TSWV.

111. Rice Dwarf Virus, Wound Tumor Virus, Maize Rough Dwarf Virus and Fiji disease of sugarcane are transmitted by plant hopper.

112. The plant Reoviruses and Rhabdoviruses closely resemble animal viruses. Viruses reported to be transmitted by beetles in India include PVX by *Epilacha ocellata*. Urdbean Leaf Crinkle by *Henosepilachna dodocastigma* (epilachna beetle), Cucumber Green Mottle Mosaic Virus by *Raphidopalpa foveicollis* (red pumpkin beetle).

113. Wheat Streak Mosaic Virus (WSMV), Agropyron Mosaic Virus (AMV) and Ryegrass Mosaic Virus are transmitted by mites. Two species of lace bugs, one is *Piesma cinereum* transmits sugar beet savoy virus in circulative manner (in North America) and other is *Piesma quadratum* transmits Beet Leaf Curl Virus in circulative and propagative manner (in Europe).

114. Non-persistent viruses include Potyviruses, Carlaviruses, Cucumoviruses and Alfalfa Mosaic Virus. Some examples of aphid borne non-persistent viruses are Alfalfa Mosaic Virus, Bean Common Mosaic, Papaya (Mosaic) Ring Spot, Potato Virus Y, Soybean Mosaic and Turnip Mosaic Virus.

115. Some examples of semi-persistent viruses and their vectors are Beet Yellows, Citrus Tristeza, Parsnip Yellow Fleck (by aphid), Maize Chlorotic Dwarf, Rice Tungro (by leafhopper) and Abutilon Yellows, Cucumber Yellows, Lettuce Infections Yellows (by whitefly).

116. Circulative (persistent) viruses are Banana Bunchy Top, Barley Yellow Dwarf, Groundnut Rosette, Lettuce Necrotic Yellows, Pea Enation Mosaic, Strawberry Crinkle (by aphid), Beet Curly Top (by leafhopper), Maize Streak, Potato Yellow Dwarf, Rice Dwarf, Wheat Striate Mosaic, Wound Tumor Maize Rough Dwarf, Rice Grassy Stunt, Rice Stripe and Hoja Blanca of Rice (by plant hopper), Bean Golden Mosaic, Yellow Vein Mosaic of Okra, Horse Gram Yellow Mosaic, Tobacco Leaf Curl, Tomato Yellow Leaf Curl (by lace bug), Tomato Spotted Wilt (by thrips), Squash Mosaic (by beetle) and Wheat Streak Mosaic (by mite).

117. Ragi (finger millet) Mosaic is caused by potyvirus, the virus is transmitted mechanically and also by aphid in non-persistent manner. Bajra Streak Virus was first reported from Delhi.

118. Chickpea Stunt Virus, Pea Mosaic Virus and Cowpea Mosaic Virus are transmitted by aphid in non-persistent manner. Chilli Mosaic is caused by at least five different viruses namely, TMV, CMV, PVY, PVX and TSWV.

119. The first sat-RNA of CMV-S strain was designated as CARNA 5 (CMV associated RNA 5). Another member of Cucumovirus group, Peanut Stunt Virus (PVS), also possesses sat-RNA which is referred as PARNA5.

120. Tobacco Yellow Dwarf Virus, Maize Streak Virus, Wheat Dwarf Virus, Bean Golden Mosaic Virus, African Cassava Mosaic Virus, Squash and Tobacco Leaf

Curl Virus, Tomato Golden Mosaic Virus, Tomato Mottle Virus, Tomato Yellow Leaf Curl Virus, Beet Curly Top Virus, Banana Bunchy Top Virus (isometric), Subterranean Clover Stunt Virus, Coconut Foliar Decay Virus and Faba Bean Necrotic Yellow Virus (isometric) belongs to geminiviruses group.

121. Campbell *et al.* (1996) have tested the hypothesis of vectorassisted seed transmission (VAST) while working with seed-transmitted Melon Necrotic Spot Carmovirus (MNSV) infection in melons (*Cucumis melo*) which is transmitted by *Olpidium radicale*.

122. Pea Enation Mosaic Virus (PEMV) is caused by a complex of two viruses, PEMV-1 is a Luteoviridae member and PEMV-2 is an Umbravirus. Two sizes of isometric particles are found, that of PEMV-1 (28 nm diameter) and that of PEMV-2 (25 nm diameter) having icosahedral and quasi-icosahedral symmetry, respectively.

Multiple Choice Questions (Choose the correct answer)

1. **"Leaf drop streak" is the main symptom of:**
   (A) Vein banding severe mosaic of potato (B) Crinkle of potato
   (C) Leaf curl of potato (D) Rugose mosaic of potato

2. **Crinkle of potato is cause by:**
   (A) Mycoplasma (B) PVX and PVY
   (C) Viroids (D) Rickettsia like organism

3. **The genome of Tobacco Rattle Virus is:**
   (A) Monopartite (B) Bipartite
   (C) Tripartite (D) Multipartite

4. **Which among the plant viruses is transmitted by the nematode *Xiphinema index*:**
   (A) Tobacco rattle virus (B) Grapevine fan leaf virus
   (C) Pea early browning virus (D) Tobacco necrosis virus

5. **Nepoviruses are transmitted from plant to plant by:**
   (A) Pollen (B) Seed
   (C) Nematode (D) None of these

6. **Who was the first to demonstrate that virus are nucleoproteins:**
   (A) Bowden (B) Corbett
   (C) Stanley (D) Beijerinck

7. **An example of a DNA containing virus is:**
   (A) Tobacco mosaic virus (B) Cauliflower mosaic virus
   (C) Rice tungro virus (D) Cassava mosaic virus

8. **A bacteriophage is a:**
   (A) Bacterium attacking a virus
   (B) Virus attacking an alga
   (C) Bacterium attacking bacteria
   (D) Virus attacking bacteria

9. **The first recorded virus diseases is:**
   (A) Tobacco mosaic
   (B) Bunchy top of banana
   (C) Root (wilt) of coconut
   (D) Tulip break

10. **What is the size of cucumber mosaic virus:**
   (A) 17 nm
   (B) 30 nm
   (C) 35 nm
   (D) 60 nm

11. **Potato spindle tuber disease is caused by:**
   (A) Potato virus X
   (B) Potato virus Y
   (C) Tobacco mosaic virus
   (D) Viroid

12. **Rice tungro virus of two types of particles. The genome of these particles consists of:**
   (A) Both particles having RNA
   (B) Both particles having DNA and RNA
   (C) Both particles having DNA
   (D) One particle having RNA and the other DNA

13. **Which of the following viruses have double stranded DNA:**
   (A) Geminivirus
   (B) Caulimovirus
   (C) Tobamovirus
   (D) Tobravirus

14. **Banana bunchy top virus is transmitted by:**
   (A) Leafhopper
   (B) Whiteflies
   (C) Mealy bug
   (D) Aphid

15. **Plant viruses with helical symmetry contain about:**
   (A) 5-40 per cent nucleic acid
   (B) 5 per cent nucleic acid
   (C) 25 per cent nucleic acid
   (D) 50 per cent nucleic acid

16. **Match cacao swollen shoot virus:**
   (A) Potyvirus
   (B) Carnavirus
   (C) Bandavirus
   (D) Geminivirus

17. **Match carnation etched ring virus:**
   (A) Caulimovirus
   (B) Geminivirus
   (C) Potyvirus
   (D) None of these

18. **First gene was cloned:**
    (A) 1973 (B) 1918
    (C) 1969 (D) 1985
19. **First monoclonal antibody diagnostic kit was produced in:**
    (A) 1973 (B) 1981
    (C) 1969 (D) 1985
20. **First expression of genes cloned from different species in bacteria in:**
    (A) 1974 (B) 1981
    (C) 1973 (D) 1985
21. **Size of potex virus is:**
    (A) 680-900 x 11 nm (B) 130 x 18 nm
    (C) 470-580 x 13 nm (D) 300 x 20 nm
22. **Size of badnavirus is:**
    (A) 680-900 x 11 nm (B) 130-18 nm
    (C) 470-580 x 13 nm (D) 300 x 20 nm
23. **Seed material infected with viruses and viroids initiate:**
    (A) After infection (B) Primary infection
    (C) Latent infection (D) Secondary infection
24. **Seed plot technique is suggested to obtain potato free from:**
    (A) Late blight infection (B) Bacterial infection
    (C) Viral infection (D) Charcoal rot infection
25. **Who made first time structure of DNA:**
    (A) J. Watson and F. Crick (1962) (B) J. Lederbeg (1952)
    (C) S.A. Waksman (1952) (D) None of these
26. **Moong bean yellow mosaic is transmitted by:**
    (A) *Bemisia tabaci* (B) Soil
    (C) Sap (D) Seed
27. **The typical symptom of potato virus Y is:**
    (A) Mottle and streak with necrosis (B) Rolling of leaves
    (C) Mosaic (D) Undulation of leaf margin
28. **The leaf drop streak symptom is showed by:**
    (A) PVX (B) PVY
    (C) Papaya ring (D) Plum pox

29. Tungro infected rice plants show symptoms of:
(A) Mottling and yellow orange discoloration of the leaves
(B) Rolling of leaves
(C) Mosaic
(D) Undulation of leaf margin

30. Tomato leaf curl virus belongs to:
(A) Nepovirus (B) Tobamovirus
(C) Geminivirus (D) Cucumovirus

31. Which of the following can multiply in its vector?
(A) Geminivirus (B) Potyvirus
(C) Comovirus (D) Rhabdovirus

32. Double stranded RNA viruses were investigated by:
(A) Diener and Raymer (B) Stanley
(C) Kassanis (D) Gierer and Schram

33. Apricot gummosis is caused by:
(A) MLO (B) Virus
(C) Fungi (D) Bacteria

34. Avocado sun blotch disease is caused by:
(A) RLO (B) Virus
(C) Fungi (D) Chlamydae

35. Tobacco ring spot virus remains viable in soybean seeds at 16 to 32°C for:
(A) 5 years (B) 10 years
(C) 15 years (D) None of these

36. Greening disease of citrus is caused by:
(A) Tristeza virus (B) Mycoplasma
(C) Copper deficiency (D) None of these

37. The particles of oryzaviruses have only one protein shell (the core) and also have:
(A) 2 spikes (B) 5 spikes
(C) 8 spikes (D) 12 spikes

38. Nucleorhabdovirus such as:
(A) Potato yellow dwarf virus (B) Potato leaf roll virus
(C) Lettuce necrotic yellow virus (D) None of these

39. Who first demonstrated virus transmission by nematode:

(A) A. Lwoff (B) Hewitt *et al.*

(C) S.E. Luria (D) R.E.F. Mathews

40. Hoja blanca of rice is caused by:

(A) Fungus (B) Nematode

(C) Virus (D) Bacterium

41. First plant gene expressed in a plant of a different species in:

(A) 1982 (B) 1983

(C) 1985 (D) 1988

42. Match cotton leaf curl:

(A) Potyvirus (B) Geminivirus

(C) Badnavirus (D) Carnavirus

43. Crystallization of viruses has done by:

(A) Stanley (B) Mayer

(C) Bejerinck (D) Ivanovsky

44. Who proved that virus could pass through filters with pores small enough to retain bacteria:

(A) Stanley (B) Ivanovsky

(C) Mayer (D) Barkley

45. The minimum period of time that virus needs for acquisition and subsequent transfer to a virus free plant is the:

(A) Transmission threshold period (B) Inoculation feeding period

(C) Inoculation threshold period (D) Inoculation access period

46. The actual period of feeding is called:

(A) Transmission threshold period (B) Inoculation feeding period

(C) Inoculation threshold period (D) Inoculation access period

47. A virus does not induce symptom development in the host:

(A) Latent infection (B) Latent virus

(C) Latent period (D) Dormant period

48. Host is infected but does not show symptoms:

(A) Latent infection (B) Latent period

(C) Infection period (D) None of these

49. **Tobacco necrosis satellite virus can replicate in the susceptible cell:**
    (A) On its own
    (B) In the presence of TMV
    (C) In the presence of TNV
    (D) In the presence of TSWY

50. **Which of the following viruses is transmitted by its vector in a non persistent manner?**
    (A) Cucumber mosaic virus
    (B) Tobacco mosaic virus
    (C) Wound tumour virus
    (D) Bhindi yellow viin mosaic virus

51. **Cucumber mosaic virus is transmitted by:**
    (A) Aphid
    (B) Leafhopper
    (C) Nematode
    (D) All of these

52. **Which of the following viruses multiplies in its vector?**
    (A) Rice tungro virus
    (B) Cowpea mosaic virus
    (C) Potato virus Y
    (D) Wound tumour virus

53. **Which of the following viruses is transmitted by its vector in semi persistent manner?**
    (A) Cowpea mosaic virus
    (B) TMV
    (C) Rice tungro virus
    (D) None of these

54. **Rice tungro virus is transmitted by:**
    (A) Leafhopper
    (B) Mechanically transmitted
    (C) Nematode
    (D) None of these

55. **Persistent virus is:**
    (A) Wound tumor virus
    (B) Rice tungro virus
    (C) Cucumber mosiaic virus
    (D) None of these

56. **Examples of wound tumor virus are:**
    (A) Rice tungro virus
    (B) Rice dwarf virus
    (C) Cowpea mosaic virus
    (D) None of these

57. **In which of the following tolerance levels allowed for a potato mosaic viruses:**
    (A) 0.5-1 per cent
    (B) 0.1-2 per cent
    (C) 1.2 per cent
    (D) 1.5 per cent

58. **Yellow vein mosaic of okra is transmitted by:**
    (A) Aphid
    (B) Whitefly
    (C) Leafhopper
    (D) Thrips

59. Bunchy top of banana virus is transmitted by:

(A) Aphid (*Pentalonia nigronervosa*) (B) Whitefly (*Bemisia tabaci*)

(C) Aphid (*Myzus persicae*) (D) None of these

60. The genome of tobacco mosaic virus is:

(A) Monopartite (B) Bipartite

(C) Tripartite (D) Multipartite

61. Cauliflower mosaic virus is famous for:

(A) The disease is causes (B) Its promoter

(C) Its transmission by seed (D) Its transmission by vector

62. Cross protection is a phenomenon in which:

(A) A mild strain of the virus prevents infection by a severe strain

(B) A severe strain of the virus causes less severe infection

(C) A virus prevents infection by another virus

(D) A virus protects another virus from losing infectivity

63. Bean golden mosaic virus is a member of group:

(A) Badnaviruses (B) Potyviruses

(C) Carmoviruses (D) Geminiviruses

64. Banana streak virus is a member of group:

(A) Badnaviruses (B) Potyviruses

(C) Carmoviruses (D) Gaminiviruses

65. Rice tungro bacilliform virus is a member of group:

(A) Potyviruses (B) Tobamoviruses

(C) Tobraviruses (D) Badnaviruses

66. Which of the following viruses is transmitted by a fungus:

(A) Tobacco mosaic virus (B) Tobacco necrosis virus

(C) Tobacco leaf curl virus (D) Tobacco ring spot virus

67. Tobacco necrosis virus transmitted by:

(A) *Alternaria* (B) *Fusarium*

(C) *Olpidium* (D) *Agrobacterium*

68. Tobacco ring spot virus transmitted by:

(A) *Olpiduum* (B) *Xiphinema*

(C) Whitefly (D) None of these

69. Nepoviruses are transmitted by the following genus of nematodes:

(A) *Meloidogyne* (B) *Trichodorus*
(C) *Tylenchus* (D) *Xiphinema*

70. Tobravirus is transmitted by:

(A) *Trichodorus* (B) *Xiphinema*
(C) *Tylenchus* (D) None of these

71. Tobacco ring spot virus is transmitted by:

(A) *Xiphinema* (B) *Tylenchus*
(C) *Trichodorus* (D) None of these

72. Which of the following viruses can be used as a vehicle/vector for useful gene is genetic engineering:

(A) Geminiviruse (B) TMV
(C) PVX (D) Caulimovirus

73. Transduction is the mediated through:

(A) Bacteriophages (B) Mycoplasma
(C) *E. coli* (D) All of the above

74. Detection range of ELISA is:

(A) 1-10 mg/ml antigen (B) 1-10 lg/ml antigen
(C) 1-10 ng/ml antigen (D) 1-10 pg/ml antigen

75. Detection range of rocket electrophoresis is:

(A) 0.2-1.0 lg/ml antigen (B) 1-10 μg/ml antigen
(C) 1-10 ng/ml antigen (D) 1-10 pg/ml antigen

76. Detection range of western blotting test is:

(A) 5-10 mg/ml antigen (B) 1-10 μg/ml antigen
(C) 1-10 ng/ml antigen (D) 1-10 pg/ml antigen

77. Detection range of double immunodiffusion technique is:

(A) 5-10 mg/ml antigen (B) 2-20 lg/ml antigen
(C) 1-10 ng/ml antigen (D) 1-10 pg/ml antigen

78. Detection range of liquid precipitin technique is:

(A) 1-10 mg/ml antigen (B) 1-10 lg/ml antigen
(C) 1-10 ng/ml antigen (D) 1-10 pg/ml antigen

79. Detection range of Radial immunodiffusion (RID) technique is:

(A) 0.5-10 lg/ml antigen (B) 1-10 μg/ml antigen
(C) 1-10 ng/ml antigen (D) 1-10 pg/ml antigen

80. Detection range of Passive hemagglutination is:

(A) 20-50 ng/ml antigen (B) 1-10 ug/ml antigen

(C) 1-10 ng/ml antigen (D) 1-10 pg/ml antigen

81. Detection range of immunoosmophoresis is:

(A) 5-20 mg/ml antigen (B) 50-100 ng/ml antigen

(C) 1-10 ng/ml antigen (D) 1-10 pg/ml antigen

82. Detection range of latex test is:

(A) 5-20 ng/ml antigen (B) 50-100 ng/ml antigen

(C) 1-10 ng/ml antigen (D) 1-10 pg/ml antigen

83. Detection range of complement fixation is:

(A) 5-20 mg/ml antigen (B) 50-100 ng/ml antigen

(C) 1-10 ng/ml antigen (D) 1-10 pg/ml antigen

84. The tungro virus of rice is transmitted by:

(A) *Nilaparvata lugens* (B) *Nephotettix virescens*

(C) *Bemesia tabaci* (D) *Aphis maydis*

85. The tungro virus of rice is:

(A) Semi persistent manner (B) Persistent manner

(C) Non persistent manner (D) None of these

86. Antigenicity of a plant virus is specified by:

(A) Coat protein (B) Lipid

(C) Carbohydrate (D) Nucleic acid

87. A System of cryptogram of virus nomenclature was introduced by:

(A) Mathews *et al.* (B) Gibbs *et al.*

(C) Johnson *et al.* (D) Smith *et al.*

88. Which of the following groups of microorganisms do not form siderophores:

(A) Virus (B) Fungi

(C) Aerobic bacteria (D) Anaerobic bacteria

89. A Lambda phage is about:

(A) 45500 bp (B) 47000 bp

(C) 48502 bp (D) 50000 bp

90. Which of the following disease of pigeon pea is caused by a virus:

(A) Wilt (B) Sterility

(C) Leaf spot (D) Blight

91. The movement of the plant viruses from cell to cell is through:

(A) Xylem (B) Phloem

(C) Plasmodesmata (D) Ectodesmata

92. Soluble and host coded proteins are:

(A) PR Protein (B) Enzyme

(C) Specific gene (D) None of these

93. The genome of caulimoviruses is:

(A) ds RNA (B) ss RNA

(C) ds DNA (D) ss RNA

94. The genome of Betacryptovirus is:

(A) ds RNA (B) ssRNA

(C) dsDNA (D) ssDNA

95. The genome of Geminiviruses is:

(A) ss DNA (B) ssRNA

(C) dsDNA (D) dsRNA

96. The genome of Alphacryptovirus is:

(A) ds DNA (B) ssDNA

(C) dsRNA (D) ssRNA

97. Fijivirus contain:

(A) dsRNA (B) ssRNA

(C) dsRNA (D) ssDNA

98. Wound tumor virus contains:

(A) dsRNA (B) ssRNA

(C) dsDNA (D) ssDNA

99. Match Alpha cryptovirus:

(A) White clover cryptic virus 1 (B) White clover cryptic virus 2

(C) Rice fiji disease virus (D) Wound tumor virus

100. Match Beta cryptovirus:

(A) White clover cryptic virus 1 (B) White Clover cryptic virus 2

(C) Rice fiji disease virus (D) Wound tumor virus

101. Tomato spotted wilt virus is:

(A) Negative ss RNA (B) Negative ds RNA

(C) Positive ss RNA (D) None of these

102. Nucleorhabdovirus is a:

(A) Positive ss RNA
(B) Positive ds RNA
(C) Negative ss RNA
(D) Negative ds RNA

103. Match Nucleorhabdovirus:

(A) Potato yellow dwarf virus
(B) Rice ragged stunt virus
(C) Rice tungro bacilliform virus
(D) Tomato spotted with virus

104. Folded filamentous virus is:

(A) Tenuivirus
(B) Cytorhabdovirus
(C) Tospovirus
(D) None of these

105. The genome of badnavirus is:

(A) ss DNA
(B) ds DNA
(C) ss RNA
(D) ds RNA

106. Single stranded negative RNA virus is:

(A) Rahabdoviridae
(B) Bunyaviridae
(C) Both (A) and (B)
(D) None of these

107. A self replicating extra chromosomal hereditary circular DNA is:

(A) Plasmid
(B) Viroid
(C) Virusoid
(D) None of these

108. Tobacco mosaic virus (TMV) was first studied by:

(A) Ivanovsky
(B) Mayer
(C) Beijerinck
(D) Fisher

109. Crystalline protein nature of plant viruses was demonstrated by:

(A) Bawden
(B) Stanley
(C) Hashimoto
(D) Flor

110. Which of the following viruses contains single stranded DNA:

(A) Tobamovirus
(B) Nepovirus
(C) Caulimovirus
(D) Geminivirus

111. Potato virus Y is transmitted by:

(A) Whiteflies
(B) Leaf hopper
(C) Nematode
(D) Aphid

112. Satellite virus is found associated with:

(A) Tobacco necrosis virus
(B) Tobacco mosaic virus
(C) Tobacco leaf curl virus
(D) Tobacco ring spot virus

113. Which of the following viruses is being found very useful in molecular biology:

(A) Cucumber mosaic virus
(B) Tobacco mosaic virus
(C) Wound tumor virus
(D) Cauliflower mosaic virus

114. Resistance to viruses in plants can be developed by:

(A) Treating seeds with chemical
(B) By spraying plants with antibiotics
(C) By incorporating by gene into plant genome
(D) By incorporating viral coat protein gene into plant genome

115. Which of the following is the most sensitive serological test:

(A) Tube precipitin
(B) Double gel diffusion
(C) Enzyme -linked assay
(D) Slide agglutination

116. Which of the following viruses also multiplies in insect vector:

(A) Tobacco mosaic virus
(B) Tobacco rattle virus
(C) Cucumber mosaic virus
(D) Wound tumor virus

117. The genome to toabacco rattle virus is:

(A) Monopartite
(B) Bipartite
(C) Tripartite
(D) Multipartite

118. Match bunyaviridae virus:

(A) Tomato spotted wilt virus
(B) Potato yellow dwarf virus
(C) Rice stripe virus
(D) All of above

119. Tobamovirus is a:

(A) Positive ss RNA
(B) Positive ds RNA
(C) Negative ss DNA
(D) Negative ds DNA

120. Match Furovirus:

(A) Barley stripe mosaic virus
(B) Soil borne wheat mosaic virus
(C) Tobacco rattle virus
(D) All of the above

121. Tobamovirus is a:

(A) Monopartite
(B) Biprtite
(C) Multipartite
(D) None of these

122. All flexuous viruses are:
(A) Monopartite (B) Biprtite
(C) Multipartite (D) None of these

123. Match multipartite virus:
(A) Comovirus (B) Nepovirus
(C) Cucumovirus (D) All of the above

124. Match Tenuivirus:
(A) Rice stripe virus (B) Tomato spotted wilt virus
(C) Brome mosaic virus (D) None of these

125. Match Geminivirus:
(A) Beet curly top virus (B) Bean golden mosaic virus
(C) Banana bunchy top virus (D) All of the above

126. Match envelope virus:
(A) Tomato spotted wilt virus (B) Cucumber mosaic virus
(C) Tobacco streak virus (D) None of these

127. Match closterovirus:
(A) Beet yellow virus (B) Ryegrass mosaic virus
(C) Potato virus Y (D) None of these

128. Who was the first to demonstrate that mosaic disease of tobacco is transmissible?
(A) Beijerinck (B) Mayer
(C) Stanley (D) Ivanovsky

129. Contagium vivum fluidum theory was proposed by:
(A) Beijerinck (B) Mayer
(C) Stanley (D) Ivanovsky

130. The only known mode of transmission through seed is in:
(A) Tobamovirus (B) Potexvirus
(C) Cucumovirus (D) Cryptovirus

131. In 1935 the first major contribution to the nature of viruses was made by:
(A) Gierrer (B) K.M. Smith
(C) Stanley (D) Raymer

Fill in the Blanks with Correct Answer:

1. Barley stripe mosaic virus (BSMV) is found in_______

   Ovules and pollen

2. Rosset disease of groundnut is caused by_______

   Virus

3. The most common insect vector are_______

   Aphid

4. Most serious viral disease of citrus _______

   Tristeza

5. _______is the smallest and simplest of all known organisms.

   Virus

6. Viruses can be_______

   Crystallized

7. All_______can pass through fine filter paper.

   Viruses

8. In addition to nuclear material of viruses also posses _______

   Protein

9. When a virus infects a host, new viruses are formed by _______

   Multiplication

10. Nucleic acid of viruses are _______

    RNA and DNA (Never both)

11. Cauliflower mosaic virus particles contain_______

    dsDNA

12. Moon bean yellow mosaic virus (MYMV) is_______

    Circular ssDNA

13. Maize streak virus is a_______

    Geminivirus

14. Geminiviruses are_______

    Circular ssDNA

15. A virus capable of supporting the replication of an additional, nonessential and unrelated nucleic acid component called________

    Helper virus

16. Radioimmunosorbent Assay (RISA) method is developed by________

    Ghabrial and Shepherd (1980)

17. Chromatography is used for purification of________

    Plant viruses

18. Grente and Sauret (1969) coined the term________

    Hypovirulence

19. The hypovirulence strain carry virus like ________

    dsRNA

20. Yellow group of virus is mostly transmitted by________

    Leafhopper

21. Mosaic group of virus mostly transmitted by ________

    Aphid

22. *Trichodorus* and *Paratrichodorus* nematode transmit red shaped virus i.e.________

    Tobbaco rattle virus, Pea early browning virus

23. The plasmids are________molecules ranging in size from 1-113x 10^6 daltons.

    Circular dsDNA

24. The term satellite virus was first used by ________in 1962.

    Kassanis

25. Satellite virus is a small polyhedral________

    ssRNA

26. Satellite virus multiply in the presence of ________

    TNV

27. Presence of SAT-RNA leads to________in severity of disease symptoms.

    Reduction

28. Cauliflower mosaic virus, tobacco mosaic virus and geminiviruses have be used as vectors for________

    Cloning of DNA segments

29. In 1898, the Dutch scientist________finally concluded that tobacco mosaic was not caused by a microorganism but by a 'Contagium Vivum Fluidum' which he called a virus.

    Beijerink

30. Star crack of apple and pear is caused by________

    Virus

31. Line pattern of peach and apricot is caused by________

    Virus

32. Actinomycin D compound is used for suppression of tobacco mosaic virus in________

    Chinese cabbage

33. Eggplant mottled crinkle virus, lily symptomless virus, potato virus M and potato virus S is inactivated by the use of chemical compound________

    Ribavirin

34. Size of cucumoviruses is________

    29 nm

35. Size of geminiviruses is________

    18 x 30 nm

36. Size of lettuce mosaic virus is________

    750 x 12 nm

37. Tomato aspermy virus belongs to________group.

    Bromoviruses

38. Rice gall dwarf virus belongs to________group.

    Phytoreoviruses

39. Watermelon silver mottle virus belongs to________group.

    Tospoviruses

40. Dahlia mosaic and carnation etched ring viruses belongs to________

    Caulimoviruses

41. Rice tungro bacilliform virus, banana streak virus and cacao swellen shoot virus are associated________

    Badnaviruses group

42. Beet necrosis virus is transmitted by________

    Polymyxa fungus

43. Subterranean clover stunt virus is transmitted by________

    Aphid

44. ________have no resting stage and are transmitted through a continuous infection chain.

    Plant viruses

45. 'Landmarks in Plant virology: Genesis of Concepts' has been discussed by________

    Markkham (1977)

46. Squash leaf curl virus is transmitted by the.

    Whitefly (Bemisia tabaci)

True or False

1. Viruses can be cultured in bacteriological media. (**False**)
2. All the plants viruses can be transmitted by insects. (**False)**
3. Some of virus are non pathogenic. (**False**)
4. All bacteriophages do not possess head and tail. (**True**)
5. Viruses can be cultured on nonliving media.(**False**)
6. Cauliflower mosaic virus is transmitted by aphid in persistent manner. (**False**)
7. In the single stranded RNA rhabdoviruses, the RNA is not infectious because it is the (-) strand. (**True**)
8. Sour cherry infected with prunus necrotic ring spot virus is transmitted from plant to plant through pollen. (**True**)
9. Barley stripe mosaic virus consist of three rigid rod shaped particles about 100-150 nm, long by 20 nm, in diameter. (**True**)
10. Popular mosaic virus is the Carlavirus. (**True**)
11. Bean yellow mosaic virus is not transmitted through the seed in beans. (**True**)
12. Potato leaf roll virus is transmitted though infected potato field by more than 10 species of aphids. (**True**)
13. Comoviruses are not transmitted by beetles. (**True**)
14. Fabaviruses are not transmitted by aphid. (**False**)
15. Most nepoviruses are transmitted by nematodes. (**True**)

16. Prunus necrotic ring spot virus cannot be transmitted by budding and grafting and mechanically by rubbing sap. (**False**)
17. The maize streake geminivirus is transmitted by several species of leafhoppers. (**True**)
18. Geminivirus contain dsDNA. (**False**)
19. Banana bunchy top virus contains single stranded DNA molecules. (**True**)
20. Raychaudhuri (1996) reviewed the development of plant virology in India. (**True**)
21. Tomato spotted wilt virus is an enveloped viruses, contains *ss*(-) RNA. (**True**)
22. Maize streak virus is a geminivirus. (**True**)
23. Wound tumour virus is a *ds*RNA and non enveloped virus. (**True**)
24. Badnavirus is a single stranded DNA virus. (**False**)
25. Rhabdoviridae is the family of wound tomour virus. (**False**)
26. Twin particles are the main characteristics of Geminiviridae geminivirus. (True)
27. Potato virus Y (PVY) is the single flexuous rod shaped particles 680-900 nm and 12 nm diameter. (**True**)
28. The size of potato virus X (PVX) is 470-580 x 11 nm. (**True**)
29. Tobraviruses can persist in the vector for weeks or months but does not multiply in the vector. (**True**)
30. Phytoreoviruses and fijiviruses consist of two concentric protein shells and oryzaviruses has only one protein shell. (**True**)
31. Lettuce necrotic yellows are cytorhabdovirus. (**True**)
32. Grape fruit is susceptible to tristeza virus. (**True**)
33. Kagzi lime is indicator plant for tristeza and is highly susceptible to it. (**True**)
34. *Carica cauliflora* is a source of resistance for papaya ring spot virus. (**True**)
35. Viruses causing stunting in plants, it shows reduce auxin concentration. (**True**)
36. Hypertrophy and hyperplasia are typical in nucleic acid in host parasite interactions. (**True**)
37. The cualimoviruses and geminiviruses have potential as vector for expression of alien genes. (**True**)
38. Alfalfa mosaic virus group and rhabdoviruses are bacilliform. (**True**)
39. Kagzi lime is a suitable species for cross protection technique in citrus for tristeza virus. (**True**)
40. Turnip mosaic, watermelon mosaic and zucchini yellow mosaic viruses belong to potyviruses group. (**True**)

41. Grapevine fan leaf, raspberry, yellow dwarf, raspberry ring spot and raspberry leaf curl viruses belongs to nepoviruses group. (**True**)
42. Tobacco streak, citrus variegation, prunus necrotic ring spot and citrus leaf rugose viruses belong to Ilaviruses group. (**True**)
43. Maize rough dwarf virus and oat sterile dwarf virus belongs to fijivirus group. (**True**)
44. Potato yellow dwarf, rice transitory yellowing and wheat striate mosaic viruses are associated with rahabdovirus. (**True**)
45. Rice grassy stunt and rice and maize stripe viruses are tenuiviruses. (**True**)
46. Egg plant mosaic disease is transmitted by beetle *Epitrix* sp. (**True**)
47. Safflower mosaic virus is transmitted by *Aphis gossypii*. (**True**)
48. Tomato mottle virus is transmitted by aphid. (**False**)
49. Potato virus Y is transmitted by white fly. (**False**)
50. Apple stem grooving virus is associated with capilloviruses group. (**True**)
51. Rice black streak dwarf virus is associated with fijiviruses group. (**True**)
52. Barley yellow mosaic virus belongs to the bymoviruses group. (**True**)
53. Rice dwarf virus belongs to the family reoviridae. (**True**)
54. Maize streak virus belongs to the family rhabdoviridae. (**False**)
55. Lettuce necrotic yellow virus belongs to the family Geminiviridae. (**False**)
56. Size of apple grooving virus is 600-7000 x 12 nm. (**True**)
57. Size of water melon mosaic virus is 760 x 12 nm. (**True**)
58. Size of zucchini yellow mosaic virus is 750 x 12 nm. (**True**)
59. Squash leaf curl virus is transmitted by white fly. (**True**)
60. African cassava mosaic virus is transmitted by aphid. (**False**)
61. Beet curly top virus is transmitted by leafhopper. (**True**)
62. Foorkey of large cardamom is caused by virus. (**True**)
63. Plant viruses are obligate intracellular parasites that do not have the molecular machinery to replicate without a host. (**True**)
64. Transgenic tobacco and potato plant expressing a gene from pokeweed (*Phytolacca* sp.) that codes for an antiviral, ribosomes-inactivating protein exhibited resistance against several potato and other viruses. (**True**)
65. The virus particles do not absorb nutrients from the cell protoplasm, but their nucleic acid induces self replication. (**True**)

66. Symptoms of bean common mosaic virus in bean appear on the first primary leaf. (**False**)
67. The tobacco necrosis virus is transmitted by *Olpidium brassicae* as surface bone on zoospores. (**True)**
68. Reovirus contains a transcriptase enzyme. (**True**)
69. Non persistant viruses persist through the molt or egg. (**False**)
70. Potyviruses consist of a single flexuous rod shaped particle having a single positive RNA species. (**True**)
71. Most seed transmitted viruses are found in the endosperm. (**False**)
72. Maize dwarf mosaic virus is not found in the embryo of mature maize kernels but occasionally found in the endosperm and pericarp. (**True**)

Match the Following

1. **Match the following viral disease transmitted by vectors.**

Host Plant/Disease	*Vector*
I. Maize mosaic virus	(A) *Aphis maidis*
II. Jowar yellowing virus	(B) *Peregrinus maidis*
III. Potato necrosis virus	(C) *Aphis rhamni*
IV. Leaf roll of potato	(D) *Myzus persicae*

Answer

I	*II*	*III*	*IV*
A	*B*	*C*	*D*

2. **Match the following viral disease transmitted by vectors.**

Host Plant/Disease	*Vector*
I. Bioottle gourd mosaic virus	(A) *Aulacophora foveicollis*
II. Filform leaf mosaic virus of cucurbits	(F) *Myzus persicae*
III. Black ring spot of tomato	(C) *Aphis craccivora*
IV. Chilli mosaic	(D) *Myzus persicae*

Answer

I	*II*	*III*	*IV*
A	*B*	*C*	*D*

3. **Match the following viral diseases transmitted by vectors.**

Host Plant/Disease	*Vector*
I. Onion yellow dwarf	(A) *Aphis gossypii*
II. Urd leaf crinkles	(B) *Myzus persicae*

III. Soybean mosaic (C) *Aceria cajani*

IV. Pigeon pea sterility mosaic (D) *Aphis craccivora*

Answer

I	*II*	*III*	*IV*
A	*D*	*B*	*C*

4. **Match the following viral diseases transmitted by vectors.**

Host Plant/Disease	*Vector*
I. Sesamum leaf curl	(A) *Pentalonia nigronervosa*
II. Large cardamom foorkey disease	(B) *Bemesia tabaci*
III. Coconut cadaong cadaong	(C) *Toxoptera citricidus*
IV. Citrus trestiza	(D) *Aphis gossypii*

Answer

I	*II*	*III*	*IV*
B	*A*	*D*	*C*

5. **Match the following mycoplasmal disease transmitted by vector.**

Host/Disease	*Vector*
I. Rice yellow dwarf	(A) *Nephotetix virescens*
II. Sandal wood spike	(B) *Jasus indicus*
III. Safflower phyllody	(C) *Neoaliturus fenestratus*
IV. Sesamum phyllody	(D) *Orosius albicinctus*

Answer

I	*II*	*III*	*IV*
A	*B*	*C*	*D*

6. **Match the following mycoplasmal disease transmitted by vector.**

Host/Disease	*Vector*
I. Eggplant little leaf	(A) *Hishimonus phycitus*
II. Citrus greening	(B) *Diaphorina citri*
III. Potato purple top	(C) *Orosius albicinctus*
IV. Sugarcane grassy shoot	(D) *Aphis* sp.

Answer

I	*II*	*III*	*IV*
A	*B*	*C*	*D*

7. Match the following viruses transmitted by aphids.

I.	Potato leaf roll Y virus	(A)	*Brevicoryne brassicae*
II.	Sugarcane mosaic virus	(B)	*Myzus persicae*
III.	Mosaic of crucifers	(C)	*M. persicae* and *A. gossypii*
IV.	Bean mosaic virus	(D)	*Aphis maidis*

Answer

I	*II*	*III*	*IV*
B	*D*	*A*	*C*

8. Match the viruses transmitted by nematodes with crop.

	A		*B*
I.	Raspberry ring spot	(A)	Lucerne
II.	Tobacco rattles	(B)	Red current
III.	Arabis mosaic	(C)	Potato
IV.	Early browning	(D)	Strawberry

Answer

I	*II*	*III*	*IV*
B	*C*	*D*	*A*

9. Match the following viral disease transmitted by suitable vector.

I.	Tomato ring spot virus nematode	(A)	*Longidorus xiphinema*
II.	Tobacco rattle virus	(B)	*Trichodorus* and *Paratrichodorus*
III.	Potato virus X	(C)	*Synchytrium endobioticum* fungi
IV.	Wheat streak mosaic virus	(D)	*Eriophyid* mite

Answer

I	*II*	*III*	*IV*
A	*B*	*C*	*D*

10. Match the fungi with the virus disease transmitted by them:

I.	*Olpidium brassicae*	(A)	Oat mosaic
II.	*Ploymyxa graminis*	(B)	Potato mop top
III.	*Spongospora*	(C)	Pea false leaf roll
IV.	*Pythium ultimum*	(D)	Lettuce big vein

Answer

I	*II*	*III*	*IV*
D	*A*	*B*	*C*

11. Match the suitable vector with virus disease transmitted by them:

I. *Xiphinema nematode*	(A) Tomato leaf curl virus
II. *Olpidium brassicae* fungi	(B) Peach mosaic virus
III. *Eriophyid* mite	(C) Tobacco necrosis virus
IV. *Bemisia tabaci* insect	(D) Grape fan leaf virus

Answer

I	*II*	*III*	*IV*
D	*C*	*B*	*A*

12. Match the virus transmitted by insects:

Viral Disease	*Vector*
I. Bunchy top of banana	(A) *Pentalonia nigronervosa* (aphid)
II. Mung bean yellow mosaic virus	(B) *Bemisia tabaci* (white fly)
III. Leaf roll of potato	(C) *Myzus persicae* (aphid)
IV. Wheat streak mosaic virus	(D) *Eriophyid* mite (Mite)

Answer

I	*II*	*III*	*IV*
A	*B*	*C*	*D*

13. Match the viral disease transmitted by suitable vector:

Viral Disease	*Vector*
I. Leaf curl of papya	(A) Aphid
II. Papaya mosaic	(B) Whitefly
III. Plum pox	(C) Budding and grafting
IV. Barley stripe mosaic virus	(D) Pollen

Answer

I	*II*	*III*	*IV*
B	*A*	*C*	*D*

14. Match the viral disease transmitted by suitable vector:

Viral Disease	*Vector*
I. Citrus tristeza	(A) Budding and Aphid
II. Beet yellow virus	(B) Aphid
III. Rice tungro virus	(C) Leafhopper
IV. Banana streak virus	(D) Mealy bug

Answer

I	*II*	*III*	*IV*
A	*B*	*C*	*D*

15. Match the viral disease transmitted by suitable vector:

Viral Disease	*Vector*
I. Prunus necrotic ring spot virus	(A) Budding
II. Tomato spotted wilt	(B) Plant hopper
III. Rice hoja blanca virus	(C) White fly
IV. Bean golden mosaic virus	(D) Thrips

Answer

I	*II*	*III*	*IV*
A	*D*	*B*	*C*

16. Correctly match in section 'A' to those given under the section 'B':

Section 'A' (Viral Disease)	*Section 'B' (Virus group)*
I. Tobacco mosaic virus	(A) Tobamovirus
II. Wheat mosaic virus	(B) Furovirus
III. Barley stripe virus	(C) Hordeivirus
IV. Pea streak virus	(D) Carlavirus

Answer

I	*II*	*III*	*IV*
A	*B*	*C*	*D*

17. Correctly match in section 'A' to those given under the section 'B':

Section 'A' (Viral Disease)	*Section 'B' (Virus group)*
I. Papaya ring spot virus	(A) Potyvirus
II. Citrus tristeza virus	(B) Closterovirus
III. Rice tungro virus	(C) Waikavirus
IV. Tomato ring spot virus	(D) Nepovirus

Answer

I	*II*	*III*	*IV*
A	*B*	*C*	*D*

18. Correctly match in section 'A' to those given under the section 'B':

Section 'A' (Viral Disease)	*Section 'B' (Virus group)*
I. Tobacco streak virus	(A) Ilavirus
II. Wound tumor virus	(B) Phytoreovirus
III. Cucumber mosaic virus	(C) Comovirus
IV. Radish mosaic virus	(D) Cucumovirus

Answer

I	*II*	*III*	*IV*
A	*B*	*D*	*C*

19. Correctly match in section 'A' to those given under the section 'B':

Section 'A' (Viral Disease)	*Section 'B' (Virus group)*
I. Wheat streak mosaic virus	(A) Tobravirus
II. Bean yellow mosaic virus	(B) Pomovirus
III. Potato mop top virus	(C) Rymovirus
IV. Tobacco rattle virus	(D) Potyvirus

Answer

I	*II*	*III*	*IV*
C	*D*	*B*	*A*

20. Correctly match in section 'A' to those given under the section 'B':

Section 'A' (Viral Disease)	*Section 'B' (Virus group)*
I. Bean golden mosaic virus	(A) Potyviridae
II. Tomato spotted wilt virus	(B) Bunyaviridae
III. Wheat streak mosaic virus	(C) Potyviridae
IV. White clover cryptic virus 2	(D) Partitiviridae

Answer

I	*II*	*III*	*IV*
A	*B*	*C*	*D*

21. Correctly match in section 'A' to those given under the section 'B':

Section 'A' (Viral Disease)	*Section 'B' (Virus group)*
I. Cowpea mosaic virus	(A) Secoviridae
II. Rice tungro bacilliform virus	(B) Caulimoviridae
III. Grape vine fan leaf virus	(C) Secoviridae
IV. Tomato bushy stunt virus	(D) Tombusviridae

Answer

I	*II*	*III*	*IV*
A	*B*	*C*	*D*

22. Correctly match in section 'A' to those given under the section 'B':

Virus/Section 'A'	*Virus size (nm)/Section 'B'*
I. Tobacco mosaic virus	(A) 300x18 nm
II. Tobacco rattle virus	(B) 80-110x22 nm diameter x 190x22 nm long
III. Carnation latent virus	(C) 730 x11 nm
IV. Potato virus X	(D) 470-580x11-13 nm

Answer

I	*II*	*III*	*IV*
A	*B*	*C*	*D*

23. Correctly match in section 'A' to those given under the section 'B':

Virus/Section 'A'	*Virus size (nm)/Section 'B'*
I. Potato virus Y	(A) 730 x 11 nm
II. Citrus trestiza	(B) 2000 x 12 nm
III. Barley yellow mosaic virus	(C) 550-750 x 12 nm
IV. Papaya ring spot virus	(D) 800 x 12 nm

Answer

I	*II*	*III*	*IV*
A	*B*	*C*	*D*

24. Correctly match in section 'A' to those given under the section 'B':

Section 'A' (ssDNA virus)	*Section 'B' (ssDNA virus vector)*
I. Banana bunchy top virus	(A) White fly (*Bemisia tabaci*)
II. Bean golden mosaic virus	(B) Aphid (*Penatalonia nigronervosa*)
III. Maize streak virus	(C) Leaf Hopper
IV. Tobacco leaf curl virus	(D) White fly

Answer

I	*II*	*III*	*IV*
A	*B*	*C*	*D*

25. Correctly match in section 'A' to those given under the section 'B':

Section 'A' (ssDNA virus)	*Section 'B' (ssDNA virus vector)*
I. Tomato mottle virus	(A) White fly
II. Tomato golden virus	(B) White fly
III. Tomato leaf curl virus	(C) White fly
IV. Coconut foliar decay virus	(D) Aphid

Answer

I	*II*	*III*	*IV*
A	*B*	*C*	*D*

26. Match the following vectors for important disease:

Disease	*Vector*
I. Bunchy top of banana	(A) Banana aphid (*Pentalonia nigronervosa*)
II. Citrus canker	(B) Citrus leaf minor (*Phyllocnistis citrella*)
III. Fig mosaic	(C) Mites (*Acaria* spp) also through grafting
IV. Greening (MLO's)	(D) Citrus psylla (*Diaphorina citri*)

Answer

I	*II*	*III*	*IV*
A	*B*	*C*	*D*

27. Match the following vectors for important disease:

Disease	*Vector*
I. Pear decline	(A) Pineapple mealy bug or lace bug (*Planococcus* spp)
II. Pine apple wilt	(B) Pear psylla (*Psylla pyricola*)
III. Tristeza	(C) Aphid (*Toxoptera citricida*)
IV. Papaya mosaic	(D) Aphid (*Myzus persicae*)

Answer

I	*II*	*III*	*IV*
B	*A*	*C*	*D*

28. Match the section 'A' to those given under section 'B':

Virus Group	*Size*
I. Tobamoviruses	(A) 300 nm long x 18 nm diameter
II. Tobraviruses	(B) 190 nm long x 22 nm diameter and 80-100 nm long x 22 nm diameter
III. Furoviruses	(C) 65-390 nm long x 18-24 nm diameters
IV. Hordeiviruses	(D) 100-150 nm long x 20 nm diameter

Answer

I	*II*	*III*	*IV*
A	*B*	*C*	*D*

29. Match the section 'A' to those given under section 'B':

Virus Group	*Size*
I. Potex viruses	(A) 470-580 nm long x 11-13 nm diameter
II. Carla viruses	(B) 610-700 nm long x 12-13 nm diameter
III. Capillo viruses	(C) 600-700 nm long x 12 nm diameter
IV. Tricho viruses	(D) 730 nm long x 12 nm diameter

Answer

I	*II*	*III*	*IV*
A	*B*	*C*	*D*

30. Match the section 'A' to those given under section 'B':

Virus Group	*Size*
I. Potyviruses	(A) 650-900 nm long x 11-15 nm diameters
II. Rymoviruses	(B) 650-900 nm long x 11-15 nm diameters
III. Bymoviruses	(C) 250-300 nm long x 11-15 nm diameters
IV. Closteroviruses	(D) 1100-2000 nm long x 12 nm diameter

Answer

I	*II*	*III*	*IV*
A	*B*	*C*	*D*

31. Match the section 'A' to those given under section 'B':

Virus Group	*Size*
I. Waikaviruses	(A) 30 nm diameter
II. Luteoviruses	(B) 25-30 nm diameter
III. Comoviruses	(C) 30 nm diameter
IV. Nepoviruses	(D) 30 nm diameter

Answer

I	*II*	*III*	*IV*
A	*B*	*C*	*D*

32. Match the section 'A' to those given under section 'B':

Virus Group	*Size*
I. Bromoviruses	(A) 26-35 nm diameter
II. Ilaviruses	(B) 20-32 nm diameter
III. Reoviruses	(C) 65-70 nm diameter
IV. Rhabdoviruses	(D) 200-500 nm long x 50-25 nm diameter

Answer

I	*II*	*III*	*IV*
A	*B*	*C*	*D*

33. Match the section 'A' to those given under section 'B':

Virus Group	*Size*
I. Tospoviruses	(A) 80-110 nm diameter
II. Tenuiviruses	(B) 290-2100 nm long x 3-12 nm dia
III. Caulimoviruses	(C) 50 nm diameter
IV. Badnaviruses	(D) 30-100 x 300 nm

Answer

I	*II*	*III*	*IV*
A	*B*	*C*	*D*

34. Match the section 'A' to those given under section 'B':

Section 'A' Virus	*Section 'B' Size*
I. Barley stripe mosaic virus	(A) 100-150 nm long x 20 nm diameter
II. Apple chlorotic leaf spot virus	(B) 730 x 12 nm
III. Bean common mosaic virus	(C) 750 x 12 nm
IV. Sugarcane mosaic virus	(D) 750 x 11nm

Answer

I	*II*	*III*	*IV*
A	*B*	*C*	*D*

35. Match the section 'A' to those given under section 'B':

Section 'A' Virus	*Section 'B' Size*
I. Turnip mosaic virus	(A) 720 x 12 nm
II. Tobacco etch virus	(B) 730 x 12 nm
III. Beet yellow virus	(C) 1250 x 12 nm
IV. Rice stripe virus	(D) 290 x 2100 nm

Answer

I	*II*	*III*	*IV*
A	*B*	*C*	*D*

36. Match the section 'A' to those given under section 'B':

Section 'A' Virus	*Section 'B' Size*
I. Rice grassy stunt virus	(A) 950-1350 nm
II. Banana streak virus	(B) 30-130 x150 nm
III. Cacao swollen shoot virus	(C) 27-142 nm
IV. Banana bunchy top	(D) 18-22 nm

Answer

I	*II*	*III*	*IV*
A	*B*	*C*	*D*

37. Match the section 'A' to those given under section 'B':

Section 'A' Virus Group	*Section 'B' Viruses*
I. Tobamoviruses	(A) Tobacco rattle virus, Pea early browning virus
II. Tobraviruses	(B) Potato mop top virus, Beet necrotic Yellow vine virus
III. Furoviruses	(C) Tobacco mosaic virus, Tomato mosaic virus
IV. Hordeiviruses	(D) Barley stripe mosaic virus

Answer

I	*II*	*III*	*IV*
C	*A*	*B*	*D*

38. Match the section 'A' to those given under section 'B':

Section 'A' Virus Group	*Section 'B' Viruses*
I. Potexviruses	(A) Pea streak virus, poplar mosaic virus
II. Carlaviruses	(B) Apple stem grooving virus, citrus tatter leaf virus

III. Capilloviruses	(C) Tobacco etch virus, sugarcane mosaic virus
IV. Potyviruses	(D) Potato virus X, Cymbidium mosaic virus

Answer

I	*II*	*III*	*IV*
D	*A*	*B*	*C*

39. Match the section 'A' to those given under section 'B':

Section 'A' Virus Group	*Section 'B' Viruses*
I. Rymoviruses	(A) Wheat streak mosaic virus, oat necrotic mottle virus
II. Bymoviruses	(B) Barley yellow mosaic virus, wheat spindle streak mosaic virus
III. Tobamovirus	(C) Pepper green mottle virus, Odenotogolossum ring spot virus
IV. Tobraviruses (D)	Pepper ring spot virus

Answer

I	*II*	*III*	*IV*
A	*B*	*C*	*D*

40. Match the section 'A' to those given under section 'B':

Section 'A' Virus Group	*Section 'B' Viruses*
I. Furoviruses	(A) Wheat mosaic viruses, peanut clump virus
II. Trichoviruses (hairlike)	(B) Apple chlorotic leaf spot virus
III. Potyvirus	(C) Bean common mosaic, bean yellow mosaic, lettuce mosaic
IV. Bymoviruses	(D) Rice necrosis mosaic virus

Answer

I	*II*	*III*	*IV*
A	*B*	*C*	*D*

41. Match the section 'A' to those given under section 'B':

Section 'A' Virus Group	*Section 'B' Viruses*
I. Closteroviruses	(A) Lettuce infectious yellow virus, beet yellows, citrus tristeza
II. Waikaviruses	(B) Rice tungro spherical virus, maize chlorotic dwarf virus

III. Luteo viruses	(C) Beet western yellow, potato leaf roll
IV. Comoviruses	(D) Cowpea mosaic virus, squash and radish mosaic virus

Answer

I	II	III	IV
A	B	C	D

42. Match the section 'A' to those given under section 'B':

Section 'A' Virus Group	*Section 'B' Viruses*
I. Potyviruses	(A) Papaya ring spot, plum pox virus
II. Luteoviruses	(B) Barley yellow dwarf virus
III. Nepoviruses	(C) Tomato ring spot virus, cherry leaf roll virus
IV. Bromo/Cucumoviruses	(D) Cucumber mosaic virus, peanut stunt virus

Answer

I	II	III	IV
A	B	C	D

43. Match the section 'A' to those given under section 'B':

Section 'A' Virus Group	*Section 'B' Viruses*
I. Ilaviruses	(A) Apple and Rose mosaic virus
II. Phytoreovirus	(B) Rice dwarf virus, wound tumor virus
III. Fijivirus	(C) Rice black streaked dwarf virus
IV. Oryzavirus	(D) Rice ragged stunt virus

Answer

I	II	III	IV
A	B	C	D

44. Match the section 'A' to those given under section 'B':

Section 'A' Virus Group	*Section 'B' Viruses*
I. Rhabdoviruses	(A) Lettuce necrotic yellow virus
II. Tospoviruses	(B) Tomato spotted virus
III. Tenuiviruses	(C) Rice hoja blanka virus
IV. Caulimoviruses	(D) Cauliflower mosaic virus

Answer

I	II	III	IV
A	B	C	D

Answers (Multiple Choice Questions)

(1)	(A)	(34)	(B)	(67)	(C)	(100)	(B)
(2)	(B)	(35)	(A)	(68)	(B)	(101)	(A)
(3)	(D)	(36)	(A)	(69)	(D)	(102)	(C)
(4)	(B)	(37)	(D)	(70)	(A)	(103)	(A)
(5)	(C)	(38)	(A)	(71)	(A)	(104)	(A)
(6)	(A)	(39)	(B)	(72)	(D)	(105)	(B)
(7)	(D)	(40)	(C)	(73)	(A)	(106)	(C)
(8)	(D)	(41)	(B)	(74)	(C)	(107)	(A)
(9)	(A)	(42)	(B)	(75)	(A)	(108)	(A)
(10)	(B)	(43)	(A)	(76)	(C)	(109)	(A)
(11)	(D)	(44)	(B)	(77)	(B)	(110)	(D)
(12)	(C)	(45)	(A)	(78)	(B)	(111)	(D)
(13)	(B)	(46)	(B)	(79)	(A)	(112)	(A)
(14)	(D)	(47)	(B)	(80)	(A)	(113)	(D)
(15)	(A)	(48)	(A)	(81)	(B)	(114)	(D)
(16)	(C)	(49)	(A)	(82)	(A)	(115)	(C)
(17)	(A)	(50)	(A)	(83)	(B)	(116)	(D)
(18)	(A)	(51)	(A)	(84)	(B)	(117)	(B)
(19)	(B)	(52)	(D)	(85)	(A)	(118)	(A)
(20)	(A)	(53)	(C)	(86)	(A)	(119)	(A)
(21)	(C)	(54)	(A)	(87)	(B)	(120)	(B)
(22)	(D)	(55)	(A)	(88)	(A)	(121)	(A)
(23)	(C)	(56)	(B)	(89)	(C)	(122)	(A)
(24)	(C)	(57)	(C)	(90)	(B)	(123)	(D)
(25)	(A)	(58)	(B)	(91)	(C)	(124)	(A)
(26)	(A)	(59)	(A)	(92)	(A)	(125)	(D)
(27)	(A)	(60)	(A)	(93)	(C)	(126)	(A)
(28)	(B)	(61)	(B)	(94)	(A)	(127)	(A)
(29)	(A)	(62)	(A)	(95)	(A)	(128)	(B)
(30)	(C)	(63)	(D)	(96)	(C)	(129)	(A)
(31)	(D)	(64)	(A)	(97)	(A)	(130)	(B)
(32)	(D)	(65)	(D)	(98)	(A)	(131)	(C)
(33)	(B)	(66)	(B)	(99)	(A)		

Chapter 4

Viroid, Virusoid, Mollicutes and Protozoa

Queries with Timeline of Viroid, Virusoid, Mollicutes and Protozoa

1. In 1971 Theodor Otto Diener, discovered viroid in potato spindle tuber disease which was small single stranded circular molecule of infectious RNA. Viroid size 250-400 bases long. T.O. Diener coined the term 'viroid'.
2. In 1982, Diener again observed small circular single stranded linear RNA and called them 'virusoid'. It is found in velvet tobacco mottle virus. Virusoids just like viroid except in smallness, virusoid size 300-400 bases long. Virusoids are circular single-stranded RNAs, dependent on plant viruses for replication and encapsidation. The genome of virusoids consists of several hundred nucleotides and only encodes structural proteins.
3. Virusoids are similar to viroids in size, structure and means of replication (rolling-circle replication). Virusoids, while being studied in virology, are not considered as viruses but as sub viral particles. Since they depend on helper viruses, they are classified as satellites.
4. In 1982, Prusiner, discovered and reproduced the term 'prions'. He found evidence of neurological diseases caused by agents that appeared resistant to the processes that normally destroy nucleic acids. Once highly controversial, the idea won Prusiner the Nobel Prize in Physiology or Medicine in 1997.
5. A prion is an infectious protein which is hypothesized to infect tissues and 'reproduce' first by changing its own conformation, and secondly by inducing

other benign proteins to do the same. Prions have both genetic and infectious etiologies. So far, viroids have not been found in animals although similar proteinaceous infectious particles (smaller type of infectious agent than viroids) known as 'prions' were reported in scrapie disease of sheep and goats. So far, no prions have been found to infect plants. This disease has been recently mentioned in connection with the mad disease in UK.

6. The term 'Viroid' was coined by Altenburg (1946) to designate hypothetical (ultra microscopic organisms which are akin to viruses but which are useful symbionts that occur universally within the cells of larger organism). Many years later, Diener (1971) redefined the term to include the case of infective nucleic acid, too small to contain the genetic information necessary for self replication and with no capsid protein such as the potato spindle tuber virus.
7. All viroids sequenced so far are single stranded, circular RNAs. Since the properties of potato spindle tuber diseases (PSTV) differed from the typical viruses in molecular and physical aspects, the redefined 'viroid' was coined for potato spindle tuber like agent (Diener, 1971).
8. Viroid replication is considered to be through a rolling mechanism (Branch *et al.*, 1984). Recently, viroids replicate by direct RNA copying, in which circular positive strand viroids RNAs are transcribed into multimeric negative strand linear RNAs. The linear (-) strand then serve as a template for replication of multimeric strand of (+) RNA. The (+) RNA is subsequently processed (cleaved) by enzyme that release linear unit length viroid (++RNA) and these circularize and produce many copies of the original viroid RNA.
9. Potato spindle tuber, chrysanthemum stunt, citrus exocortis, chrysanthemum chlorotic mottle, coconut cadang cadang, pear blister canker, apple scar skin, citrus bent leaf, hop latent, hop stunt, avocado sun blotch and peach latent mosaic diseases are caused by viroids.
10. Potato spindle tuber viroid is transmitted by pollen and seed and contaminated mouthparts of the insects such as grasshopper, flea beetles and bugs. The viroid is mechanically transmissible mostly.
11. Virusoid RNA is structurally similar to viroid RNA it is incapable of independent replication and relies for this on its helper virus. The virusoid associated with lucerne transient streak virus (LTSV) shares no sequence homology with the genome of its helper virus LTSV can however replicate independently of the virusoid RNA. Some other viruses for example tobacco velvet mottle virus appear in contrast to be entirely dependent on the presence of virusoid RNA without which the virus cannot replicate.
12. Virusoid are encapsidated viroid like RNAs. The virusoid was coined by Haseloff and Symons (1982). Velvet mottle virus, solanum nodiflorum mottle virus, lucern transient streak virus and subterranean clover mottle virus are caused by virusoids.

13. The distinguishing features of viroids which separate them from virusoids are (i) Naked, low molecular weight, circular and linear RNA and absence of bifurcates (ii) Absence of sequence homology with virusoids and (iii) Absence of code for nucleoprotein.
12. E. Nocard and E.R. Roux in 1898, had discovered another agent of diseases caused in animal which was similar to viruses in size but which could be cultured on artificial media. This group was known as (*Mycoplasma mycoides* var. *mycoides*) mycoplasma.
13. In 1967, Doi *et al.*, and Ishii *et al.* Japanese scientist (in Japan) found/observed that mycoplasma like bodies/organisms in the phloem of plants infected with several leafhopper transmitted could be responsible for most of the diseases of yellows types caused by viruses in plants. These scientists discovered pleomorphic MLOs in the phloem cells affected by four different yellows type diseases. These were aster yellows, mulberry dwarf, potato witches broom and paulownia witches broom.
14. Davis *et al.*, (1972) first noticed a motile helical wall-less microorganism associated with corn stunt disease, they called it 'spiroplasma' and this organism could be cultured and characterized.
15. Mycoplasma like organism (MLO) and Spiroplasma are phloem inhabiting bacteria. Edward *et al.* (1967) proposed a new class which they called 'mollicutes' (bacteria without cell wall). Mollicutes ribosome's are mostly 72 S.
16. Pear decline, grape yellows, coconut lethal yellowing, x-disease of peach, apple proliferation, aster yellows of vegetables and ornamentals, elm yellow (phloem necrosis), witches broom and stolbur of tomato are caused by mollicutes (Mycoplasma like organism) and citrus stubborn and corn stunt disease are known to be caused by helical mollicutes (*Spiroplasmas*).
17. Mycoplasma like organism (MLO) belongs to the division Tenericutes, class Mollicutes, family Spiroplasmataceae (genus *Spiroplasma*) and other phytoplasmas (unidentified genus) family still unknown.
18. Properties of true *Mycoplasmas,* are prokaryotic organism that have no cell walls and belongs to class mollicutes, order mycoplasmatales, the order has three families each with one genus Mycoplasmataceae, genus *Mycoplasma*; Acholeplasmataceae, genus *Acholeplasma* and Spiroplasmataceae, genus *Spiroplasma*.
19. True *Mycoplasma* are bounded by unit membrane, lack a true cell wall, 175 to 250 nm in diameter, no flagella, produces no spore, reproduces by budding and by binary fission, gram negative and all true *Mycoplasma* parasitic to humans and animals and all saprophytic ones can be grown on more or less complex nutrient media.
20. Phytoplasma observed in the plants and insect vectors, which do not include the spiroplasmas, resemble the mycoplasmas of the genera *Mycoplasma* or

Acholeplasma in all morphological aspects. Genetically, phytoplasma are more related to *Acholeplasma* than to *Mycoplasma*.

21. Phytoplasma are prokaryotic organisms, lack of cell walls, are bonded by unit membrane, have cytoplasm, ribosome and strands of nuclear material, pleomorphic shape (spherical to ovoid, tubular to filamentous, 10 nm diameter).
22. Phytoplasma and spiroplasma are generally present in the sap of a small number of phloem sieve tubes. Most plant mollicutes are transmitted by leafhopper and some are transmitted by psyllids and plant hoppers.
23. The vector cannot transmit the mollicutes immediately after feeding on the infected plant, but it begins to transmit them after an incubation period of 10 to 45 days, depending on the temperature, the shortest incubation period occurs at about 30°C, the longest at about 10°C.
24. All the mollicutes are prokaryotic cells without cross walls, a few of them have a helical structure and are called *Spiroplasmas*. Most, however, are round to elongate but are not spiral and they are now called '*Phytoplasmas*'.
25. Phytoplasmas (pear decline, grape yellow, coconut lethal yellowing, X-diseases of peach, apple proliferation etc) cannot be grown on artificial nutrient media and so far no plant disease has been reproduced on healthy plants inoculated directly with phytoplasmas obtained from diseases plants.
26. Spiroplasmas are helical mollicutes. Spiroplasmas are cells that vary in shape from spherical or slightly ovoid, 100 to 240 µm or larger in diameter, to helical and branched, 2 to 4 µm long during active growth and spiroplasma can be cultured on nutrient media. It's produce mostly helical forms in liquid media and multiplies by fission. They lack true cell wall and are bounded by unit membrane. The helical filaments are motile, moving by a slow undulation of the filament and probably by a rapid rotary or screw motion of the helix. Spiroplasmas are gram positive.
27. The disease caused by spiroplasma spp includes citrus stubborn (*Spiropasma citri*) and corn stunt (*S. kunkelii*). *Spiroplasma citri* was obtained in pure culture by Sanglio *et al.* in 1971and by Fudl-allah *et al.* in 1972.
28. Aster yellows are transmitted by budding or grafting and by several leafhoppers. Lethal yellowing of coconut palms is transmitted by plant hopper *Myndus crudus*. Elm yellows (phloem necrosis) are transmitted from diseased to healthy trees by the leafhopper *Scaphoideus luteolus*.
29. X. disease of peach is transmitted by several species of leafhoppers of the genera *Colladonus* and *Scaphytopius* and budding and grafting also. Pear decline is transmitted by pear psylla (*Psylla pyricola*).
30. *Spiroplasma citri* (citrus stubborn) is found in the phloem sieve tubes. The pathogen is gram positive and is insensitive to penicillin but is highly sensitive to tetracycline. This disease is transmitted by several leafhoppers such as *Circulifer tenellus*, *Scaphytopius nitridus* and *Neoaliturus haemoceps*.

31. *Spiroplasma kunkelii* (corn stunt disease) is transmitted in nature by leafhoppers such as *Dalbulus elimatus, D. maidis* and incubation period of 2 to 3 weeks.

32. Grassy shoot disease (GSD) of sugarcane is caused by mycoplasma like organism (MLO)/mycoplasma like bodies (MLB) and in India it was first detected in 1949. The pathogen is transmitted by several aphids such as *Aphis maydis, Aphis sacchari* and *Aphis idiosacchari* and transmitted by dodder (*Cuscuta campestris*) has also reported.

33. Little leaf of brinjal is found throughout India. In India it was first reported from Coimbatore (Tamil Nadu). The disease is caused by MLO and is transmitted by leafhopper (*Hishimonus phycitis*). The same MLO occurs on *Datura fastuosa* and *Vinca rosea*. Terramycin, aureomycin, ledermycin, cultural practices and vector control are effective for the management of the disease.

34. Purple top roll (PTR), marginal flavescence (MF), potato stolbur (stolbur), purple top wilt (PTW) and witches brooms (PWB) are phytoplasmal disease of potato and are transmitted by leafhopper. Potato phyllody (PP) is caused by phytoplasma and is readily transmitted through tubers and graft. Insect vector of the pathogen is not known.

35. Sandle spike disease is caused by MLO and was first reported by Mc Carthy from Coorg district of Karnataka in 1899. The disease is transmitted by leafhoppers such as *Nephotettix virescens* and *Moonia albimaculata* and through dodder also. The MLOs infect a large number of collateral hosts such as *Eucalyptus grandis, Vinca rosea, Zizyphus oenoplia* and *Dodonia viscosa*.

36. Sesamum phyllody disease is caused by MLO and is transmitted by leafhopper (*Orosius albicinctus*). The incubation period of the pathogen in leafhoppers is found to be 15 to 63 days and 13 to 61 days in the host depending upon the atmospheric conditions (temperature). Coconut (root) wilt is caused by MLO and is transmitted by lace bug (*Stephanitis typicus*).

37. Flagellate protozoa were first found to be associated with plants in 1909, in Mauritius, when Lafont, reported that they parasitize the latex bearing cells of *Euphorbia* (Dugdhika). Plant protozoa were placed in a new genus, *Phytomonas*, and one described by Lafont was named *P. davidi*. Several other species of phytomonas have been reported from plants, *e.g. P. elmassiani* on milkweed, *P. bancrofti* on ficus, *P. leptovasorum* on coffee and *P. francai* on cassava.

38. In 1931, Stahel found flagellate infecting the phloem of coffee trees and causing abnormal phloem formation and wilting of the trees.

39. In 1963, Vermeulen presented additional and more convincing evidence of the pathogenicity of flagellates to coffee trees and in 1976 flagellates were also reported to be associated with several diseases of coconut and oil palm trees in South America and in Africa.

40. All plant flagellates belong to the order Kinetoplastida, family Trypanosomatidae. Phytomonas species live in the phloem sieve tubes of non-laticiferous plants

such as coconut and oil palms and coffee. All phytomonas can be grown on specialized nutrient media.

41. Phloem necrosis of coffee is caused by the pathogen *Phytomonas leptovasorum*, it is a trypanosomatid flagellate and is transmitted through root grafts and pentatomid insects of the genus *lincus*. Hartrot of coconut palms is caused by Phytomonas protozoa and is transmitted by pentatomid insects of the genera *Lincus* and *Ochlerus*.

42. Sudden wilt (Marchitez) of oil palm and empty root of cassava is caused by flagellate protozoa Phytomonas, the organism is transmitted by *Lincus* and *Ochlerus* and empty root disease can be transmitted by grafting.

43. Antignus *et al.* (2007) have confirmed that bumblebees (*Bombus terrestris*) transmit the tomato apical stunt viroid from infected to healthy tomato plants in the form of secondary spread.

Multiple Choice Questions (Choose the correct answer)

1. **Citrus stubborn disease of citrus is caused by:**
   (A) *Xanthomonas* (B) *Colletotrichum*
   (C) *Spiroplasma* (D) *Mycoplasma*

2. **Citrus greening disease is caused by:**
   (A) Mycoplasma like organism (B) Rickettsia like organism
   (C) Fungal like organism (D) Virus like organism

3. **A derivative of gram positive bacterium without a cell wall is called:**
   (A) Protoplast (B) Spheroplast
   (C) L-Form (D) Mycoplasma

4. **Who coin the term viroid for hypothetical symbionts?**
   (A) Altenburge (1946) (B) Diener (1971)
   (C) De Barry (1807) (D) None of these

5. **Single stranded circular RNA, sub viral pathogen is:**
   (A) Virus (B) Viroids
   (C) Virion (D) None of these

6. **Viroids consist of:**
   (A) 246-375 nucleotide (B) 400-525 nucleotide
   (C) 450-550 nucleotide (D) None of these

7. **Viroids replication is considered to be through a:**
   (A) Binary fission (B) Budding
   (C) Rolling circle mechanism (D) None of these

8. **Lacking a central conserved region (CCR) in:**
   (A) Group A Viroid (B) Group B Viroid
   (C) Group C Viroid (D) Group D Viroid

9. **Potato spindle tuber viroids are transmitted by:**
   (A) Aphid (B) Leafhopper
   (C) Mite (D) Pollen

10. **Avocado sunblotch disease is caused by:**
    (A) Virus (B) Viroid
    (C) Virusoid (D) None of these

11. **MLO lack:**
    (A) Protein (B) Cell wall
    (C) RNA (D) DNA

12. **Coconut cadang-cadang is caused by:**
    (A) Viroid (B) Virus
    (C) MLO (D) RLO

13. **Grapevine yellow speckle-1 is caused by:**
    (A) Viroid (B) Virus
    (C) Bacterium (D) MLO

14. **Chrysanthemum stunt disease is caused by:**
    (A) Viroid (B) Virus
    (C) Fungi (D) Bacteria

15. **Peach latent mosaic disease is caused by:**
    (A) Bacteria (B) Viroid
    (C) Virus (D) None of these

16. **A derivative of gram positive bacterium without a cell wall is called:**
    (A) Spheroplast (B) Protoplast
    (C) Mycoplasma (D) L-Form

17. **The causal agent of sandal spike disease is:**
    (A) BLO (B) MLO
    (C) RLO (D) BC

18. **Velvet tobacco mottle virus is caused by:**
    (A) Virus (B) Viroids
    (C) Virusoid (D) None of these

19. The casual agent of tobacco mottle disease is:

(A) RLO (B) Virusoid

(C) Viroid (D) MLO

20. Which group of bacteria has defective cell wall?

(A) Gracellicutes (B) Firmicutes

(C) Mendosicutes (D) Tenericutes

21. The viroid as a casual agent of a plant disease was first discovered in:

(A) Tubers of potato (B) Leaves of sugarcane

(C) Flowers of wheat (D) Stem of mango

22. The phytoplasma grow in phloem cells of the plant because:

(A) The phloem cells rich in sterol

(B) The phloem cells are rich in protein

(C) The phloem cells contain vitamin

(D) The phloem cells are easy to penetrate

23. Rickettsia differs from Phytoplasma in:

(A) Presence of 70 S ribosomes

(B) Presence of nuclear membrane

(C) Presence of double unit membrane

(D) Absence of cell wall

24. Match the discovery of Asuyama *et al.*:

(A) Hybridoma technology (B) Viroid

(C) Forest pathology (D) Mycoplasma discovery

25. Match the discovery of Diener *et al.*:

(A) Viroid (B) Hybridoma technology

(C) Forest pathology (D) Mycoplasma discovery

26. Who coined the term Virusoids:

(A) Haseloff and Symons (1982) (B) Kassanis (1966)

(C) Gierrer and Schramm (1956) (D) None of these

27. Rotary or screw like motion appears in:

(A) Spiroplasma (B) Mycoplasma

(C) Virus (D) Viroids

28. Match helical organism:

(A) Virus (B) Fungi

(C) Mycoplasma (D) Spiroplasma

29. Who first time noticed Spiroplasma:

(A) Devis *et al.* (1972) (B) Saglio *et al.* (1973)

(C) Townsend *et al.* (1977) (D) None of these

30. Size of helical organism spiroplasma is:

(A) 15-18 x 0.88 μm (B) 3-12 x 0.22 μm

(C) 20-25 x 1.20 μm (D) None of these

31. Sterols are not required for growth of:

(A) Mycoplasmatacease (B) Acholeplasmataceae

(C) Virus (D) None of these

32. The casual agent of pierce's disease of grape:

(A) Mycoplasma (B) Spiroplasma

(C) Bandavirus (D) Rickettsia like organism

33. Which of the following disease is caused by *Spiroplasma*:

(A) Citrus greening (B) Citrus rubbery wood

(C) Citrus canker (D) Citrus stubborn

34. Viroids, the smallest plant pathogens consist of:

(A) Protein only (B) DNA only

(C) DNA and RNA (D) RNA only

35. Spindle tuber disease of potato is caused by:

(A) Viroid (B) Virus

(C) Virusoid (D) Mycoplasma

36. Rickettsia grows readily on:

(A) Yolk sac of embryonated egg (B) Potato dextrose agar

(C) Waksman medium (D) Corn meal agar medium

Fill in the Blanks with Correct Answer

1. Ratoon stunning diseases (RSD) is caused by ________gram positive xylem inhabiting fastidious bacteria.

   Clavibacter xyli

2. Potato spindle tuber disease is caused by________

   Viroid

3. Citrus greening is caused by________previously known as Rickettsia like organism (RLO).

   Phloem inhabiting fastidious bacteria

4. Pierce's disease of grape is caused by________

   Xylella fastidiosa

5. Helical mollicutes are________

   Spiroplasma

6. Most of the mycoplasma is resistant to________

   Penicillin

7. Phloem necrosis of coffee is caused by________it is flagellate protozoa.

   Phytomonas leptovasorum

8. Heart rot of coconut palm and sudden wilt (Marchitez) of oil palm are caused by________

   Phytomonas sp

9. ________is a just like virus but differing from shell which it has devoid off.

   Viroid

10. ________is a naked nucleic acid which is able to cause the disease in the plant.

    Viroid

11. Tetracycline has recently been found to be effective against plant disease caused by________

    Mycoplasma

12. Rubber wood of apple and pear is caused by________

    Phytoplasma

13. Mycoplasmal disease coconut lethal yellowing is transmitted by________

    Slow flying insect

14. ________ proposed a new class of which they called mollicutes

    Edward et al. (1967)

15. Lack of protein coat and apparently exist and free RNA is call________

    Viroid

16. Phony peach disease, plum leaf scald, almond leaf scoarch and elm leaf scoarch are caused by ________

    fastidious vascular bacteria

17. In 1967, Japanese scientists found that was mycoplasma-like-organism could be responsible for most of the ________supposed to be caused by virus in plants.

    Yellow disease

18. Citrus stubborn disease is caused by ________and can be grown o artificial media.

    Spiroplasma citri

19. Fastidious vascular bacteria are sensitive to________ *i.e.* Tetracyclin, Penicillium.

    Antibiotic

20. The two families of viroids are the________

    Pospiviroidae and Avsunviroidae

21. In Japan, Tomato chlorotic dwarf viroid is also transmitted from infected tomato plants to neighbouring healthy plants through________ during pollination activities (Matsuura *et al.* 2010).

    Bumblebees (Bombus ignites)

True or False

1. MLO of plants cannot be grown on artificial media. (**True**)
2. Viruses and viroids reproduce within living host cells. (**True**)
3. Spiroplasmas are helical mollicutes, moving by a slow undulation of the filament and probably by rapid rotary or screw motion of helix. (**True**)
4. *Spiroplasma citri* is gram positive and found in the phloem. (**True**)
5. Viroids are circular, single standard RNA molecules with extensive base paring in parts of the RNA stand. (**True**)
6. Spiroplasmas have helical cell shape. (**True**)
7. Plasma membrane is also known as cytoplasmic membrane. (**True**)
8. Gumming disease of cereal and grasses such as yellow ear rot of wheat is caused by *Clavibacter tritici*. (**True**)
9. Fastidious vascular bacteria are sensitive to antibiotic *e.g.* Tetracycline, Penicillin and sensitive to high temperature. (**True**)
10. Aster yellow phytoplasma is limited primarily to the phloem of infected plants. (**True**)
11. Grassy shoot stunt disease was first observed in Maharashtra on sugarcane variety CO 419 in 1942 (Vasudeva). (**True**)
12. Mycoplasmal disease sugarcane white leaf is transmitted by *Epitettix hiroglyphicus*. (**True**)
13. Exocortis in citrus is caused by viroid. (**True**)
14. Mollicutes have only cell membrane and lack cell wall. (**True**)

Match the Following

1. **Match disease and host with the parasite**

Disease Host	*Parasite*
I. Potato spindle tuber	(A) Viroids
II. Velvet tobacco mottle	(B) Virusoids
III. Phloem necrosis of coffee	(C) Spiroplasma
IV. Citrus stubburn	(D) Phytomonas protozoa

Answer

I	*II*	*III*	*IV*
A	*B*	*C*	*D*

2. **Match the following some fruits and their maladies:**

Malady	*Cause*
I. Bunchy top of banana	(A) Viral
II. Cadang cadang in coconut	(B) Viroids
III. Canker in acid lime	(C) ***Xanthomonas campestris***
IV. Greening in citrus	(D) Fastidious vascular bacteria

Answer

I	*II*	*III*	*IV*
A	*B*	*C*	*D*

3. **Match the following crop disease and their casual organism.**

Crop, Disease	*Casual organism*
I. Potato spindle tuber	(A) Viroid
II. Chrysanthemum stunt	(B) Viroid
III. Tomato bunchy top	(C) Viroid
IV. Avocado sun blotch	(D) Viroid

Answer

I	*II*	*III*	*IV*
A	*B*	*C*	*D*

Answers (Multiple Choice Questions)

(1)	(C)	(10)	(B)	(19)	(B)	(28)	(D)
(2)	(B)	(11)	(B)	(20)	(D)	(29)	(A)
(3)	(D)	(12)	(A)	(21)	(A)	(30)	(B)
(4)	(A)	(13)	(A)	(22)	(A)	(31)	(B)
(5)	(B)	(14)	(A)	(23)	(D)	(32)	(D)
(6)	(A)	(15)	(B)	(24)	(D)	(33)	(D)
(7)	(C)	(16)	(C)	(25)	(A)	(34)	(D)
(8)	(A)	(17)	(B)	(26)	(A)	(35)	(A)
(9)	(D)	(18)	(C)	(27)	(A)	(36)	(A)

Chapter 5
Molecular Plant Pathology

Queries with Timeline of Molecular Plant Pathology

1. Detection and diagnostic techniques for plant pathogenic bacteria are microscopical observation, isolation, biochemical characterisation, serology (mainly through immunofluorescence and Enzyme-Linked Immunosorbent Assay (ELISA) using polyclonal and/or monoclonal antibodies), bioassays and pathogenicity tests etc.
2. Biological indexing, electron microscopy and some biochemical and staining tests have been used for testing pathogens of the genus *Spiroplasma* and phytoplasmas.
3. For detection of viruses and viroids, biological indexing (using herbaceous and/or woody indicator plants), electrophoresis, electron microscopy and ELISA based techniques have been the choice.
4. The development of hybridization based methods and nucleic acid based methods which depends on PCR and real-time PCR protocols, the advances in microarray, microchip or biochip technology allows to test simultaneously, the prospect of a wide variety of pathogenic microorganisms (nematodes, fungi, bacteria, phytoplasmas, viruses and viroids).,
5. For detection and diagnosis of fungal, bacterial, and nematodal pathogens, ribosomal RNA genes and intervening sequences are common targets for PCR amplification. The internal transcribed spacer (ITS) regions of ribosomal genes are especially useful targets for species-specific primers and commonly used in fungal identification. The ITS region of ribosomal RNA genes (rDNA) region consists of multiple copies (up to 200 copies per haploid genome) arranged

in tandem repeats comprising the 18 S small subunit, the 5.8 S, and the 28 S large subunit genes separated by internal transcribed spacer regions (ITS1 and ITS2). This region contains highly conserved areas adequate for genera- o species-consensus primer designing (RNA ribosomal genes), alternate with highly variable areas that allow discrimination over a wide range of taxonomic levels (ITS region).

6. The intergenic spacer sequence (IGS) placed between the 28 S and 18 S rRNA genes are the region with the greatest amount of sequence variation in rDNA. It is frequently used in PCR-based methods when there are not enough differences available across the ITS. Other housekeeping genes with higher variability, including nuclear genes such as β-tubulin, translation elongation factor 1 alpha (*TEF* 1α), calmodulin, avirulence genes and mitochondrial genes such as cytochrome oxidase (co x I and cox II) genes and their intergenic region and mating type genes are also suitable targets for distinguishing pathogen species.

7. Molecular hybridization technique for the detection of viruses is non-isotopic dot-blot hybridization using digoxigenin-labelled probes. This technique has been employed for detection of Apple mosaic virus (ApMV), Prunus necrotic ringspot virus (PNRSV), Prunedwarf virus (PDV), PPV, and Apple chlorotic leaf spot virus (ACLSV).

8. According to Bonants *et al.* (2005), molecular approaches developed over the last ten years to detect many bacteria, *Spiroplasma*, phytoplasmas, viruses, and viroids in plant or environmental samples can be grouped as follows, a) RNA level: RT-PCR, NASBA or AmpliDet RNA; and b) DNA level: hybridisation, FISH, and PCR variants (conventional PCR, nested PCR, cooperative PCR, multiplex PCR, real-time PCR).

9. Fluorescence *in-situ* hybridisation (FISH) combines microscopical observation of bacteria and the specificity of hybridisation and is dependent on the hybridisation of DNA probes to species-specific regions of bacterial ribosomes.

10. Hybridization-based techniques are based on the Watson-Crick complementary rules of base pairing and require the use of labeled nucleic acid molecules as hybridization probes.

11. Probes can be short single-stranded nucleic acid segments (oligonucleotides) of synthetic origin or cloned DNA segments that bind to complementary nucleic acids, forming hybrid molecules. In conventional analyses, probes are hybridized to target DNA molecules that have been previously cut (digested) with restriction endonuclease enzymes and transferred (blotted) to a membrane support according to the Southern procedure probes represent single copy genomic segments, the procedure generates restriction fragment length polymorphic (RFLP) markers. More advanced applications involve the use of oligonucleotide arrays ("chips") that confine individual oligonucleotides to defined physical addresses in solid supports such as glass or silicon

(Southern, 1996). These hybridization arrays are very popular as they can be used successfully in both sequencing and genotyping applications.

12. Restriction fragment length polymorphisms (RFLP) of DNA representing selected genes can be used to identify pathogen species. This approach to species identification depends on having an excellent database on the variability in fragment length polymorphisms that may be found among isolates of individual species, since conspecific isolates may differ in the presence or absence of particular restriction sites, consequently changing the RFLP banding profile.
13. Amplified fragment length polymorphism (AFLP), a modification of the RFLP technique, has been used for species identification and more commonly to examine genotypic diversity within a population. It is useful in discerning the geographic origin of pathogens associated with new disease outbreaks.
14. Arrays or oligonucleotide arrays consist of discrete spots of pathogen-specific oligonucleotide sequences immobilized onto a solid surface such as a nylon membrane or glass slide. The sample DNA is amplified by PCR, labeled, then hybridized to the array.
15. Amplification-based techniques or Nucleic acid amplification methods use oligonucleotides to drive the exponential accumulation of specific sequences from defined regions in a genome or transcriptome (Landegren, 1993). Thermostable DNA polymerase enzymes that copy (replicate) the accumulating nucleic acid segments with efficiency and high fidelity usually mediate these experimental strategies.
16. The choice of the appropriate molecular markers for this purpose depends on the degree of diversity (polymorphism) that is anticipated. Markers producing multilocus profiles (*e.g.*, RAPD and DAF) are good choices when analyzing closely related genotypes.
17. Polymerase Chain reaction (PCR) allows the enzymatic amplification of millions of copies of specific DNA sequences by repeated cycles of denaturation, polymerisation and elongation at different temperatures using specific oligonucleotides (primers), deoxyribonucleotide triphosphates (dNTPs) and a thermostable *Taq* DNA polymerase in the adequate buffer.
18. Primers, which are small strands of DNA containing may be 15-20 nucleotides, are used to adhere to the now separate, single stranded, DNA.
19. A polymerase is an enzyme required for DNA elongation, or essentially, filling in the "holes". Primers are designed to anneal to the single strands (they have a melting temperature higher than the actual DNA and thus are able to bind.) Because the primers are so small, they only fill in a small "hole" in the single stranded DNA. This is when the polymerase comes in and completes the rest of the strand, creating now two separate, double stranded pieces of DNA. The polymerase used in PCR reaction is called the Taq polymerase, as it was

isolated from a thermophillic bacteria, *Thermus aquaticus* and is able to function at higher temperatures than other enzymes (the enzymes within our body are only able to maintain stability at a vary narrow range of temperatures, *i.e.* 98.6 F).

20. PCR based markers have been widely used in the detection and identification of pathogens of potential regulatory concern or of high economic consequence. For *e.g.* Pathogens where molecular detection is a fundamental tool include *Phytophthora ramorum*, (sudden oak death); *Xanthomonas citri* strains (citrus canker); and *Tilletia indica* (Karnal bunt of wheat). Discrimination of the Asian soybean rust pathogen, *Phakopsora pachyrhizi*, from the related species *Phakopsora meibomiae* by PCR is central to soybean rust management programs in the United States.

21. *Co-operational PCR is a* new PCR concept, based on the simultaneous action of four or three primers, it can be performed easily in a simple reaction increasing the sensitivity level and using ten times less reagent than in conventional PCR. The technique was first developed and used successfully for the detection of plant RNA viruses, such as CTV, PPV, *Cucumber mosaic virus* (CMV), *Cherry leaf roll virus* (CLRV) and Strawberry latent ringspot virus (SLRSV) and then for the bacterium *R. solanacearum* in water and in *Pelargonium* spp

22. SSR/ISSR primers designed to target sequences flanking simple sequence-repeat (SSR) regions to produce powerful nucleic acid markers, because these microsatellite sequences are highly variable and reveal many allele variants. However, one important limitation is the prerequisite for partial or total knowledge of the target sequence.

23. Dubbed quantitative real-time PCR (qPCR), this technique makes it possible to measure the amount of DNA produced during each PCR cycle. This refinement involves the use of fluorescent dyes or probes that label double-stranded DNA molecules. These fluorescent markers bind to the new DNA copies as they accumulate, making "real-time" monitoring of DNA production possible. As the number of gene copies increases with each PCR cycle, the fluorescent signal becomes more intense. Plotting fluorescence against cycle number and comparing the results to a standard curve (produced by real-time PCR of known amounts of DNA) enables scientists to determine the amount of DNA present during each step of the PCR reaction.

24. RT-PCR can be used to determine how gene expression changes over time or under different conditions.

25. All nucleic acid synthesis in a cell occurs in the 5′-3′ direction, because new monomers are added via a dehydration reaction that uses the exposed 3′ hydroxyl as a nucleophile.

26. PCR/electrospray ionization-mass spectrometry (PCR/ESI-MS; previously known as "TIGER", an acronym referring to Triangulation Identification for the Genetic Evaluation of Risks) is a new technique for rapid identification

of microorganisms in environmental samples. It is based on the use of mass spectrometry to characterize the base compositions of PCR products generated using a suite of conserved primers. For species determinations, the resulting base compositions are compared to a database of microbial basecomposition signatures.

27. Expressed sequence tags (ESTs) are generated by large scale single pass sequencing of randomly picked cDNA clones, have been cost effective and valuable resource for efficient and rapid identification of novel genes and development of molecular markers. ESTs have been employed in bioinformatic analysis to identify the genes that are differentially expressed in various tissues, cell types or developmental stages of the same or different genotypes.
28. The DNA sequences from which the primers are designed for bacteria come from three main origins: pathogenicity/virulence genes, ribosomal genes, and plasmid genes.
29. Avirulence gene codes for an elicitor molecule or protein controlling the synthesis of an elicitor. The Resistance gene codes for a receptor molecule which 'recognises' the elicitor. A plant with the Resistance gene can detect the pathogen with the Avirulence gene. Once the pathogen has been detected, the plant responds to destroy the pathogen. Both the Resistance gene and the Avirulence gene are dominant.
30. Resistant genes are functional genes that regulate the functioning of defense genes. Structure of resistance gene refers to the protein structure motifs recognizable in the derived amino acid sequences of R genes. Resistance genes have been divided into 6 classes: Three of these classes contain: leucine rich repeats (LRRs), of which the class of nucleotide binding site (NB)-LRRs proteins is most abundant. NBS LRR resistance proteins generally contain one of two types of N-terminal domains. On the basis of the deduced N-terminal features of R proteins NB –LRR class can be subdivided into TIR-NB-LRR and CC-NB-LRR proteins. These are either a domain that has homology with the Toll and Interleukin-1 Receptor proteins (TIR) or a predicted coiled-coil domain (CC). The other two classes of LRR proteins involve the LRR-transmembrane-anchored (LRR-TM) proteins and the LRR-TM kinase proteins. Fourth class of R genes represents protein kinases. Fifth class of R genes contain cytoplasmic coiled coil domain. Sixth class of R gene encode TM-surface glycoproteins having an extracellular LRR domain, endocytosis like signals, and leucine zipper (LZ) or Pro-Glu-Ser-Thr (PEST) sequences. These structures form homo- or hetero- oligomeric associations facilitating interactions between proteins and possibly playing a role in the interaction of R-proteins with molecules downstream in the signal transduction pathway.
31. NBS-LRR proteins (inside the cell) recognize specific elicitors produced by Avr and Hpr genes, activates resistance to obligate and hemibiotrophs.

32. Quantitative trait locus (QTL) is a region of the genome that is associated with an effect on a quantitative trait. Conceptually, a QTL can be a single gene or may be a cluster of linked genes that affect the trait.

33. In the case of polygenic resistance, several genes contribute to a quantitative trait loci (QTL) that show additive effect against different races of the pathogen. With the use of DNA markers flanking the QTL of interest, it is possible to facilitate breeding for complex traits such as downy mildew resistance in pearl millet.

34. Pathogen associated molecular patterns (PAMPs/MAMPs) recognised by Pattern recognition receptors (PRRs), are pathogen derived general elicitors (non specific elicitors), evolutionary conserved structures which are functonally important. Flagellin, LPS, Chitin, initiates basal defense.

35. Elicitor is a molecule which induces any plant defence response. Proteins made by the pathogen avirulence genes, or the products of those proteins. Elicitor can be a polypeptide coded for by the pathogen; it may be avirulence gene, a cell wall breakdown product or low-molecular weight metabolites. Not all elicitors are associated with gene-for-gene interactions.

36. Programmed cell death or apoptosis is a physiological process in which unwanted and damaged cells are removed from the plant body. Events takes place during PCD are: Loss of cell-to-cell contact, Cell shrinkage, Condensation of chromatin, Membrane "blebbing", mitochondrial dysfunction, Systematic DNA degradation or "laddering", Activation of caspases (cysteine-aspartic proteases).

37. Events takes place during Hypersensitive Reaction includes: Oxidative reactions (production of hydrogen peroxide), Deposition of callose (related to cellulose), Opening of ion channels, Apoptosis (programmed cell death), Cessation of cell cycle, Induction of genes that promote resistance, induction of Phenylpropanoid pathway which stimulates production of salicylic acid (secondary inducer: induces other pathogenesis-related proteins), lignins (cell wall), and flavonoids, Pathogenesis-related (PR) proteins, increased production of Phytoalexins, Fortification of cell walls with lignin, hydroxyproline-rich glycoproteins (HRGPs), etc.

38. Potential sources of ROS in plants are some reactions of normal aerobic metabolism, such as photosynthesis and respiration, while others belong to pathways enhanced during abiotic stresses, such as photorespiration. Newly identified sources of ROS in plants are NADPH oxidases, amine oxidases, and cell wall-bound peroxidases. These are tightly regulated and participate in the control of processes such as programmed cell death, stress response, and pathogen defense.

39. Major ROS scavenging mechanisms of plants include superoxide dismutase (SOD), ascorbate peroxidase (APX), and catalase (CAT). The balance between SOD, and APX (and/or CAT) activity in cells is considered to be crucial for determining the steady-state level of O_2^- and H_2O_2.

40. A number of components that include the MAPKKK, AtANP1 (also NPK1), the MAPKs, AtMPK3/6, and Ntp46MAPK, and calmodulin may be involved in the ROS signal transduction of plants.

41. H_2O_2 plays multiple roles during host pathogen interactions: it induces defense-related genes, induces apoptosis, causes cross-linking of cell wall proteins (more resistant to wall-degrading enzymes) and are toxic so it may directly kill pathogens.

42. Ca^{+2} also plays important role as a signaling molecule and required for subsequent steps: may mediate phosphorylation-dephosphorylation events involved in transcriptional or post- transcriptional gene regulation (there are a number of genes whose transcription increases, and some decrease).

43. Plants have evolved a number of inducible defense mechanisms against pathogen attack. Recognition of a pathogen often triggers a localized resistance reaction, known as the hypersensitive response (HR), which is characterized by rapid cell death at the site of infection.

44. When the plant turns on its defense mechanism in every part of the plant in response to a chemical or pathogen, it is known as Systemic Acquired Resistance (SAR).

45. The induction of systemic resistance by rhizobacteria is referred as Induced Systemic Resistance. ISR was analytically established by Ku´c *et al.* (1959) and Ross (1966).

46. SAR is dependent on the production of salicylic acid and the level and activity of NPR1, a protein that interacts with transcripton factors that regulate the expression of defense related genes. The activated state of SAR is characterized by broad spectrum resistance against viruses, bacteria, and fungi, and by a set of biochemical responses; both spectrum of protection and biochemical responses vary according to plant species.

47. Unlike race-specific (vertical) resistance or pesticides, induced resistance does not appear to apply selective pressure to pathogen or parasite populations on the basis of any single genetic determinant or specific mode of action, but rather is quantitative because of the cumulative effects of numerous plant defense mechanisms.

48. Salicylic acid in turn triggers both local and systemic responses. These include HR (programmed cell death at the site of infection), local resistance to the pathogen and Systemic Acquired Resistance (SAR). SAR is a generalized state of increased resistance to pathogen attack and occurs at sites distant from the original site of infection.

49. Treatment with synthetic elicitors, PGPR bacteria and fungi have been reported to sensitize plants to defend themselves against pathogen attack by triggering various defense mechanisms including production of phytoalexins, synthesis of phenolics, accumulation of pathogenesis related proteins and deposition

of structural barriers. There are several commercial sprays that will turn on genes in the plant, producing a systemic acquired resistance to disease.

50. System biological approaches also referred as Proteomics are utilized extensively to characterize the interplay between the various components of the cells in order to fully understand cellular processes.
51. Proteome refers to a complete set of proteins that are specified by the genome, and analogous to genomics, Proteomics describes the study and characterization of the complete set of proteins present in the cell, organ or organism at a given time.
52. The term proteomics was coined by Marc Wilkins, back during the 1994 Siena Meeting, to simply refer to the "PROTein complement of a genOME". The proteome can be defined as being the total set of protein species present in a biological unit (organule, cell, tissue, organ, individual, species, and ecosystem) at any developmental stage and under specific environmental conditions. By using proteomics one aim to know how, where, when, and what for are the several hundred thousands of individual protein species produced in a living organism, how they interact with one another and with other molecules to construct the cellular building, and how they work with each other to fit in with programmed growth and development, and to interact with their biotic and abiotic environment.
53. In the case of fungi, a new area has also be defined as Secretomics (the secretome is defined as being the combination of native proteins and cell machinery involved in their secretion), since many fungi secrete a vast number of proteins to accommodate their saprotrophic lifestyle; this would be the case of proteins implicated in the adhesion to the plant surface, host-tissue penetration and invasion, effectors, and other virulence factors.
54. Genomics or genome level studies reveals or suggest what could theoretically happen, whereas the proteomics or proteome level investigations provide insights into the actual players involved in mediating specific cellular responses.
55. Genomics is research by means of large scale characterization of genes and gene products into the elucidation of the way genes, RNA, proteins and metabolites interact in the functioning of cells, tissues, organs and the complete organism and its environment, both in an individual or in populations of species, as well as between species'.
56. Agricultural genomics may reveal what genes or combinations of genes do in plants and livestock. Genomics of plant pathogens may identify the causes of plant diseases and indicate new ways of fighting them, suggest strategies for minimizing resistance development. Genomics will prevent pollution by reducing the use of pesticides and weed-killers in agriculture, and it may stimulate the use of plants in the cleaning up of soils contaminated with heavy metals or other undesired compounds.

57. Gene silencing is generally used to describe the "switching off" of a gene by a mechanism other than genetic modification. That is, a gene which would be expressed (turned on) under normal circumstances is switched off by machinery in the cell.
58. RNA silencing is a nucleotide sequence-specific process that induces mRNA degradation or translation inhibition at the post-transcriptional level (named PTGS in plants) or epigenetic modification at the transcriptional level, depended on RNA-directed DNA methylation (a process named RdDM in plants).
59. Post-transcriptional gene silencing is the result of mRNA of a particular gene being destroyed or blocked. The destruction of the mRNA prevents translation to form an active gene product (in most cases, a protein). A common mechanism of post-transcriptional gene silencing is RNAi. Genes may be silenced by DNA methylation during meiosis, as in the filamentous fungus *Neurospora crassa*.
60. The expression of genes encoded in DNA begins by transcribing the gene into RNA, a second type of nucleic acid that is very similar to DNA, but whose monomers contain the sugar ribose rather than deoxyribose.
61. Genes that encode proteins are composed of a series of three-nucleotide sequences called codons, which serve as the *words* in the genetic *language*. The genetic code specifies the correspondence during protein translation between codons and amino acids.
62. All genes have regulatory regions in addition to regions that explicitly code for a protein or RNA product. A regulatory region shared by almost all genes is known as the promoter, which provides a position that is recognized by the transcription machinery when a gene is about to be transcribed and expressed. A gene can have more than one promoter, resulting in RNAs that differ in how far they extend in the 5′ end.
63. In general, the genetic code specifies 20 standard amino acids; however, in certain organisms the genetic code can include selenocysteine—and in certain archaea—pyrrolysine.
64. Variation caused during plant tissue culture is called *somaclonal variation.* Somaclonal variations arise as a result of chromosome structural changes, gene mutations, plasma gene mutations, gene amplification, transposable elements etc. There are three steps of somatic hybridization: (i) Isolation of protoplasts, (ii) Fusion of protoplasts of desired species, (iii) Culture of hybrid protoplast to produce whole plants.
65. Two most common methods used for resistance gene isolation are positional (or map-based) cloning, and transposition tagging. Isolation of several disease-resistance genes from different crop plants and their common structural features has led to the development of a third and popular method of isolation of disease resistance genes by PCR and primers, which is popularly known as candidate gene approach.

66. Isolation of resistance genes based on map-based cloning is dependent on their position on a genetic map. The process of gene tagging relies on the transposable elements to inactivate, and at the same time molecularly tag, gene *of interest.*

67. *In PCR bas*ed cloning, the consensus nucleotide (and amino acid) sequence in the (R) genes has led to the development of polymerase chain reaction (PCR) based approach to isolate disease resistance genes using oligonucleotide primers.

68. In serological techniques, antibodies are raised in animals against specific antigens from the pathogen. These antibodies are then purified and used in diagnostic tests. IgG is the most commonly used soluble antibody in vertebrates. It is a Y-shaped molecule with two antigen-binding sites.

69. Variations in the amino acid composition of antigen-binding site determine the specificity, while the other regions of IgG can be labelled, for example by attachment of enzymes, for the diagnostic detection system.

70. Polyclonal antibodies (Pabs) are made by injecting extracts from the pathogen into an animal (usually a rabbit), and then collecting blood from it. The blood samples are allowed to clot, and the serum collected from it can contain antibodies to the test extract. These antibodies can be used directly or after further purification.

71. Monoclonal antibodies made by fusing antibody-producing cells (lymphocytes) from the spleens of an inoculated animal (usually mice or rats) with cultured myeloma cells. This generates many hybrid cell lines (hybridomas) which each produce a different single (monoclonal) antibody. These individual cell lines are propagated and single monoclonal antibodies are harvested from the culture medium. These antibodies can be more specific and the method provides unlimited supplies of standardized reagents with homogeneous binding behavior.

72. Phage Display is a new method of producing antibodies, uses libraries of functional fragments of antibody molecules which have been amplified by PCR from a number of different animal species, including humans.

73. The main objective of an immunodiagnostic assay is to detect or quantify the binding of the diagnostic antibody with the target antigen.

74. Immunofluorescence (IF) microscopy is another method that allows *in situ* localisation of the pathogen. Plant samples are applied to microscope slides in thin tissue sections and fixed. Detection is achieved by conjugating a fluorescent dye to the specific antibody (direct IF) or to a molecule that detects the specific antibody (indirect IF). This method has been used to identify specific plant pathogen spores on microscope slides.

75. Genetically modified organisms are in which one or more traits has been introduced by incorporation of novel gene constructs with molecular biology

tools to avoid any recombination barrier and keeping intact normal genetic blueprint of the organism.

76. Recombinant DNA is a tool in understanding the structure, function, and regulation of genes and their products. For *e.g.*, Identifying genes, Isolating genes, Modifying genes, Re-expressing genes in other hosts or organisms. Recombinant technology begins with the isolation of a gene of interest. The gene is then inserted into a vector and cloned. Commonly used vectors are bacterial plasmids and viral phages. The gene of interest (foreign DNA) is integrated into the plasmid or phage, and this is referred to as recombinant DNA. Before introducing the vector containing the foreign DNA into host cells to express the protein, it must be cloned. Cloning is necessary to produce numerous copies of the DNA since the initial supply is inadequate to insert into host cells. Once the vector is isolated in large quantities, it can be introduced into the desired host cells such as mammalian, yeast, or special bacterial cells. The host cells will then synthesize the foreign protein from the recombinant DNA. When the cells are grown in vast quantities, the foreign or recombinant protein can be isolated and purified in large amounts.

77. A transgenic plant contains a gene or genes that have been artificially inserted. The inserted gene sequence is known as the transgene, it may come from an unrelated plant or from a completely different species. The purpose of inserting a combination of genes in a plant is to make it as useful and productive as possible. This process provides advantages like improving shelf life, higher yield, improved quality, pest resistance, tolerant to heat, cold and drought resistance, against a variety of biotic and abiotic stresses.

78. Genetically engineered plants are generated in a laboratory by altering the genetic-make-up, usually by adding one or more genes of a plant's genome. The nucleus of the plant-cell is the target for the new transgenic DNA.

79. Gene transfer techniques *viz.* Agrobacterium mediated transformation, biolistic gene gun or microprojectile bombardmnet approach, direct gene transfer to protoplasts. Others methods of gene transfer are Micro injection, micro targetting, electroporation, electrophoresis, sonication assisted transformation, silicon carbide mediated and in planta gene transfer.

80. The "Gene Gun" method, also known as the "Micro-Projectile Bombardment" or "Biolistic" method is most commonly used in the species like corn and rice. In this method, nucleic acid is bound to the tiny particles of Gold or Tungsten, which is subsequently shot into plant tissue or single plant cells, under high pressure using gun. The accelerated particles are penetrating both into the cell wall and membranes. The DNA separates from the coated metal and it integrates into the plant genome inside the nucleus.

81. Gene Gun method has been applied successfully for many crops, especially monocots, like wheat or maize, for which transformation using *Agrobacterium tumefaciens* has been less successful

82. Agrobacterium method involves the use of soil-dwelling bacteria, known as *Agrobacterium tumefaciens*. It has the ability to infect plant cells with a piece of its DNA. The piece of DNA, that infects a plant, is integrated into a plant chromosome, through a tumor inducing plasmid (Ti plasmid). The Ti plasmid can control the plant's cellular machinery and use it to make many copies of its own bacterial DNA. The Ti plasmid is a large circular DNA particle that replicates independently of the bacterial chromosome. Agrobacterium method works especially well for the dicotyledonous plants like potatoes, tomatoes and tobacco plants.

83. Transgenic plants are used to express proteins, like the cry toxins from *Bacillus thuringiensis*, herbicide resistant genes and antigens for vaccinations.

84. Glyphosate is an active ingredient of many broad spectrum herbicides. Glyphosate resistant transgenic tomato, potato, tobacco, cotton etc are developed by transferring aro A gene into a glyphosate EPSP synthetase from *Salmonella typhimurium* and *E. coli*. Sulphonylurea resistant tobacco plants are produced by transforming the mutant ALS (acetolactate synthetase) gene from *Arabidopsis*. QB protein of photo system II from mutant *Amaranthus* hybrids is transferred into tobacco and other crops to produce atrazine resistant transgenic plants.

85. *Bacillus thuringiensis* is a bacterium that is pathogenic for a number of insect pests. Its lethal effect is mediated by a protein toxin it produces. Through recombinant DNA methods, the toxin gene can be introduced directly into the genome of the plant, where it is expressed and provides protection against insect pests of the plant.

86. TMV resistant tobacco and tomato plants are produced by introducing viral coat proteins. Other viral resistant transgenic plants are (A) Potato virus resistant potato plants, (B) RSV resistant rice, (C) YMV resistant black gram and (D) YMV resistant green gram etc.

87. Nano-phytopathology is a cutting-edge science which uses nanotechnology for detecting, diagnosing and controlling plant disease and their pathogens at an early stage, owing to crop protection from epidemic diseases. Nanoparticles such as nanosized silica-silver have recently been applied as antimicrobial and antifungal agents. Additionally, nanomaterials can be used for mycotoxin detection and detoxication, increasing plant resistance, plant disease forecasting and nano-molecular diagnostics of plant pathogens.

88. Ethical and safety concerns have been raised around the use of genetically modified food. A major safety concern relates to the human health implications of eating genetically modified food, in particular whether toxic or allergic reactions could occur. Gene flow into related non-transgenic crops, off target effects on beneficial organisms and the impact on biodiversity are important environmental issues. Ethical concerns involve religious issues, corporate control of the food supply, intellectual property rights and the level of labeling needed on genetically modified products.

List of Pathogen Recognition Receptors

Class	Pattern Recognition Receptor (PRR)	Plant species	Predicted features of MAMPs protein	Molecule/protein recognized	Pathogen Species
1	LeEix1, LeEix2	Tomato	LZ-eLRR-TM-ECS	EIX an ethylene induced xylanase	*Trichoderma viride*
2	FLS2	Arabidopsis	eLRR-TM-kinase	flg22, a 22 amino acid peptide derived from the N-terminal fragment of the flagellin protein	Multiple bacteria species
3	EFR	*Arabidopsis*	eLRR-TM-kinase	EF-Tu – acetylated N terminus of the elongation factor Tu	Bacteria
4	GBP – 75-kDa b-glucan binding protein	Soybean and Fabaceae species	Soluble, cell wall located protein with intrinsic endo-glucanase activity	HG heptaglucoside	*Phytophthora*cell-wall derived
5	N-glycoproteins of 162 and 50 kDa	Tobacco, Arabidopsis and Acer pseudo-platanus	Plasma membrane localized	Lipid-transfer proteins–elicitins which bind sterols	Oomycetes (*Phytophthora* species and *Pythium* species)
6	CEBiP	Rice	Plasma membrane localized glycoprotein with two extracellular LysM motifs	Chitin oligomers (chitooligo-saccharide)	Fungi
7	100-kDa Pep-13 binding protein	Parsley	Plasma membrane localized	Pep13, a surface exposed 13 amino acid sequence present within a cell wall (transglutaminase)	*Phytophthora sojae* and other *Phytophthora* species
8	Not known	Tobacco	Plasma membrane localized	RNP-1 cold shock inducible RNA-binding protein	Gram-negative and Gram-positive bacteria
9	Not known	Arabidopsis and rice	Not known	LPS- lipopolysaccharides	Gram-negative bacteria
10	HrBP1	Arabidopsis	Not stated	Harpin	*Erwinia amylovora*

Online Primer Design Tools

Name of Tool	Function
Auto prime	Design primers for real time PCR, measurement of eukaryotic gene expression
Batch Primer 3	Design primers
CODEHOP	Consensus Degenerate Hybrid Oligonucleotide Primers designed from protein multiple sequence alignments
EXON primer	Design intronic primers for amplification of exons
Genefisher	Genefisher interactive PCR primer design
IDT Antisense Design	Antisense oligo design and selection tool
IDT Oligo Analyzer	Online calculation of oligonucleotide parameters such as melting temperature
IDT Primer Quest	Design and select primer and probe
MEDUSA	A tool for automatic selection and visual assessment of PCR primer pairs
mPrimer 3	Modified primer 3
Methprimer	Design primers for methylation PCR
Netprimer	Java applet for primer design
Oligonucleotide analyzer	Generates Tm, free energy, molecular weight and hairpin and dimer formation structure
Osprey	Oligonucleotide design software for sequencing and gene expression
PCR suite	Design primers for exon amplification, SNP and cDNA flanking regions
PRIDE	Less automated web version of PRIDE
Primaclade	Accepts a multiple species nucleotide alignment file as input and identifies a set of PCR primers that will bind across the alignment
Primer 3	Utility for locating oligonucleotide primers for PCR amplification of DNA sequences. A common software used for designing primers
Primer 3 Plus	Use primer 3 to pick primers for specific tasks
Primer X	Automated design of mutagenic primers for site-directed mutagenesis
Primer Blast	A web based application from NCBI
Primer Quest	Primer design at IDT
Primer Pro	PCR primer design
Primer Generator	Automated generator of primers for site directed
Primique	Automated design of specific PCR primers for each sequence in a family
Primo UniqueProbe Wiz Server	Finds multiple primer pairs, each uniquely amplify one gene in a family.The CBS probe Wiz WWW server predicts optimal PCR primer pairs for generation of probes for cDNA arrays
PUNS	Primer Uni Gene Selective Testing compares primer sequences against both the genome and transcriptome to assess the potential for multiple amplicons
Quant Prime	Design high throughput primer pairs and test specificity of realtime qPCR on any organism
RNAi design	Design primer for RNAi at IDT
Site Find	Design primers for site directed mutagenesis that include a novel restriction for use as a marker for successful mutation

Name of Tool	Function
SNP cutter	SNP PCR RFLP Assay Design. Design primer for restriction analysis of SNPs
SPADS	Specific primers and amplicon design software for amplification of individual members of gene family
UCSC In Silico PCR	In silico PCR searches a genome sequence database with a given pair of PCR primers
Web primer	Design primers and sets for amplifying yeast ORFs

List of Molecular Markers

Acronym	Full Form	Year
RFLP	Restriction Fragment Length Polymorphism	1974
VNTR	Variable Number Tendem Repeat	1985
ASO	Allele Specific Oligonucleotide	1986
AS-PCR	Allele Specific Polymerase Chain Reaction	1988
OP	Oligonucleotide Polymorphism	1988
A PCR	Asymmetric PCR	1988
IPCR	Inverse PCR	1988
M-PCR	Multiplex PCR	1988
SSCP	Single Stranded Conformational Polymorphism	1989
STS	Sequence Tagged Site	1989
RAPD	Random Amplified Polymorphic DNA	1990
AP-PCR	Arbitrarily Primed– Polymerase Chain Reaction	1990
STMS	Sequence Tagged Microsatellite Sites	1990
RLGS	Restriction Landmark Genome Scanning	1991
TD-PCR	Touch down PCR	1991
CAPS	Cleaved Amplified Polymorphic Sequence	1992
DOP-PCR	Degenerate Oligonucleotide Primer –PCR	1992
SSR	Simple Sequence Repeat	1992
MAAP	Multiple Arbitrary Amplicon Profiling	1992
IS-PCR	***In-situ*** or Slide PCR	1992
SCAR	Sequence Characterized Amplified Region	1993
Q PCR	Quantitative or Real time PCR	1993
ISSR	Inter Simple Sequence Repeat	1994
SAMPL	Selective Amplification of Microsatellite Polymorphic Loci	1994
SNP	Single Nucleotide Polymorphism	1994
SP PCR	Solid Phase PCR	1994
RT PCR	Reverse Transcriptase PCR	1995
AFLP (SRFA)	Amplified Fragment Length Polymorphism (Selective Restriction Fragment Amplification)	1995

Acronym	Full Form	Year
ASAP	Allele Specific Associated Primers	1995
BIO-PCR	Biological Polymerase Chain Reaction	1995
CFLP	Cleavase Fragment Length Polymorphism	1996
ISTR	Inverse Sequence Tagged Repeats	1996
N-PCR	Nested PCR	1996
DAMD-PCR	Directed Amplification of Minisatellite DNA PCR	1997
S-SAP	Sequence Specific Amplified Polymorphism	1997
MSAP	Methylation Sensitive Amplification Polymorphism	1997
IC-PCR	Immuno-capture PCR	1998
RBIP	Retrotransposon Based Insertional Polymorphism	1998
IRAP	Inter Retrotransposons Amplified Polymorphism	1999
REMAP	Retrotransposon Microsatellite Amplified Polymorphism	1999
MITE	Miniature Inverted Repeat Transposable Element	2000
TE-AFLP	Three Endonuclease AFLP	2000
LAMP	Loop-mediated isothermal amplification	2000
IMP	Inter MITE Polymorphism	2001
SRAP	Sequence Related Amplified Polymorphism	2001
Co-PCR	Co-operational Polymerase Chain Reaction	2002
TRAP	Target Region Amplification Polymorphism	2003
—	Hot start PCR	2003
SCAR	Sequence Characterized Amplified Region	2003
ITS	Internal Transcribed Spaces	2004
TBP	Tubulin Based Polymorphism	2004
SQ PCR	Semi-quantitative PCR	2004
LATE PCR	Linear After The Exponential PCR	2004
—	Assembly PCR	2005
C PCR	Colony PCR	2008

Multiple Choice Questions (Choose the correct answer)

1. **In RAPD, the resulting amplified DNA markers are:**
   (A) Random polymorphic segments with band sizes from 100 to 3000 bp
   (B) Non-arbitrary segments
   (C) None of the above
   (D) All of the above

2. **AFLP is a combination of:**
   (A) Hybridization and amplification based strategies
   (B) Hybridization only

(C) Amplification only

(D) None of the above

3. **To generate insect resistant plants, gene for the insect Helicoverpa resistance has been transferred from soil-borne bacterium** *Bacillus thuringiensis* **in a transgenic of:**

(A) Cotton (B) Corn

(C) None of the above (D) Both (A) and (B)

4. **The role of ROS in plant biology is**

(A) Toxic byproducts of aerobic metabolism

(B) Key regulators of metabolic and defense pathways

(C) None of the above

(D) Both (A) and (B)

5. **In response to stress signals, evidence suggests that infected cells produce large quantities of extra-cellular superoxide and hydrogen peroxide which may**

(A) Damage the pathogen

(B) Strengthen the cell walls

(C) Trigger/cause host cell death through Oxidative Burst

(D) All of the above

6. **Which of the following is not an ethical concerns over genetically modified (GM) foods:**

(A) Their potential to trigger allergies or disease in humans

(B) Eco-friendly

(C) Challenge of cross-pollination, which could create overzealous weeds

(D) Risk of entry in the food chain

7. **What is the acronym "BLAST" stands for?**

(A) Basic Local Alignment Search Tool

(B) Better local assessment search tool

(C) Basic local alignment search technique

(D) Better localized assessment search tool

8. **PCR is a :**

(A) DNA degradation technique

(B) DNA sequencing technique

(C) DNA amplification technique

(D) All of the above

9. **Asymmetric PCR is used to generate:**
   (A) Single stranded copies for DNA sequencing
   (B) Double stranded copies for DNA sequencing
   (C) All of the above
   (D) None of the above

10. **CT value (cutoff threshold) in the real time PCR experiments mean:**
   (A) The lower the value, the higher the transcript amount
   (B) The higher the value, the higher the transcript amount.
   (C) No transcript, if the value is lower than 30
   (D) No transcript, if the value is larger than 30

11. **In tissue culture disease resistance can be obtained by:**
   (A) Soma clonal variation (B) Meristem culture
   (C) Somatic hybridization (D) Anther culture

12. **ELISA plate is made up of:**
   (A) Plastic (B) Nitrocellulose
   (C) Polystyrene (D) Polypropylene

13. **RNA induced silencing complex was produced by the combination of:**
   (A) Si (ssRNA) + protein (B) Si (dsRNA) + protein
   (C) Si (dsRNA) + fatty acid (D) Si (ssRNA) + fatty acid

14. **SSRs are:**
   (A) Codominant and multiallelic markers
   (B) Diallelic and dominant markers
   (C) Dominant markers
   (D) None of the above

15. **Which of the following viruses have been used as a vector for gene transfer in genetic engineering:**
   (A) CMV (B) TMV
   (C) PVX (D) CaMV

16. **Chemical gene transfer methods are :**
   (A) Poly-ethylene glycol (PEG)-mediated
   (B) Diethyl amino ethyl (DEAE) dextran-mediated
   (C) Calcium phosphate precipitation
   (D) All of the above

17. A region on a DNA or RNA which is recognized by RNA polymerase in order to initiate transcription known as:

(A) Promoter
(B) Proliferation
(C) Activator
(D) Precursor

18. Reverse transcriptase PCR uses:

(A) mRNA as a template to generate cDNA
(B) RNA as a template to generate DNA
(C) DNA as a template to generate ssDNA
(D) All of these

19. Tn-5 generally used in molecular Plant Pathology as:

(A) Mutagenic agent
(B) Curing agent
(C) Nucleotide sequencer
(D) Replicating enzyme

20. Avirulence (avr) genes, first identified by:

(A) H.H. Flor
(B) Robinson
(C) Vonder plank
(D) None of these

Fill in the Blanks with Correct Answer

1. Liquid nitrogen boils at________ As such, it presents a severe frostbite hazard.

   77 Kelvin (-194 Celsius)

2. The lyophilizer, when in good working order, should have a pressure of________ (normal atmospheric pressure is 760 torr).

   5-50 millitorr

3. ________is a chemical used for the visualization of nucleic acids. Its appearance is dark red to purple.

   Ethidium Bromide (EtBr)

4. Real time PCR or quantitative or kinetic PCR is used to quantify gene expression to confirm differential expression of genes and also to measure the abundance of particular________ in different samples.

   DNA or RNA sequences

5. Real time PCR uses commercially available________ to amplify specific nucleic acid sequences.

   fluorescence detecting thermocycler

6. In RAPD, multiple________ primers are added to each individual sample of DNA (which is subjected to PCR).

   10 base pair (bp) oligonucleotide

7. In tobacco, insect resistance has been achieved by transferring________ from cow pea.

   Trypsin inhibitor gene

8. After electrophoresis, silver staining is used for________

   Polyacrylamide gels

9. ________are placed in the electrophoresis gels to made wells.

   Combs

10. ________is a method for probing for the presence of a specific DNA sequence within a DNA sample.

    Southern blot

11. The________is used to study the expression patterns of a specific type of RNA molecule as relative comparison among a set of different samples of RNA.

    Northern blot

12. R-avr interaction triggers a series of signalling responses which results in the Biosynthesis of.which acts as a central signalling intermediate in plant defence.

    Salicylic acid

13. SAR takes________to start, can last for months.

    24-48 h

14. ________, which is very rapid and takes place within 24 hours. It is not always needed for resistance.

    Local response or Hypersensitive Response

15. Jasmonates are ubiquitously occurring lipid-derived compounds with signal functions in plant responses to________, as well as in plant growth and development.

    Abiotic and biotic stresses

16. Jasmonic acid and its various metabolites are members of the oxylipin family. These molecules may alter________positively or negatively in a regulatory network with synergistic and antagonistic effects in relation to other plant hormones such as salicylate, auxin, ethylene and abscisic acid.

    Gene expression

17. ________control and regulate biological processes such as programmed cell death, hormonal signaling, stress responses, and development.

    ROS or ROI (reactive oxygen species or reactive oxygen intermediates)

18. The steady state level of ROS in the different cellular compartments is determined by interplay between multiple________

    ROS-producing pathways and ROS-scavenging mechanisms

19. One dramatic hallmark of the resistance response is the induction of a________ response at the site of the infection.

    Localized cell death

20. AOS are rapidly produced by plant cells during the HR in a phenomenon termed as the________

    Oxidative burst

21. The HR is likely to be important for limiting a pathogen's nutrient supply, since the dying tissue rapidly becomes________

    Dehydrated

22. 1st GM food in the world was________ consumed in USA. Developed by companies namely Bayer crop sciences, Monsanto, Dupont/Pioneer, Syngenta.

    Calgene's delayed ripening tomato, Flavr savr,1994

23. Herbicide tolerant soybean came in the market in________

    2006

24. While phytoalexins are mainly characteristic of the local response, PR proteins occur both________

    Locally and systemically

25. .involves an enzyme-mediated colour change reaction to detect antibody binding.

    Enzyme linked immunosorbent assay (ELISA)

26. Used for detecting pathogens in plant tissue.

    Dot immuno- (binding) assay (DIA/DIBA)

27. The transmission of genetic material from parent to offspring is referred to as________

    Vertical gene transfer

28. ________is referred to a process in which genetic material is gained from a non-parental organism.

    Horizontal gene transfer (or lateral gene transfer)

29. ________provides the genetic makeup of a genome, that is, the entire genetic complement of a living organism.

    DNA

30. ________provides the expression vehicle of the entire complement of genes, that is, the transcriptome.

    RNA

31. ________is the use of nanobiotechnology to diagnose plant diseases and this can be termed as nanodiagnostics?

    Nanomolecular diagnostic

32. ________has been employed to obtain primers for X. *fragariae* and X. *hortorum* pv. *pelargonii.*

    REP-PCR

33. ________is a powerful nonsequencing approach to find genetic differences between bacterial strains.

    Genomic subtraction

34. PCR can be used for bacterial detection because of it's relatively high sensitivity. and good specificity.

    (1 – 10^3 cells/ml of plant extract)

True or False

1. In Real time PCR, fluorescent DNA intercalating dyes such as SYBR GREEN or DNA sequence specific TaqMan probes and Molecular Beacons are used instead of Taq DNA polymerase enzymes. **(True)**
2. PCR amplification product is visualized by gel electrophoresis. **(True)**
3. Semi quantitative PCR uses the properties of a thermo stable polymerase (Taq polymerase) to amplify small segments of DNA through several rounds of synthesis. **(True)**
4. RAPD is a modified PCR technique which involves the amplification of whole genomic DNA extracted from plant pathogens. **(True)**
5. AFLP is a technique in which a selected restricted fragment of total genomic DNA is digested and detected by PCR amplification. **(True)**
6. Mantel matrix correspondence test is used to find out the correlation between the two matrices obtained with different types of markers (AFLP, RAPD ISSR and SSR). **(True)**
7. Transgenic virus resistant genotypes have been developed in crops like tobacco, tomato, potato and cucumbers. **(True)**
8. Gene barnase from *Bacillus amyloliquefaciens* and genes rol B, rol C from *A. rhizogenes* are used to induce male sterility in plants. **(True)**
9. In indirect ELISA method, two sets of complete antibodies are used. **(True)**

10. While preparing PCR master mix DNA template is added along with sterile water, Taq polymerase, DNTPs, Primers and Polymerase buffer. **(False)**
11. The Avirulence genes (*avr* genes) are very diverse and perform different functions in various organisms. In bacteria, they code for cytoplasmic enzymes involved in the synthesis of secreted elicitor and in fungi, some code for secreted proteins, some for fungal toxins. **(True)**
12. Resistance gene code for a receptor for a specific elicitor associated with the interacting *avr* gene in a particular host pathogen interaction. **(True)**
13. Unlike the terms "avirulence", "elicitor", "toxin", and "virulence", the term effector is neutral and does not imply a negative or positive impact on the outcome of the disease interaction. **(True)**
14. Theoritically, FISH can detect single cells but in practice, the detection level is near 103 cells/ml of plant extract. **(True)**
15. Macroarrays may be more useful in plant disease diagnosis than microarrays because of their greater sensitivity and freedom from the need for highly specialized equipment. **(True)**
16. In Plant Pathology, the most frequently used molecular techniques have been, first, molecular hybridisation and, afterwards, the polymerase chain reaction (PCR). **(True)**

Match the Following

1. **Match the following with appropriate group.**

I. RPW8- Arabidopsis, powdery mildew species	(A) Membrane protein with CC domain
II. Cf-2, Cf-4, Cf-5, Cf-9- Tomato, *Cladosporium fulvum*	(B) Extracellular LRR
III. Xa-21- Rice, *Xanthomonas oryzae*	(C) Extracellular LRR/kinase domain
IV. Hm-1 – Maize, *Helmintosporium maydis*	(D) Detoxifying enzyme HC-toxin reductase

Answer

I	*II*	*III*	*IV*
A	*B*	*C*	*D*

2. **Match the following.**

I. *L6*- Flax, *Melampsora lini*	(A) TIR/NBS/LRR
II. Pto- Tomato, *Pseudomonas syringae* (*avrPto*)	(B) Intracellular ser/thr protein kinase

III. N – Tobacco, *Tobacco mosaic virus* — (C) TIR/NBS/LRR

IV. Pi-ta- Rice, *Magnaporthe grisea*

Answer

I	*II*	*III*	*IV*
A	*B*	*C*	*D*

3. **Match the following.**

I. G-protein — (A) Guanine nucleotide-binding *proteins* (involved in transmitting signals from a variety of stimuli outside a cell to its interior)

II. NADPH oxidase — (B) Nicotinamide adenine dinucleotide phosphate *oxidase* (catalyzes the production of superoxide– a reactive free radical)

III. Protein Kinase — (C) Pto (a tomato serine-threonine)

IV. SCAR — (D) Sequence Characterized Amplified Region

Answer

I	*II*	*III*	*IV*
A	*B*	*C*	*D*

4. **Match the following.**

I. DAMPs — (A) Damage-associated molecular patterns (Endogenous elicitors)

II. NBS-LRR — (B) Nucleotide binding site leucine rich repeats proteins

III. AFLP — (C) Amplified Fragment Length Polymorphism

IV. SSR — (D) Simple sequence-repeat

Answer

I	*II*	*III*	*IV*
A	*B*	*C*	*D*

5. **Match the function/use of following chemicals.**

I. Sodium Dodecyl Sulphate — (A) Anionic detergent, disrupts the lipid layers (helps to dissolve membranes and binds positive charges of chromosomal proteins (*histones*) to release the DNA into the solution)

II. EDTA	(B) Chelates the magnesium ions required for DNase activity
III. CTAB	(C) Cationic detergent, useful for isolation of DNA from tissues containing high amounts of polysaccharides
IV. β mercaptoethanol	(D) A reducing agent, helps in denaturing proteins by breaking the disulfide bonds

Answer

I	*II*	*III*	*IV*
A	*B*	*C*	*D*

6. Match the following:

I. SYBR Green probe	(A) binds to double stranded DNA and emits light on excitation
II. TaqMan® probes (also called Double-Dye Oligonucleotides, Double-Dye Probes, or Dual Labelled probes)	(B) consist of two fluorescent tags, reporter dye (6-carboxyfluorescein (6-*FAM*) or FAM), and quencher dye (Carboxytetramethylrhodamine or TAMRA)
III. Molecular Beacons	(C) Probes that contain a stem-loop structure, with a fluorophore and a quencher at their 5′ and 3′ ends, respectively.
IV. The rDNA-ITS or β-tubulin sequences (of fungal isolates)	(D) Can be used to design primers for RT-PCR.

Answer

I	*II*	*III*	*IV*
A	*B*	*C*	*D*

7. Match the following:

I. Nested PCR	(A) Instead of one pair of primers, two pairs of primers are used to amplify desired fragment.
II. Semi quantitative RT-PCR	(B) Is used when the copy number is low or small amounts of sample are available for analysis.
III. InterSequence-Specific (ISSR)	(C) PCR- Primers from a commonly repeated sequence of DNA (called microsatellites) are used to generate a unique fingerprint of amplified product lengths.

IV. Multiplex-PCR — (D) Multiple primer pairs to various target sequences are used to enable simultaneous analysis of more than one sequence of interest.

Answer

I	*II*	*III*	*IV*
A	*B*	*C*	*D*

8. **Match the following elicitors with their producers and function.**

I. Glucan (cell wall components)	(A) By oomycetes, induce phytoalexins (in rice and soybean)
II. Chitin oligomers (fungal cell wall)	(B) By higher fungi, induce phytoalexins in rice and lignification in wheat.
III. Harpins (Involved in type III secretion system)	(C) By gram-ve bacteria; cause callose formation and defense gene in tobacco.
IV. Flagellin	(D) By gram-ve bacteria; cause HR, ROS and defense response

Answer

I	*II*	*III*	*IV*
A	*B*	*C*	*D*

9. **Match the following elicitors with their producers and function.**

I. Glycopeptide fragments) invertase (enzyme in yeast metabolism	(A) By yeast, activate defense genes and ethylene production in tomato
II. Syringolids (acyl glycosides, Signal compound for the bacterium)	(B) By *Pseudomonas syringae* pv. HR in soybean (Rpg4 gene)
III. Nod factors (lipochitooligosaccharides, Signal in symbiosis communication)	(C) By *Rhizobium* and other Rhizobia, Nod formation in legumes
IV. FACs (fatty acid amino acid conjugates, Emulsification of lipids during digestion)	(D) By Various *Lepidoptera*, Monoterpenes in tobacco-"indirect defence"

Answer

I	*II*	*III*	*IV*
A	*B*	*C*	*D*

10. Match the following elicitors with their producers and function.

I. Mycotoxins (*e.g.* fumonisin B1, Toxin in necrotrophic interaction; disturb metabolism)	(A) by *Fusarium moniliforme,* PCD and defence genes in tomato, *Arabidopsis*
II. Victorin (toxin)	(B) By *Helminthosporium victoriae* (rust), Programmed cell death (PCD) in oat
III. Glycoproteins	(C) By *Phytophtora sojae;* Phytoalexin defense genes in parsley
IV. Viral coat protein	(D) By TMV; HR in tomato, tobacco

Answer

I	*II*	*III*	*IV*
A	*B*	*C*	*D*

11. Match the following

I. To improve protein quality of alfalfa	(A) transfer of ovalbumin gene from chicken
II. Improved protein quality of potato	(B) transfer of serum albumin gene from human
III. Improved protein quality of tobacco	(C) transfer of glutenin gene of wheat.
IV. Vitamin A rich (increased production of beta-carotene) golden or yellow rice	(D) transfer of 3 genes, two from daffodils and one from a microorganism.

Answer

I	*II*	*III*	*IV*
A	*B*	*C*	*D*

12. Match the following

I. Cold resistance in tomato	(A) Transfer of anti-freezing gene from fish (winter flounder)
II. Cold resistance in tobacco	(B) Transfer of anti-freezing gene from fish and *Arabidopsis thaliana*
III. Male sterility in rapeseed	(C) By transferring a gene barnase from *Bacillus amyloliquefaciens.*
IV. An effective fertility restoration gene in plants	(D) Gene barstar from *B. amyloliquefaciens* (is an intracellular inhibitor of barnase).

Answer

I	*II*	*III*	*IV*
A	*B*	*C*	*D*

13. Match the following components of signal transduction pathways

I.	Stimulus	(A)	Are Hormones, physical environment, pathogens
II.	Receptor	(B)	Present on the plasmamembrane, or are situated internally
III.	Secondary messengers	(C)	Are Ca^{2+}, G-proteins, Inositol Phosphate
IV.	Effector molecules	(D)	Are Protein kinases or phosphatases and Transcription factors

Answer

I	*II*	*III*	*IV*
A	*B*	*C*	*D*

14. Match the following examples where plants with heritable resistance have been obtained from unselected tissue culture:

	Plant disease/Host		*Tissue culture*
I.	*Fusarium oxysporum* f. sp. *medicaginis*/alfalfa	(A)	Callus
II.	*F. oxysporum* f. sp. *cubense*/banana	(B)	Shoot tip
III.	*Phoma lingam/Bassica napus*	(C)	Callus, stem, embryo culture
IV.	*F. oxysporum* f. sp. *apii*/ Race-2/celery	(D)	Apical meristem

Answer

I	*II*	*III*	*IV*
A	*B*	*C*	*D*

15. Match the following examples where plants with heritable resistance have been obtained from unselected tissue culture:

	Plant disease/Host		*Tissue culture*
I.	*Alternaria solani*/potato	(A)	Mesophyll protoplast
II.	*Helminthosporium maydis* Race-T and HMT Toxin/ Maize cyptoplasm (CMS-T)	(B)	Callus from immature embryo
III.	*F. oxysporum* f. sp. *fragariae*/ strawberry	(C)	Leaf callus
IV.	Fiji disease of sugarcane	(D)	Callus

Answer

I	*II*	*III*	*IV*
A	*B*	*C*	*D*

16. Match the following important transgenic developed for the control of plant disease:

Disease/Pathogen	*Gene involved*
I. Tobacco mosaic virus	(A) Coat protein expression
II. Tobacco brown spot	(B) Chitinase gene
III. Tomato aspermy virus	(C) Satellite RNA expression
IV. Tobacco wildfire	(D) Toxin resistant gene (Acetyl transferase gene)

Answer

I	*II*	*III*	*IV*
A	*B*	*C*	*D*

17. Match the following important transgenic developed for the control of plant disease:

Disease/Pathogen	*Gene involved*
I. Tobacco, Rhizoctonia solani	(A) Chitinase gene
II. Tomato spotted wilt virus	(B) Nucleic capsid gene
III. Rice, *Helminthosporium oryzae*	(C) Expression antisense RNAs
IV. Potato leaf roll virus	(D) HS toxin gene

Answer

I	*II*	*III*	*IV*
A	*B*	*D*	*C*

18. Match the following important transgenic developed for the control of plant disease:

Disease/Pathogen	*Gene involved*
I. Bean, *Rhizoctonia solani*	(A) Chitinase gene
II. Potato, late blight (*Phytophthora infestans*)	(B) Osmotin gene
III. Tobacco brown spot (*Alternaria longipes*)	(C) Chitinase gene
IV. Cucumber mosaic virus	(D) Coat protein expression

Answer

I	*II*	*III*	*IV*
A	*B*	*C*	*D*

19. Match the following:

I.	WIPO	(A)	World Intellectual Property Organisation
II.	TRIPS	(B)	Trade related Intellectual Property Rights
III.	PCT	(C)	Patent Co-operation Treaty
IV.	GURT	(D)	Genetic use restriction technology.

Answer

I	*II*	*III*	*IV*
A	*B*	*C*	*D*

Answers (Multiple Choice Questions)

(1)	(A)	(6)	(B)	(11)	(A)	(16)	(D)
(2)	(A)	(7)	(A)	(12)	(C)	(17)	(A)
(3)	(D)	(8)	(C)	(13)	(A)	(18)	(A)
(4)	(D)	(9)	(A)	(14)	(A)	(19)	(A)
(5)	(D)	(10)	(A)	(15)	(D)	(20)	(A)

Chapter 6
Seed Pathology

Queries with Timeline of Seed Pathology

1. The relationship of seed to disease in the crop produced has been reflected by the observations of Remnant (1637). Hellwig (1699) first time reported *Claviceps purpurea* of rye carried by seed. However, the work on seed health was initiated when Tillet (1755) with his field trials proved conclusively that stinking or hill bunt of wheat was contagious and carried through seed and said this disease is caused by a 'poisonous substances'.

2. The term seed pathology was first used by Paul Neergaard and Mary Noble in the 1940s. Paul Neergaard (1907-1987) is considered the father of seed pathology. He established the seed borne nature of several hundred host-pathogen combinations. He demonstrated the etiology of seed borne of *Colletotrichum, Bipolaris, Drechslera, Fusarium, Myrothecium* and other fungi.

3. In 1807, Prevost proved that stinking bunt was caused by aparasitic fungus '*Tilletia caries*'. Frank (1983) demonstrated the internally seed borne nature of a fungus *Colletotrichum lindemuthianum* causing anthracnose of bean (*Phaseolus vulgaris*).

4. In 1892, Beach, in New York proved the seed borne nature of bacterial plant pathogen *Xanthomonas campestris* pv. *phaseoli* in common bean seeds. Smith (1909) the first bacterial pathogen recorded as seed borne was *Xanthomonas stewartii* on corn. Rolfs (1915) reported the internal transmission of *Xanthomonas campestris* pv. *malvacearum* in cotton seeds. Clayton in 1929 reported *Xanthomonas campestris* pv. *campestris* on cauliflower.

5. Seed transmission of *Xanthomonas phaseoli* causing common bean blight was reported by Orton (1931). External transmission of bacterium was shown first by Stewart in 1897 studying *Erwinia stewartii* on maize. Virus seed transmission was studied by Mayer in 1886 that tobacco seeds from tobacco mosaic infected plants yielded diseased seedlings. J.A. McClintock reported in 1916 that cucumber mosaic virus (CMV) was seed transmitted.

6. Stewart and Reddick (1917) were first to demonstrate the internal transmission of a virus causing mosaic (bean common mosaic on bean) of *Phaseolus vulgaris*. Doolittle and Gilbert (1919) demonstrated seed transmission of CMV in wild cucumber (*Echinocystis lobata*). Chen published a monograph on internal fungal parasites of agricultural seeds in 1920. In 1931 Alcock published a list of seed borne mycoflora of forage crops, ornamental plants and vegetables in Scotland.

7. In 1869, the first official seed testing station was established by Nobbe in Tharandt (Germany) with its major function to test seeds for germination and purity. The step toward international cooperation in seed testing was taken at the First International Seed Testing Congress (ISTC) in Hamberg in 1906. At the third ISTC in Copenhagen in 1921, the European Seed Testing Association was founded.

8. The first seed testing station was established at Wageningen in Netherland and the first seed health testing laboratory was setup in 1918.The 'Manual for the Determination of seed borne Diseases' was published by L.C. Doyer in 1938. She was the first official seed pathologist and was the first chairperson of the ISTA Plant Disease Committee for Seed Health Committee. Her contribution published in the 1930 ISTA Proceedings, was a systematic survey of diseases and pests on weeds involving bacteria, fungi, insects, viruses and other injurious organisms.

9. The blotter method is used as incubation method tostudy seed mycoflora, it was developed by Doyer in 1938 which was later included in the ISTA rules of 1966. 2, 4-D method was also used for seed health, it was first used by Hagborg *et al.* (1950) in testing bean seeds for the presence of *Colletotrichum lindemuthianum*. Deep freeze method is used to detect slow growing pathogens; this method was developed by Limonard (1976).

10. In Northern Ireland, Muskett and Malone (1941) first time used 'Agar Plate method' for seed health testing of flax seeds. Mangan (1971) modified agar plate method to isolate *Phoma betae* and named the method as 'Hold fast method'. Rolled towel paper method is used for the detection of *Xanthomonas campestris* pv. *oryzae* on paddy seed, it was developed by Singh and Rao (1977). Hiltner brick stone method is used for field performance test giving information on seedling symptoms and was developed by Hiltner in 1917. Standard soil method was developed by Karlberg (1974) for detection of mycoflora.

11. Test tube agar method developed by Khare, Mathur and Neergaard in 1977 for the detection of *Septoria nodorum* in wheat grains. The embryo count method developed by Hewett in 1970 for detection loose smut of wheat and barley.

Phage method was used first to detected and identified seed borne *Xanthomonas campestris* pv. *phaseoli* by Katznelson and Sutton in 1951. Guthrie *et al.* (1965) developed serological technique for detection of halo blight in bean seed samples.

12. An international rule for seed testing was published in 1966. Two ISTA hand books on testing methods have also been published by De Tempe (1961) and de Tempe and Binnerts (1979). In 1957, the plant disease committee (PDC) of ISTA established a comparative seed health testing programe to standardize techniques for detection of seed borne pathogens. The first PDC worshop was held at the seed testing station, Cambridge, UK in 1958. In India, first training was organized on seed pathology in 1971 in New Delhi.
13. The first plant quarantine regulation was perhaps promulgated in France in 1660 to eradicate the barbery plants which have been known since early times to harbor black rust disease of wheat. The first Federal Plant Quarantine Act was passed in USA in 1912. In India, the plant protection services started with the enforcement of Destructive Insects and Pests Acts (DIPA), 1914. Plant Quarantine Act (PQA) was enforced in 1912. Directorate of Plant Protection Quarantine and Storage (DPPQS) was established by Government of India in 1946 at New Delhi.
14. During 2002, Khan *et al.* have reported the transmission of phytoplasma in both seed and seedling progeny of alfalfa plants affected by witches' broom disease, indicating the seed transmission in certain plant host–phytoplasma pathosystem.
15. Necas *et al.* (2008) have reported the preliminary information of possibility of European stone fruit yellows phytoplasma (ESFY) transmission through apricot seeds.
16. Common bunt, stinking smut or hill bunt of wheat is of two types one is rough spored bunt (low smut) caused by *Tilletia caries* and the other is smooth bunt (high bunt) is caused by *Tilletia foetida*. Primary inoculum consists of the spores carried on the seed or present in the soil however, under Indian conditions the disease is mainly seed borne. A temperature ranging from 5-15°C is highly favourable for infection.
17. Karnal bunt of wheat (*Neovosia indica* Mundukur, 1940) was first reported in India from karnal (Haryana) by Mitra in 1931 under the name *T. indica*. Karnal bunt is a soil borne disease but later it was shown to be air borne. The air borne sporidia germinates on the glumes.
18. False smut of paddy (*Claviceps oryzae sativae*, conidial stage *Ustilaginoidea virens*) was first reported in India by Cooke in 1818 from Tinnevally in south India.
19. Bunt of rice (Syn. black smut, kernel smut, rice smut) is caused by *Neovossia horrida* (Tak) Padwick and Khan. The disease is soil borne and externally seed borne.

20. Grain smut of jowar (syn. covered smut, kernel smut short smut) is caused by *Sphacelotheca sorghi*. It is an externally seed borne. Optimum temperature for spore germination is only 20-30°C.

21. Loose smut of sorghum (*Sphacelotheca cruenta*) is mainly externally seed borne, optimum temperature for spore germination is only 18-32°C. The spores germinate to form a 4- celled promycelium with laterally borne sporidia.

22. Head smut of sorghum and maize is caused by *Sporisorium reilianum*. The pathogen may be externally seed borne, the major source of infection is soil borne inoculum.

23. Leaf bight of wheat is caused by *Alternaria triticina* and it was described by Prasada and Prabhu in 1962 for the first time.The disease develops most at 25° C and more than 80 per cent relative humidity. The pathogen is both externally as well as internally seed borne.

24. Formaldehyde and nitrogen trichloride is used to disinfect lemon storage boxes and SO_2 is recommended for figs packing boxes.

25. Some growth regulator like IAA and MH is used for the control of postharvest pathogens such as *Aspergillus* and *Rhizopus* (rots of papaya fruits).

26. Mustard, castor, groundnut and paraffin oils are effective against *Rhizopus* rot of mango and *Aspergillus* rot of papaya. Leaf extract of tuli reduces spore germination, growth, pectolytic and cellulolytic enzymes of various rot pathogens. Salt solutions (sodium lignin sulphenate), and hot water dip at 50°C for 5 minutes controlled various postharvest rots caused by *Trichothecium reseum*,*Glomerella cingulata*, *Penicillium expansum*, *Rhizopus stolonifer* and *Monilinia laxa*.

27. Udabatta disease of paddy (*Ephelis oryzae* PS: *Balansia oryzae*), loose smut of wheat (*Ustilago segatum* var. *tritici*), stripe diseare of barley (*Drechslera graminea* PS: *Pyrenophora graminea*), loose smut of barley (*Ustilago nuda*), whip smut of sugarcane (*Ustilago scitaminea*) and sugarcane red rot (*Colletotrichum falcatum* PS: *Glomerella tucumanensis*) are (internally seed borne) seed born diseases.

28. *Fusarium moniliforme* (foot rot of paddy), *Tilletia tritici* and *T. laevis* (common bunt of wheat), *Drechslera oryzae* (brown spot of paddy), *Ustilago hordei* (covered smut of barley), *Ustilago kolleri* (covered smut of oats), *Sphacelotheca sorghi* (grain smut of jowar) and *Rhizoctonia solani* and *R. bataticola* (black scurf of potato) are externally seed borne pathogens.

29. *Cochliobolus sativus* (foot rot of wheat), *Alternaria triticina* (leaf blight of wheat), *Peronospora pisi* (downy mildew of pea) and *Phoma rabie* (blight of gram) are both externally as well as internally seed borne pathogens.

30. *Phytophthora infestans* (late blight of potato), *Pyricularia grisea* (blast of rice), *Trichoconis padwickii* (stack burn of rice), *Urocystis agropyri* (flag smut of wheat), *Helminthosporium sativum* (foot rot and seedling blight of varley), *Colletotrichum capsici* (die back and fruit rot of chillies) and *Protomyces macrosporus* (stem gall of coriander) are both soil as well as externally seed borne pathogens.

31. *Cercospora nicotianae* (frog eye leaf spot of tobacco) and *Cephalosporium sacchari* (wilt of sugarcane) (seed sett), *Ascochyta rabiei* (ascochyta blight in chickpea), *Alternaria brassicae* (grey leaf spot of crucifers), *Alternaria brassicicola* (black spot of crucifers), *Botryodiplodia theobromae* (black kernel rot of maize), *Fusarium moniliforme* (cob rot of maize), *Phomopsis sojae* (pod and stem blight of soybean), *Cercospora kikuchii* (purple seed stain of soybean), *Phomopsis vexans* (foot rot of brinjal), *Thielaviopsis basicola* (black root rot of carrot), *Alternaria radicina* (black rot of carrot) are seed borne pathogens.
32. Fungi may establish themselves as dormant mycelium in the various layers which envelope the seed, such as seed coat or testa, or the pericarp (as in *Graminiae* and *Compositae*). Some genera *i.e. Cercospora, Colletotrichum, Fusarium, Phoma, Septoria* and many others are frequently carried in this manner. Bacteria are often harboured in the seed coat but only rarely within the embryo. *Xanthomonas malvacearum*, causing bacterial blight of cotton, infects some parts of the embryo as well as the seed coat, *Corynebacterium insidiosum*, the cause of bacterial wilt of lucern, invades the endosperm and *Xanthomonas campestris*, the cause of black rot of crucufers, the seed coat.
33. *Xanthomonas vesicatoria* on tomato and tobacco mosaic virus is an example of a bacterium carried on the surface of seeds. The latter may be passed to the seedling from an infected seed coat remaining on the cotyledons after germination. Lettuce mosaic virus in lettuce, alfalfa mosaic virus in alfalfa and yellow dwarf virus in onion are important examples of pollen and seed transmission.
34. Polymerase chain reaction (PCR), double antibody sandwich-enzyme linked immunosorbent assay (DAS-ELISA), isothermal amplification methods, fingerprinting, DNA hybridization technology and sequencing (Massive sequencing techniques, DNA barcoding) are molecular tools for detection of pathogenic fungi.
35. Certain seed-transmitted plant viruses through pollen are prevalent in bromovirus, alfamovirus, cucumovirus and ilarvirus groups. Pollen and seed transmission are closely related factors in virus epidemiology.

Multiple Choice Questions (Choose the correct answer)

1. **Loose smut of wheat is:**

   (A) Externally seed borne (B) Internally seed borne

   (C) Both of them (D) Soil borne

2. **Two most important diseases of potato whose spread in India has been successfully restricted through plant quarantine include:**

   (A) Wart and golden nematode (B) Ring rot and mosaic

   (C) Brown rot and late blight (D) Bacterial wilt and charcoal rot

3. **Seed transmitted viruses are generally carried:**
   (A) In embryo
   (B) In seed coat
   (C) On seed surface
   (D) All of the above

4. **Which of the following viruses are carried on the seed coat (testa) as surface contaminant:**
   (A) TMV in tomato seed
   (B) Citrus tristeza virus
   (C) Bean common mosaic virus
   (D) Tobacco ring spot virus

5. **Latex flocculation test is used for the detection of seed borne:**
   (A) Fungi
   (B) Viruses
   (C) Nematode
   (D) None fo these

6. **Seed transmission is of great important where:**
   (A) Disease is endemic
   (B) Disease is pandemic
   (C) Disease is absent in an area
   (D) Disease is sporadic

7. **The large majority of seed-borne viruses are carried:**
   (A) In the embryo
   (B) In seed coat
   (C) One seed surface
   (D) None of these

8. **Tobacco mosaic tobamovrus (TMV) is carried:**
   (A) In embryo
   (B) In seed coat
   (C) On seed surface
   (D) All of the above

9. **A majority of seed transmitted viruses persist in:**
   (A) Endosperm
   (B) Embryo
   (C) Cortex
   (D) None of these

10. **Plant pathogenic *Alternaria* spp overwinter in infected plant debris, on seeds, tubers etc. as:**
    (A) Conidia
    (B) Mycelium
    (C) Both A and B
    (D) None of these

11. ***Colletotrichum orbiculare* causing anthracnose of cucurbits overwinters:**
    (A) In infected debris in soil
    (B) On or in the seed
    (C) Both A and B
    (D) None of these

12. **International Seed Testing Association was founded in________during the 4th International Seed testing Congress at Cambridge, UK:**
    (A) 1923
    (B) 1926
    (C) 1925
    (D) 1924

Fill in the Blanks with Correct Answer

1. The first seed health testing laboratory in the world was established in ________at the government seed testing station Wageningen.

   1918

2. When the dormant mycelium is present in the embryo, the disease in known as________

   Internally seed borne

3. Seed borne microorganisms never demonstrated to cause________

   Disease

4. Pathogens that infect seed in fields or in storage, and reduce________

   Seed quality

5. Pea early browning virus is transmitted by________

   Seed

6. Lettuce mosaic virus is transmiited by________

   Seed

7. The first seed transmission of tobacco mosaic virus through tomato seed was suspected by________

   Westerdijk in 1910 and later by Allard (1914)

8. Seed transmission of virus disease of Lima bean mosaic was studied by________

   Mc Clintock (1917)

9. Seed transmission of Cucumber mosaic virus (CMV) in wild cucumber (Echinocystis lobata) was reported by________

   Doolittle and Gilbert (1919)

10. Cryptoviruses are transmitted only through________and consist of a doublestranded RNA genome encapsidated in protein, without lipid envelope (Fraser 1989).

    Seed and pollen

11. Viroid diseases are effectively transmitted through________(Mink 1993, Diener 1999, Hadidi *et al.* 2003, Flores *et al.* 2005, Hammond and Owens 2006).

    Seed and pollen

12. Viroid diseases are effectively transmitted________through pollen and ovule to the seed and seedling.

    Vertically

13. Apple dapple viroid and apple scar skin viroid is transmitted by________

    Seed

14. Chrysanthemum stunt viroid, coleus viroid, grapevine viroid and hop stunt viroid diseases are transmitted by________

    Seed

15. Cucumber pale fruit, avocado sunblotch and chrysanthemum stunt viroid diseases are transmitted by________

    Seed

16. Red clover vein mosaic viruses (RCVMV) in certain legume cultivars, pea seed-borne mosaic (PSbMV), soybean stunt, and cucumber mosaic (CMV) are transmitted by________

    Seed

17. Tomato apical stunt viroid and grapevine yellow speckle viroid 1 and 2 are transmitted through seed as in________

    High degree

18. Keller *et al.* (1998) have reported that potyvirus genome-linked protein (VPg) determines the________

    Pea seed-borne mosaic virus pathotype

19. ________ (1995) have studied the viral genetic basis of symptom severity and seed transmission of two pathotypes of TSV, which differ in physico–chemical properties and RNA secondary structure.

    Walter et al.

20. Potato spindle tuber viroid survive in poato seed for________years (Singh *et al.*, 1991).

    21

21. Seed-transmitted viruses possess a unique property which enables them to invade even ________

    Reproductive tissues

22. Seed transmission virus *i.e.* tobacco ring spot virus discovered by Yang and Hamilton in________

    1974

23. Soybean mosaic in soybean, bean southern mosaic in bean, bean common mosaic in bean, and peanut mottle viruses inheritance pattern are________

    Monogenic

24. The level of the pathogen in seeds which gives rise to an unacceptable risk of disease is often referred to as________

    Inoculum threshold

25. Banana bunchy top virus, banana streak mosaic virus and peanut stripe virus introduced into India from________

    Srilanka in 1940

True or False

1. Seed coat transmission of plant viruses is common. (**False**)
2. Bean common mosaic virus transmitted by pollen and the first experimental evidence was demonstrated by Reddick and Stewart. (**True**)
3. Seed infection of many viruses occurs through introduction of male gametophytes (pollen) into the embryo sac. (**True**)
4. Barley stripe mosaic virus is transmitted less through pollen than ovules. (**True**)
5. Alfalfa mosaic virus is transmitted at a much higher frequency through pollen than ovules. (**True**)
6. Tobacco mosaic virus in tomato is carried in the gelatinous matrix on tomato seed testa during seed extraction. (**True**)
7. *Corynebacterium tritici* is carried as a contaminant on the surface of galls of *Anguina tritici* in barely. (**False**)
8. Tobacco ring spot virus remains viable in soybean seeds for 15 years at 16 to 32°C. (**False**)
9. *Ascochyta pisi* survives in broad bean seeds for 9 years and in pea seeds for 7 years. (**False**)
10. *Xanthomonas campestris* pv. *malvacearum* survives in cotton seeds for 20 years. (**False**)
11. *Xanthomonas campestris* pv. *lycopersici* survives in tomato seeds for 20 years. (**True**)
12. Bean common mosaic virus survives in bean seeds for 38 years. (**True**)
13. Barley strips mosaic virus survives in the barley seed for 19 years. (**True**)
14. *Xanthomonas campestris* pv. *malvacrearum* in cotton seeds survives in the basal end of the chalaza, as a contaminant on the seed surface or in the lint. (**True**)
15. Externally seed borne organisms (mycelium, spore, bacterial cell) remain in the seed. (**False**)
16. The plant and plant organs works as means of autonomous dispersal. (**True**)
17. Karnal bunt of wheat (*Neovossia indica*) is soil and externally seed borne. (**True**)

18. Seed transmitted viruses *i.e.* alfalfa mosaic in alfalfa, barley stripe mosaic in barley, blackeye cowpea mosaic in soyabean, pea seed borne mosaic in lentil, blackeye cowpea mosaic in cowpea and cucumber mosaic viruses in mungbean, their inheritance pattern is monogeneic resistance. (**True**)
19. Solid phase radioimmunoassay (SPRIA) technique is being widely used for seed-transmitted virus detection. (**True**)
20. *Alternaria brassicae* and *A. brassicicola* in crucifers, *A. longipes* in tobacco, *A. radicina* in carrot, *Ascochyta pinodella* in pea, *Drechslera sorokiniana* and *D. oryzae* in rice are transmitted on seed coat surface. (**True**)

Match the Following

1. **Match the following phytotoxin with seed borne fungi.**

Phytotoxin	*Seed borne fungi/crop*
I. Piricularin, picolinic, pyriculol	(A) *Pyricularia oryzae*/rice
II. Ophiobolin A, B, C and D	(B) *Cochiliobolus miyabeanus/rice*
III. Helminthosporil	(C) *Helminthosporium sativum*/wheat
IV. HMT-toxin	(D) *H. maydis*/maize

Answer

I	*II*	*III*	*IV*
A	*B*	*C*	*D*

2. **Match the following phytotoxin with seed borne fungi:**

Phytotoxin	*Seed borne fungi/crop*
I. HC-toxin	(A) *Helminthosporium carbonum*/maize
II. Ten-toxin	(B) *Alternaria brassicae*/rapeseed
III. Zinniol	(C) *A. zinnia*/sunflower
IV. Dextruxin	(D) *Alternaria alternata*/chilli

Answer

I	*II*	*III*	*IV*
A	*D*	*C*	*B*

3. **Match the longevity of some important seed transmitted viruses in different crops:**

Virus/Crop	*Viability (Years)*
I. Alfalfa mosaic (*Medicago sativa*)	(A) 3-7
II. Barley stripe mosaic (*Hordeum vulgare*)	(B) 3-19

III. Bean western mosaic (*Phaseolus vulgaris*) — (C) 3

IV. Bean southern mosaic (*Phaseolus vulgaris*) — (D) 0.6

Answer

I	*II*	*III*	*IV*
A	*B*	*C*	*D*

4. Match the longevity of some important seed transmitted viruses in different crops:

Virus/Crop	*Viability (Years)*
I. Cucumber mosaic (*Phaseolus vulgaris*)	(A) 2.25
II. Dodder latent mosaic (*Cuscuta campestris*)	(B) 1
III. Muskmelon mosaic (*Cucumis melo*)	(C) 3-5
IV. Soybean mosaic (*Glycine max*)	(D) 1-2

Answer

I	*II*	*III*	*IV*
A	*B*	*C*	*D*

5. Match the longevity of some important seed transmitted viruses in different crops:

Virus/Crop	*Viability (Years)*
I. Eggplant mosaic (*Solanum melongena*)	(A) 0.6
II. Raspberry ring spot virus (*Stellaria media*)	(B) 6
III. Squash mosaic (*Cucurbita pepo*)	(C) 5
IV. Tobacco ring spot virus (*Glycine max*)	(D) 5

Answer

I	*II*	*III*	*IV*
A	*B*	*C*	*D*

6. Match the following externally pollen borne viruses with appropriate host:

Virus	*Host*
I. Tobacco mosaic	(A) *Vigna unguiculata*
II. Potato virus X	(B) *Petunia hybrida*

III. Squash mosaic	(C) *Cucurbita pepo*
IV. Potato virus T	(D) *Datura stramonium*

Answer

I	*II*	*III*	*IV*
A	*B*	*C*	*D*

7. **Match the maximum per cent virus disease certification standards in foundation seed crops:**

Virus/crop	*Standard in foundation seed (per cent)*
I. Common mosaic/cowpea	(A) 0.10
II. Jute chlorosis/jute	(B) 1.0
III. Potato leaf roll/potato	(C) 0.5
IV. Sweet potato/sweet potato	(D) 0.05

Answer

I	*II*	*III*	*IV*
A	*B*	*C*	*D*

8. **Match the maximum per cent virus disease certification standards in certified seed crops:**

Virus/crop	*Standard in certified seed (per cent)*
I. Common mosaic/cowpea	(A) 0.20
II. Jute chlorosis/jute	(B) 2.0
III. Potato leaf roll/potato	(C) 1.0
IV. Sweet potato/sweet potato	(D) 0.10

Answer

I	*II*	*III*	*IV*
A	*B*	*C*	*D*

9. **Match the virus and virus-like diseases transmission through vegetative propagules:**

Virus/crop	*Vegetative propagule*
I. Apple stem pitting/apple	(A) Grafting
II. Colocasia bobone disease/ colocasia	(B) Corm
III. Dahlia mosaic/dahlia	(C) Bulb, bulblets
IV. Lily mottle/lily	(D) Root tubers

Answer

I	*II*	*III*	*IV*
A	*B*	*D*	*C*

10. Match the virus and virus-like diseases transmission through vegetative propagules:

Virus/crop	*Vegetative propagule*
I. Rose leaf curl/rose	(A) Runners
II. Strawberry crinkle/ strawberry	(B) Corm
III. Sweet potato latent/ sweet potato	(C) Tuberous roots
IV. Sugarcane mosaic/sugarcane	(D) Stem cutting

Answer

I	*II*	*III*	*IV*
D	*A*	*C*	*B*

Answers (Multiple Choice Questions)

(1)	(B)	(4)	(A)	(7)	(A)	(10)	(C)
(2)	(A)	(5)	(B)	(8)	(C)	(11)	(C)
(3)	(D)	(6)	(C)	(9)	(B)	(12)	(D)

Chapter 7

Phanerogamic Plant Parasites and Non-parasitic Diseases

Queries with Timeline of Phanerogamic Parasites and Non-parasitic Diseases

1. Mistletoes were one of the earliest parasites of plants recognised by Albertus Magnus around 1200 A.D. (Horsfall and Cowling 1977). They cause sometimes major problems for the horticulturists as well as the foresters. *Striga* sp. is recognized as parasitic weed problems before 1910 in both India (Barber, 1954) and Soth Africa (1905).

2. Some important literature on parasitic higher plants is that of Hawksworth and Wiends (1970) '**Biology and Taxonomy of Dwarf Mistletoes'**, Musselman (1980) **'The Biology of Striga, Orobanche and other Root Parasitic Weeds'**, and Musselman (1987) **'Parasitic Weeda in Agriculture'.** Thody (1991) has dicussed the histological relation between Cuscuta and its host.

3. In ozone injures, the leaves of plants exposed for even a few hours at concentration of 0.1 to 0.3 ppm, sulphur dioxide (SO_2) at 0.3 to 0.5 ppm, peroxyacetyl nitrate (PAN) at 0.01 to 0.02 ppm, hydrogen fluoride (HF) at 0.1 to 0.2 ppm, chlorine and hydrogen chloride (HCl) at 0.1 ppm ethylene (CH_2CH_2) at 0.05 ppm and nitrogen dioxide (NO_2) at 2 to3 ppm.

4. Peroxyacetyl nitrate (PAN) causes 'silver leaf' on plants *i.e.* bleached white to bronze spots on lower surface of leaves and ozone (O_3) causes stippling, mottling and chlorosis of leaves.

5. Little leaf of apple, stone fruits and grape, sickle leaf of cacao, white tip of corn, rosette of apple, motle loaf of citrus and khaira disease of rice symptoms is caused due to deficiency of zinc.
6. Copper (Cu) deficiency causes dieback of citrus, pome and stone fruits. Calcium (Ca) deficiency causes blossom end rot of tomato, pepper, apple, black heart of celery and wither tip of flax. Black heart of potato is caused by unfavourable oxygen relation.
7. Tip burn of paddy (pansukh, dakhia, chatra, ukra, kadamora) is caused by reduced supply of oxygen to the root. Heat canker of linseed caused by high soil temperature near the soil surface.Iron (Fe) is essential for the synthesis of chlorophyll but is not constituents of chlorophyll and it causes chlorotic symptoms in citrus are termed as 'green netting'.
8. Mgnesium (Mg) dificieney causes 'sand drawn'in tobacco and manganese (Mn) causes chlorosis and shows other diseases *i.e.* gray speck of oat, speckled yellows in sugarbeet, little leaf of apple and marsh spot of garden peas.
9. Molydbenum (Mo) deficiency causes whiptail disease of cauliflower and it exhibit severe yellowing and stunting. Its deficiency also affected on corn, sugarbeet, legumes, flax, tomato and melons.
10. Boron (B) increases the mobility of sugars and calcium and is important in cell division and protein synthesis. It is essential for pollination and also affects flower formation, fruit set and seed production. Apical growing points of plants are stunted and die due to boron deficiency. In case of excess of boron in plants, leaf tips and margins are turned brown and die. It's deficiency causes hallow stem of brassica, fruit pitting and dieback of olive, rosetting and anthocynescence of alfalfa, heart rot of beet, internal cork of apple, brown heart of cabbage and turnip, terminal bud breakdown of tobacco, bitter pit in apple and black tip of mango.
11. *Cuscuta* (love vine, dodder, amarbel, *Ccuscuta gronovii*) are entirely dependent (holo parasite) stem parasites. It attacks on clover, berseem, flax, onion, alfalfa, and many oil seed crops and also attacks on garden or ornamental and hedge plants.
12. Loranthus (banda, giant mistletoe, virkshabhaksh), *Dendrophthoe* spp are semiparasite or hemi stem parasite (partially dependent) of tree trunks and branches of mango. *Dendrophthoe falcata* is the common species reported in India. It also attacks citrus, apple and guava.
13. Spraying of 2, 4-D, PCP (pentachlorophenol) or diesel oil fortified with DNBP (4, 6-dinitro-o-butylphenol) for the control of dodder and glyphosate is effective even after dodder has established connection with the host.
14. The most common host of loranthus is the mango tree in northern India and other host such as *Ficus* spp, *Albizzia* spp, *Dalbergia* spp and in the common host is *Madhuca latifolia*. The parasite is disseminated through its seeds carried by birds.

15. Injection of $CuSO_4$ and 2, 4-D into infected branches is effective for the control of loranthus, diesel oil emulsion in soap waeter is quite efficacious in eradication of the parasite from mango trees and catterpillers of butterfly *Delias eucharis* is potential biological agent for the control of loranthus on mango and cotton plants. They feed on *D. falcata* leaves and tender branches causing heavy defoliation.

16. *Orobanche* spp (broom rape or tokra is a total root parasites/entirely dependent) affecting tomato, brinjal, tobacco, cabbage, cauliflower, turnip and many other solanaceous and crucifer plants. It is especially destructive to tobacco. *O. cernua* var. *desertorum* affects on tobacco, *O. aegyptiaca* parasitizes many crop plants in India.

17. *Orobanche* contaminated lots of seeds are treated with 25 per cent $CuSO_4$ solution and spraying of 25 per cent $CuSO_4$ solution in the soil is effective for the control of *Orobanche*.

18. *Striga* spp (Witchweed) is semi-root parasites (partially dependent) of sugarcane, corn, maize, cereals and millets. 2, 4-D, Agroxone (McPA), Fernoxone (2-4 D) and methylbromide as soil fumigants are used for the control of *Striga* spp

19. Dr. Y.L. Nene and his associates first discovered Khaira disease of rice in the Tarai region of the Uttaranchal Pradesh.

20. Mango necrosis or black tip of mango disease was first reported from Bihar, toxic gases sulphur dioxide is the major causal factor.

21. Excess manganese is known to cause a crinckle leaf disease in cotton and internal bark necrosis in apple.

22. Nitrogen is a constituent of such compound as proteins, amino acids, enzymes, harmones, chlorophyll and vitamins. Plants deficient in nitrogen are stunted and woody and the leaves become yellow or light brown. Sugar accumulates in the tissues because deficiency of nitrogen affects respiration. Plants excess in nitrogen is appeared in dark green colour, succulent and blossom abortion and lack of fruit set occurred.

23. Phosphorous is a constituent of phospholipids, nucleic acids and many proteins. It is essential for carbohydrtate transformations and respiration. It is invoved in high energy bonding as in the conversion of high energy bonding (i.e. in ATP and ADP).The phosphorous deficient plants become bluish green with purple tints. High phosphorous interfere with normal Ca nutrition, and typical Ca deficiency symptoms occurred.

24. Potassium is essential as a catalyst in nitrate reduction and inreactions that convert starch to sugar. Potash deficient plants are appeared as mild chlorosis and browning of the tips of leaves. Older leaves edges seem as burned, a symptoms known as scoarch.

25. Calcium is indirectly involved in cell division. It acts as an activator of the enzyme phospholipase arginine kinase and several other enzymes. The growing tips of roots and leaves are turned brown and die and also leaves curl back with irregular ragged margins.

Multiple Choice Questions (Choose the correct answer)

1. ***Striga* is:**
   (A) A partial parasite (B) A total stem parasite
   (C) A total root parasite (D) A partial root parasite

2. ***Cuscuta* is:**
   (A) A partial stem parasite (B) A entirely stem parasite
   (C) A total root parasite (D) A total stem parasite

3. ***Orobanche* is:**
   (A) A total root parasite (B) A partial stem parasite
   (C) A entirely stem parasite (D) None of these

4. ***Loranthus* is:**
   (A) A partial stem parasite (B) A total stem parasite
   (C) A total root parasite (D) A portial root parasite

5. **Black tip of mango is also known as:**
   (A) Chimney disease (B) Soft nose of mango
   (C) Seed jelly of mango (D) Withch's broom

6. **Older leaves of rice are resistance to blast pathogen due to greater absorption of:**
   (A) Nitrogen (B) Silicon
   (C) Phosphorus (D) Magnesium

7. **Common deficiency of micronutrient noticed in:**
   (A) Zinc (B) Iron
   (C) Copper (D) Manganese

8. **Endoxerosis a disorder associated with water deficiency. It is associated with which of the following fruit crop:**
   (A) Apple (B) Pear
   (C) Lemon (D) Guava

9. **Which of the following is a holoparasitic phanerogam on root?**
   (A) *Orobanche* (B) *Loranthus*
   (C) *Striga* (D) *Cuscuta*

10. Black heart of potato is a:

(A) Fungal disease (B) Viral disease

(C) Bacterial disease (D) None parasitic disease

11. The most susceptible citrus species for cold injury:

(A) Lime (B) Lemon

(C) Sweet orange (D) Trifoliate orange

12. Soft nose is a disorder of:

(A) Mango (B) Banana

(C) Apple (D) Guava

13. The most susceptible fruit crop to waterlogging:

(A) Papaya (B) Guava

(C) Mango (D) Banana

14. Granulation is a physiological disorder of:

(A) Apple (B) Guava

(C) Citrus (D) Mango

15. Interveinal chlorosis is caused by the deficiency of:

(A) Nitrogen (B) Potassium

(C) Calcium (D) Magnesium

16. Leaf scorching in mango:

(A) Nitrogen deficiency (B) Chloride toxicity

(C) Chloride deficiency (D) Potassium deficiency

17. Sulphur dioxide is not injurious at:

(A) 500-200 ppm (B) 100-50 ppm

(C) 10-5 ppm (D) 1-2 ppm

18. Black heart disease of potato is caused by:

(A) Extremely low temperature (B) Darkness/unfavorable light

(C) Low oxygen and high storage temp. (D) All of these

19. Match *Striga*:

(A) Bacteria (B) Fungi

(C) Mycoplasma plant parasites (D) Phanerogamic pl. parasites

20. Root parasite of bajra is:

(A) *Striga* (B) *Cuscuta*

(C) *Orobanche* (D) *Loranthus*

21. Sulphur dioxide is injurious to plant after its absorption in the form of:
(A) Sulphite ion (B) Free sulphur
(C) Sulphate (D) None of these

22. Sulphur acted against pathogen in the form of:
(A) Sulphite (B) Sulphide
(C) H_2S (D) None of these

23. Mistletoe haustoria are located in host:
(A) Root (B) Branches
(C) Leaves (D) Fruits

24. *Orobanche* spp are:
(A) Semi root parasite (B) Total root parasite
(C) Non root parasite (D) Only stem parasite

25. Temporary wilting in plants result due to lack of:
(A) O_2 (B) SO_2
(C) H_2O (D) CO_2

26. Khaira disease of rice is caused by:
(A) Iron deficiency (B) Boron deficiency
(C) Potassium deficiency (D) Zinc deficiency

27. Black tip of mango is caused by:
(A) B deficiency (B) *Fusarium moniliforme*
(C) *Diplodia natalensis* (D) None of these

28. Fasciation/flatted fruit multiple crown and sun scald are the disorder of:
(A) Pine apple (B) Grape
(C) Guava (D) Mango

29. Litchi leaf scorching is associated with the deficiency of:
(A) N (B) P
(C) Ca (D) K

30. Little leaf of mango is due to the deficiency of:
(A) Zn (B) Cu
(C) K (D) Fe

31. Which of the following is the chief cause of spreading citrus decline:
(A) Citrus nematode (B) Burrowing nematode
(C) Fungal disease (D) Nutritional deficiency

32. Which of the following gases is involved with black tip in mango:

(A) CO (B) SO_2

(C) Ethyulene (D) All of these

33. Yellow spot a disorder in citrus due to deficiency of:

(A) Mo (B) S

(C) Ca (D) B

34. Bronzing in Guava is associated with the deficiency of:

(A) Zn (B) Mn

(C) Cu (D) Mg

35. Khaira can be controlled with the spray of ________in 1000 litre water per hectare:

(A) 2.5 kg $ZnSO_4$+5.0 kg $CaCO_3$ (B) 5 kg $ZnSO_4$+2.5 kg $CaCO_3$

(C) 10 kg $ZnSO_4$+5.0 kg $CaCO_3$ (D) None of these

36. Mark the aboitic elicitors :

(A) Glucon

(B) Salicylic acid

(C) Jasomic acid

(D) 2, 2, dichloro-3, 3-demethylcyclopropane carboxylic acid

37. Black tip disease of mango is caused by:

(A) Fungi (B) Bacteria

(C) Boron deficiency (D) Nematode

38. Ozone injures, the leaves of plants exposed for even a few hours at concentrations of:

(A) 1.2 ppm (B) 3.4 ppm

(C) 0.1-0.3 ppm (D) 0.5-1 ppm

39. Air pollutants sulphur dioxide (SO_2) is toxic at:

(A) 0.3-0.5 ppm (B) 0.8-1 ppm

(C) 1-1.5 ppm (D) 2.3 ppm

Fill in the Blanks

1. Mottling in citrus is due to deficiently ________

   Zn

2. ________of citrus is due to molybdenum deficiency.

   Yellow spot

3. Leaf scorch in mango is due to excess of________

   MOP

4. Deformed flowers are a common symptom of________deficiency.

   Boron

5. Due to________deficiency leaves may develop tan gray spots.

   Manganese

6. Black tip of mango is controlled by________and ________

   Caustic soda, lime spray

7. 'Choke throat' is a physiological disorder of________

   Banana

8. ________is the responsible pollutant causing tip burn and necrosis of conifers.

   Fluoride

9. In________deficient plants chlorosis is most severe at the top of the plant.

   Manganese

10. In________deficient plants yellowing near leaf margins develop in a V shaped pattern.

    Manganese

11. Water berry, shot berry and mummification is a physiological disorder of________

    Grape

12. Pink berry is serious physiological disorders in grape variety of ________

    Thompson seedless

13. Collar of slips is physiological disorder of________

    Pineapple

14. Witche's broom of peach and apricot is caused by________

    Mn deficiency

15. Broom rape of brinjal is caused by________

    Orobanche cernua and O indica

16. Nitrogen dioxide (NO_2) is toxic at ________

    2 to 3 ppm

17. Deficiency of highly mobile nutrients is first seen in the________leaves.

    Lower

18. Hollow stem of *Brassica* is also caused by________deficiency.

    Boron

19. Strangle weed is another name for the stem parasite________

    Cuscuta

20. Hydrogen fluoride (HF) cause injury to the plants at concentrations as low as ________

    0.1 to 0.2 ppm

21. .________deficiency in banana causes bunchy top crowns, narrow pointed and chlorotic young leaves.

    Znic

22. Acis soils, sandy soils and the soils containing excess of magnesium or potassium may show deficiency of________

    Calcium

23. Angiosperm parasitic host plants mainly by________

    Mechanical pressure

24. Corky spots on fruits and rusty cracks on stems are symptoms of________

    Boron

25. ________deficiency in maize may lead to bleached white blind grains in the year.

    Copper

26. ________is also known as hair weed, devils hair and love vine.

    Cuscuta

27. Internal necrosis of mango is caused due to________ deficiency.

    Boron

28. Finger tip disease is severe in________ density planting.

    High

29. Guava wilt is most common in________________

    Alkaline soils

30. Stem segment of dodder (*Cuscuta*) are carried by________to prepare nest and is thus transported to new sites.

    Birds

31. ________deficiency in plants result in chlorotic leaves with green main vein.

    Iron

32. ________deficiency is generally observed in alkaline soils and soil having excess phosphorus or aluminium concentration.

    Iron

33. Leaves turn brilliant________in colour due to molybdenum toxicity.

    Orange

34. ________deficient plants first turn yellow, then ornage and finally brown.

    Magnesium

35. ________is essential for symbiotic nitrogen fixation in the nodules due to *Rhizobium* spp

    Molybdenum

True or False

1. Roots of litchi show association with mycorrhizal fungi in acidic soil. (**True**)
2. Bitter pit in apple is due to zinc deficiency. (**True**)
3. Calcium deficiency in pear causes black and cork spot. (**True**)
4. Fruit cracking in pear is due to boron deficiency. (**True**)
5. Satellite virus can multiply by itself. (**False**)
6. Satellite virus is a very small virus measuring only 17 nm in diameter. (**True)**
7. *Rhizopus* spp produce soft rot of fleshy parts which spread rapidly at high temperature. (**True**)
8. Hen and chicken disorder caused by B deficiency. (True)
9. Fasciation/multiple crown physiological disorder caused by Genetically/ Nutritional.(**True**)
10. Gamboge physiological disorder of mangosteen caused by excess water or heavy rainfall prior to pre-harvest. (**True**)
11. Shot berry physiological disorder in grape is caused by B deficiency. (**True**)

12. Collar slips physiological disorder in pineapple is caused by high N and high rainfall with relatively low temperature. (**True**)
13. Little leaf of grape is caused by Zn deficiency. (**True**)
14. Heart rot of beet is caused by B deficiency. (**True**)
15. Fruit necrosis in aonla is due to deficiency of boron. (**True**)
16. Black tip, a disorder of mango fruit is caused by smoke of brick kiln. (**True**)
17. Little leaf in mango is due to calcium deficiency. (**False**)
18. Black tip of mango was first reported by Woodhose in 1909. (**False**)
19. The alkalinity necrosis and acidity is not involved in predisposition. (**True)**
20. Excess of calcium promotes wilt incidence in tomato. (**False**)
21. Deficiency of zinc predisposes maize plant to attack of downy mildew (*Peronosclerospora sacchari*). (**True**)
22. At high temperature sulphur is toxic to plants. (**True**)
23. Black heart is caused by excess of oxygen which kills the tissues. (**False**)
24. *Dendrophthoe* spp is semi parasitic on tree trunk and branches. (**True**)
25. *Striga* spp is total root parasites affecting brinjal and tomato. (**False**)
26. *Orobanche* spp is partial root parasites. (**False**)
27. *Striga* spp are obligate parasites, do not obtain all their nutrients from their host's root. (**True**)
28. Sodium and chlorine ions cause symptoms of poor growth and decline in trees. (**True**)
29. Blue berries grow well in acidic soils, while alfalfa grows best in alkaline soils. (**True)**
30. Zinc deficiency leads to poor nitrogen formation in legumes. (**True**)
31. Zinc deficient plants have chlorotic and necrotic leaves and stunted internodes. (**True**)
32. The root nodule bacteria may become parasitic on the legumes is generally due to deficiency of boron. (**True**)
33. The leaves of phosphorus deficient plants become bluish green with purple tints. (**True**)
34. The herbicide glyphosate effectively controls the parasite Orobanche. (**True**)
35. *Orobanche* is also known as broomrape. (**True**)

Match the Following

1. **Match the following physiological disorders:**

Disorder	*Cause*
I. Black tip of mango	(A) SO_2, CO_2 and acetylene
II. Blossom end rot	(B) Ca deficiency
III. Bud killing	(C) Excess of N_2
IV. Fruit cracking	(D) Dry spell followed by rains

Answer

I	*II*	*III*	*IV*
A	*B*	*C*	*D*

2. **Match the following physiological disorders:**

Disorder	*Cause*
I. Interveinal chlorosis	(A) Mg deficiency
II. Leaf scorching	(B) Cl toxicity
III. Spongy tissue	(C) Connective heat
IV. Soft suture of peach	(D) Fluoride toxicity

Answer

I	*II*	*III*	*IV*
A	*B*	*C*	*D*

3. **Match the following disorders of fruit:**

Disorder	*Cause*
I. Mottle leaf	(A) Cu deficiency
II. Exanthema	(B) Ca deficiency
III. Bitter pit	(C) Excess of B
IV. Cork spot	(D) Excess N_2 and K

Answer

I	*II*	*III*	*IV*
D	*A*	*B*	*C*

4. **Match the following disorders of fruit:**

Disorder	*Cause*
I. Yellow spot in citrus	(A) Mo deficiency
II. Skin russeting	(B) B deficiency
III. Internal fruit necrosis	(C) B deficiency
IV. Little Leaf	(D) Zn deficiency

Answer

I	*II*	*III*	*IV*
A	*B*	*C*	*D*

5. Match the following disorders in fruit crops:

Disorder	*Fruit crop*
I. Greening	(A) Apple
II. Bitter pit	(B) Mango
III. Sun scalds	(C) Pineapple
IV. Leaf scorching	(D) Citrus

Answer

I	*II*	*III*	*IV*
D	*A*	*C*	*B*

6. Match the following disorders in fruit crops:

Disorder	*Fruit crop*
I. Fruit cracking	(A) Litchi
II. Spongy tissue	(B) Alphanso
III. Dry neck	(C) Avocado
IV. Hard end	(D) Pear

Answer

I	*II*	*III*	*IV*
A	*B*	*C*	*D*

7. Match the following disorders in fruit crops:

Physiological disorder	*Fruit crop*
I. Core spot	(A) Apple
II. Fasciation/Multiple crown	(B) Pineapple
III. Little leaf	(C) Cashew/Mango
IV. Water berries	(D) Grape

Answer

I	*II*	*III*	*IV*
A	*B*	*C*	*D*

8. Match the following disorders in fruit crops:

Physiological disorder	*Fruit crop*
I. Stoning (Stone fruit)	(A) Custard apple
II. Leaf bronzing	(B) Guava

III. Calyx end cracking (C) Persimmon

IV. Core break down (D) Pear

Answer

I	*II*	*III*	*IV*
A	*B*	*C*	*D*

9. **Match the following disorders in fruit crops:**

Physiological disorder	*Cause in fruit crop*
I. Black tip	(A) SO_2, CO, ethylene (Mango)
II. Pink berry	(B) Unfavorable environment (Grape)
III. Exanthema	(C) Cu deficiency in citrus
IV. Granulation	(D) High Ca and Mn (Jaffa, Valencia) in citrus

Answer

I	*II*	*III*	*IV*
A	*B*	*C*	*D*

10. **Match the following disorders in fruit crops:**

Physiological disorder	*Cause in fruit crop*
I. Fruit cracking	(A) Delay in harvesting (Pomegranate)
II. Jonathan spot	(B) Cause unknown (Apple)
III. Hen and chicken	(C) B deficiency (Grape)
IV. Calyx cavity	(D) Excess of N (Persimmon)

Answer

I	*II*	*III*	*IV*
A	*B*	*C*	*D*

11. **Match the following physiological disorders:**

Physiological disorder	*Real cause*
I. Wither tip of lax	(A) Ca deficiency
II. Back heart of celery	(B) Ca deficiency
III. Blossom end rot of tomato	(C) Ca deficiency
IV. Leaf bluish green in cotton	(D) Phosphorus deficiency

Answer

I	*II*	*III*	*IV*
A	*B*	*C*	*D*

12. Match the following physiological disorders:

Physiological disorder	*Real cause*
I. Red leaf of cotton	(A) N_2 deficiency
II. Rosetting and Anthracnase of alfalfa	(B) B deficiency
III. Fruit pitting and die back of olive	(C) B deficiency
IV. Hollow stem of brassica	(D) B deficiency

Answer

I	*II*	*III*	*IV*
A	*B*	*C*	*D*

13. Match the following physiological disorders:

Physiological disorder	*Real cause*
I. Whiptail disease of cauliflower	(A) Mo deficiency
II. Marsh spot or garden pea	(B) Mg deficiency
III. Grey speck disease of oat	(C) Mg deficiency
IV. Sand drawn in tobacco	(D) Mg deficiency

Answer

I	*II*	*III*	*IV*
A	*B*	*C*	*D*

14. Match the following physiological disorders:

Physiological disorder	*Real cause*
I. Green netting in citrus	(A) Iron deficiency
II. Khaira disease of rice	(B) Zn deficiency
III. Rosette of apple	(C) Zn deficiency
IV. Mottle leaf of citrus	(D) Zn deficiency

Answer

I	*II*	*III*	*IV*
A	*B*	*C*	*D*

15. Match the following physiological disorders:

Physiological disorder	*Real cause*
I. Heat canker of linseed	(A) High soil temperature
II. Tip burn of paddy/ pansukh of rice	(B) Reduced supply of O_2

III. Black heart of potato	(C) Unfavorable O_2 reaction
IV. Citrus decline	(D) Insufficient soil moisture and nutrition

Answer

I	*II*	*III*	*IV*
A	*B*	*C*	*D*

Answers (Multiple Choice Questions)

(1)	(D)	(11)	(B)	(21)	(A)	(31)	(D)
(2)	(B)	(12)	(A)	(22)	(C)	(32)	(B)
(3)	(A)	(13)	(A)	(23)	(B)	(33)	(A)
(4)	(A)	(14)	(C)	(24)	(B)	(34)	(A)
(5)	(A)	(15)	(D)	(25)	(C)	(35)	(B)
(6)	(B)	(16)	(B)	(26)	(D)	(36)	(D)
(7)	(A)	(17)	(D)	(27)	(A)	(37)	(C)
(8)	(C)	(18)	(C)	(28)	(A)	(38)	(C)
(9)	(A)	(19)	(D)	(29)	(D)	(39)	(A)
(10)	(D)	(20)	(A)	(30)	(A)		

Chapter 8

Management of Plant Diseases

Queries with Timeline of Management of Plant Diseases

1. In 1000 BC, Homer (the Greek poet) spoke of 'the pest-averting sulphur', means he mentioned sulphur dust as pest controlling properties. 470 BC. Democbttus recommended sprinkling pure amurca of olives on plants to control blight. This is the first case of therapy. In Surapala's '*Vriksha Ayurveda*' (1000 AD) reference of tree wound dressing, tree fumigation and seed treatment with milk, honey, cowdung, cowurine and ashes for contro of plant diseases and pests of various plants like sesamum has been made.

2. Parkinson (1629) recommended urine for cankers of orchard tree management. Remnant (1637) mentioned seed treatment (probably sodium chloride) for wheat bunt disease management. This is the first case of protection. Homberg (1705) recommended mercuric chloride ($HgCl_2$) as a wood preservative.

3. In 1733, Jethro Tull described salt brine treatment for wheat seed and in 1755, Aucante mentioned arsenic and corrosive sublimate for wheat bunt control. Schulethess (1761) first time suggested the use of copper sulfate ($CuSO_4$) on wheat seed against stinking smut or bunt of wheat.

4. Prevost (1807) of France was first exponent of laboratory testing; he showed effectivity of copper sulfate in the control of stinking smut and also showed adverse effect of copper sulfate on spore germination. In 1824, J. Robertson of England reported that sulphur is effective against peach mildew. Kenrick (1833) of USA proposed boiled lime sulphur for grape mildew management.

5. In 1834, Knight, in England, recommended sprinkling with sulphur and lime to control peach leaf curl in early spring. Morren (1845) recommended lime and

copper sulfate as soil treatment for potato blight control. In 1880-81, Marshall Ward elucidated the theory of plant protection by spraying.

6. Pierre Marie Alexis Millardet (1882) a professor of botany at the University of Bordeaux, France, made his first report on the control of downy mildew of grape with Bordeaux mixture. In May 1885, he published his works and gave details of spraying with a mixture of copper sulphate and hydrated lime, this mixture now known as Bordeaux mixture. He developed sulfatine dust, Bordeaux plus sulphur in 1884.
7. In 1888, Hale in California used lime sulphur for peach leaf curl control. In the same year, Trillat first recognized fungicidal properties of formaldehyde. C.M. Weed of Ohio in 1889, first time mixed fungicides and insecticides together. In 1894, W. Saunders, S.A. Bedford and A. Mackay introduced theory of seed protection.
8. Ozanne (1885) was the first to try copper sulfate in India. At Poona, he used copper sulfate and common salt to pickle, sorghum seed against smut control before sowing and obtained fairly satisfactory results. In 1887, Mason of France first time introduced burgundy mixture in which quick lime of Bordeaux was substituted by sodium carbonate (Na_2CO_3). Its formula is 4:5:50 [4 kg copper sulfate, 5 kg washing soda (Na_2CO_3) and 200 liter of water]. Burgundy mixture is slightly less effective than Bordeaux mixture.
9. Cheshnut compound was discovered by Bewley in 1921. It contains 2 parts copper sulfate ($CuSO_4$), 11 part ammonium carbonate $(NH4)_2CO3$. It is used to drench nursery bedsfor the control of 'damping off'.
10. H.L. Bolley in 1897 first used formalin for wheat smut. A.D. Selby in 1900 introduced formaldehyde soil treatment for onion smut. In 1904, Lawrence used Bordeaux mixture for the first time in India against leaf spot of groundnut (*Cercospora* spp). In 1913, Riehm introduced organic mercuriales for seed treatment of wheat to smut control.
11. Hiltner, a German worker, in 1910 very successfully used mercuric chloride ($HgCl_2$) for the control of *Fusarium* disease of rye (*Calonectria graminicola*). In India, it was used in 1914 by Burns for steeping potato tubers against *Rhizoctonia* (black scurf of potato).Generally concentration used in 1:1000 $HgCl_2$ solutions for few minutes and then planted.
12. Tisdale and Williams in 1934 reported fungitoxic activity of dithiocarbamates and patented. The beginning of the scramble for organic fungicide. Cunningham and Sharvelle introduced chloranil as a practical organic seed protectant in 1940. Kittleson introduced captan as a fungicide in 1952.
13. The introduction of systemic fungicide in 1966 was a major landmark in the history of fungicidal management of plant diseases. It started with the discovery of oxathiins in 1966 by Von Schmeling and Marshal Kulka. It was soon followed by confirmation of systemic activity of pyrimidines (1968) and benzimidazoles (1968-69).

14. E.J. Butler (1906) recommended Bordeaux mixture brushing for the control of koleroga disease of arecanut (*Areca catechu* L.). Uppal (1929) reported the control of brown spot of rice in upper side by seed treatment with organomercurial. J.F. Dastur (1931) claimed success in controlling foot rot of betelvine by soil drenching with 2:2:50 Bordeaux mixture.

15. Thirumulachar and his coworker (1964) reported the development of aureofungin, a new antifungal antibiotic for use in plant disease control. Chatrath and his coworker for the first time in India reported successful control of loose smut of wheat through seed dressing with carboxin.

16. Metalaxyl, effective against oomycetes was developed by Ciba-Geigy in 1973 and came in use as fungicide in 1977. In the same period, organic phosphate fungicides Fosetyl-AL was also developed and used against oomycetous fungi.

17. Organomercurial was used for the first time in India by Hilson (1925) for the control of sorghum smut. Kuhn (1858) reported seed borne behaviour on nematode *Ditylenchus dipsaci*.

18. William Roberts in 1874 demonstrated the antagonistic action of micro-organisms in action of micro-organisms in liquid culture between *Penicillium glaucum* and bacteria and introduced the term antagonism. The term biological control as a feasible preposition of plant disease management was coined for the first time by C. F. Von in 1914. Sanford (1926) observed that the potato scab was suppressed by green manuring antagonistic activities.

19. In 1932, Weindling first time noted the potential of *Trichoderma* and *Gliocladium* in controlling plant pathogenic fungi. It was due to the toxic metabolites secreted by these two fungi. Initially parasitic fungus locates the host mycelium or spores, using chemical signals originating from the host fungus. Control of Rhizoctonia damping off citrus seedlings has also reported by Weindling and Emerson (1936). B. Rai and D.B. Singh (1980) investigated the antagonistic activities of some phylloplane fungi (of mustard and barley) against *Alternaria brassicae* and *Drechslera gramineae*.

20. Grossbard (1948-1952), Wright (1952-1957), and others demonstrated that antibiotics were produced in soil by *Penicilium, Aspergillus, Trichoderma, Streptomyces* sp. Kloepper (1980), demonstrated the importance of siderophores produced by *Erwinia carotovora*.

21. Campbell (1956) reported that *Myrothecium verrucaria*, can parasitize hyphae of *Cochliobolus sativus* and this high chitinase producing fungus, was also successfully used for the control of *Drechslera teres* infection of barley leaves by the seed treatment.

22. Biocontrol of cocoa pod diseases with mycoparasite mixtures has been suggested by Krauss and Soberanis (2001). Cocoa pod diseases *viz.* moniliasis caused by *Moniliophthora roreri*, witche's broom caused by *Crinipellis perniciosa* and black pod caused by *Phytophthora palmivora*.

23. During 1880s, *Metarhizium* sp. was used to control wheat chafer, *Anisoplia austriaca* and the sugarbeet curculio, *Cleonis puncitiventris*. The genera such as *Metarhizium, Beauveria, Nomuraea, Verticillium, Entomophthora* and *Neozygites* are commonly encountered in nature. Recently, entomopathogenic fungi such as *Metarhizium, Beauveria, Nomuraea, Verticillium* and *Paecilomyces* have been studied on a large scale in the field of pest control.

24. *Trichoderma harzianum*, parasitize mycelia of *Rhizoctonia, Sclerotium, Pythium, Phytophthora, Fusarium* and *Heterobasidion* (fomes). *Rhizoctonia* and *Pythium* are also parasitized by *Laetisaria arvalis* (*Corticium* sp.). Mc Rae (1909) claimed control of blister blight of tea (*Exobasidium vexana*) with 6:4:50 Bordeaux mixture in Darjeeling.

25. *Sclerotinia sclerotiorum* are parasitized by *Sporidesmium sclerotivorum, Gliocladium virens*, and *Coniothyrium minitans*. *Talaromyces flavus*, which parasitizes *Verticillium*, wilt of egg plant and also some species of *Pythium* parasitizes species of *Phytophthora*.

26. Several yeasts such as *Pichia gulliermondii* also parasitize and inhibit the growth of plant pathogenic fungi like *Botrytis* and *Penicillium*.

27. Nematophagous fungi such as *Catenaria auxiliaris, Hirsutella* sp., *Verticillium, Chlamydosporium* and *Nematophthora gynophila* are parasitic on dagger nematode *Xiphenema* and the cyst nematode *Heterodera* and *Globodera*.

28. Root knot nematode (*Meloidogyne* sp.) is parasitized by the fungi *Dactylella, Arthrobotrys, Paecilomyces* and *Hirsutella* sp.

29. *Tuberculina maxima* parasitize the white pine blister rust fungus caused by *Cronartium ribicola*.

30. *Darluca filum* and *Verticillium lecani* parasitize several rusts such as wheat leaf rust (*Puccinia recondita*) and carnation rust (*Dianthus caryophyllus*), respectively.

31. *Ampelomyces quisqualis* parasitizes several powdery mildews while *Tilletiopsis sp.* parasitizes the cucumber powdery mildew fungus *Sphaerotheca fuligena*.

32. *Nectria inventa* and *Gonatobotrys simplex* which are parasitize pathogenic species of *Alternaria*.

33. *Peniophora gigantea* controls *Heterobasidion annosum* (*Fomes annosus*), the root and butt rot of pine stumps.

34. Keeping sweet potatoes at 28-32°C for two weeks helps the wounds to heal and prevents infection by *Rhizoctonia* and soft rotting bacteria.

35. *Alternaria, Botrytis* and *Stemphylium* sporulate at 360 nm (below) ultraviolet range, it is possible to control by eliminating certain light wave lengths.

36. Synthetic chemicals reported as effective inducers of systemic acquired resistance (SAR) against several pathogens such as salicylic acid, acetylsalicylic acid (aspirin) dichloroisonicotinic acid (INA).

37. *Trichoderma harzianum, Chaetomium globosum* and *Pseudomonas fluorescens* as potential biocontrol agent to reduce downy mildew.

38. Benzothiadiazole (CGA 245704) a new synthetic chemical plant activators that have no antimicrobial activiy but mimic the biological inducfion of SAR in both dicot and monocot crops against many diverse type of, not all pathogens.

39. *Trichoderma harzianum* sold as F-stop for control of several soil borne pathogenic fungi and *T. harzianum/T. polysporum* sold as BINAB-T for control of wood decays.

40. Control of wound of fruit trees caused by *Chondrostereum purpureum* by *Trichoderma viride.*

41. Chestnut blight (*Cryphonectria parasitica*) is controlled naturally and artificially through inoculation of hypovirulent strain of the same fungus.

42. The hypovirulent strains (*Cryphonectria parasitica*) carry virus like double stranded RNAs (ds RNAs). The ds RNAs apparently pass throngh mycelial anastomoses from the hypovirulent to the virulent strains and reduced the development of canker.

43. In 1969, Grente and Sauret coined the term hypovirulence to identify the avirulent strains of *Cryphonectria* (*Endothia*) *parasitica* found in healing cankers of chestnut trees. Subsequent research conducted by Mc Donald *et al.* (1978) suggested that a ds RNA (mycovirus) was responsible for the avirulence.

44. Resistance to anthracnose in cucumber (caused by *Colletotrichum lagnearum*) was induced systemically by localized infection with tobacco necrosis virus (TNV).

45 Nonpathogenic strains of *Fusarium oxysporum* f. sp. *batatas* control the pathogenic strains of *F. oxysporum* f. sp. *batatas* (wilt of sweet potato).

46. The mycophagus nematode *Aphelenchus avenae* parasitizes *Rhizoctonia* and *Fusarium* and the amoeba *Vampyrella* parasitizes the pathogenic fungi *Cochliobolus sativus* and *Gaeumannomyces graminis* and *Thielaviopsis basicola.*

47. Fungi *Pythium nunn* and *P. oligandrum,* protect potted orramentals and vegetables from pathogenic species of *Pythium* and *Talaromyces flavus* and binucleate *Rhizoctonia* porotect plants from the pathogenic *Rhizoctonia solani.*

48. *Phytophthora cinnamomi* (root and stem rot on pine), *Fusarium oxysporum* (wilt of tomato and Douglas fir seedlings), *Verticillium* wilt (cotton wilt) and *Phytophthora megasperma* and *Fusarium solani* (root and stem rot of soybesn) is cotrolled by ectomycorrhizal association of mycorrhizae.

49. Root and stem rot of pine (*Phytophthora cinnamomi*) is controlled by ectomycorrhizal association of *Pisolithus tinctorius. Phytophthora parasitica* root rot of citrus is also controlled by *Glomus fasciculatum.*

50. A fungal parasite *Coniothyrium minitans* parasitize *Sclerotinia sclerotiorum* on sunflower, *S. trifoliorum* on bean and *S. cepivorum* on onion. *Gliocladium roseum* effectively controls black root rot of cucumber caused by *Phomopsis sclerotioides*.
51. Conidia of *Cladosporium herbarum* or *Penicillium* sp. almost completely suppressed subsequent infection of developing fruits by *Botrytis cinerea*.
52. Citrus green mold (*Penicillum digitatum*) and *Botrytis* rot of strawberries are reduced by *Trichoderma viride*.
53. *Alternaria cassia* to be sold as CASST, is effective against the weed sicklepod (*Cassia obtusifolia*) growing in peanut and southern bean crops.
54. *Colletotrichum gloeosporioides* f. sp. *malvae* to be sold as Bio Mal, is effective against the weed round leaf mallow (*Malva pusilla*) growing in small grain fields.
55. *Phomopsis amaranticola* to be used as biocontrol agents against pigweed (*Amaranthus spp*), *Colletotrichum dematium* f. sp. *crotalariae* against showy crotolaria weed, *Alternaria helianthi* against cocklebur weed, *Alternaria macrospora* against spurred anoda weed and *Colletotrichum coccodes* against velvet leaf weed.
56. Widely used fungicides benomyl, thiabendazole and imazalil for the control of post harvest diseases.
57. Plant defense activators such as salicylic acid (SA), isonicotinic acid (IND), phenolic acid and commercially available benzothiadiazole (known as compound GA 245704) which activate the natural defense of the host (SAR).
58. Flutolanil sold as Monocut or Prostar is used against the basidiomycetes *Rhizoctoina, Sclerotium rolfsii, Corticium* and *Typhula*.
59. Vinclozolin sold as Ornalin, Ronilan or Vorlan is used against sclerotial producing ascomycetes *Botrytis, Monolinia, Sclerotinia*.
60. The most important acylalanine fungicide is metalaxyl, is effective against oomycetes *Pythium, Phytophthora* and several other downy mildews, it is sold as Ridomil.
61. Benomyl is sold as Benlet, Tersan 1991 and others. Most of the benzimidazoles are converted at the plant surface to methyl benzimidazole carbamate (MBC carbendazim) and this compound interferes with nuclear division of sensitive fungi.
62. Benomyl is particularly effective for powdery mildew of all crops, scab of apples, peaches and pecans, cherry leaf spot, blast of rice, black spot of roses and various *Sclerotinia* and *Botrytis* diseases. It is highly effective against *Rhizoctonia, Thielaviopsis, Ceratocystis, Fusarium* and *Verticillium*.
63. Thiabendazole is sold as Mertect. It is also a broad spectrum fungicide and is effective against post harvest diseases such as storage rots of citrus, apple, pears, bananas, potatoes and squash.

64. Thiophanate is sold as Topsin is effective against several root and foliage fungi effective turf grasses and vegetable crops.
65. Oxathiins were the first systemic fungicides to be discovered in 1966, it inhibit succinic dehydrogenase, an enzyme important in mitochondrial respiration.
66. Oraganophosphates include primarily Fosetyl - Al sold as Aliette. It is very effective against oomycetes such as *Phytophthora*, *Pythium* and downy midews.
67. Fosetyl-Al stimulates defense reactions and the synthesis of phytoalexins against oomycetes.
68. Kitazin (IBP) and Edifenphos (Hinosan) is the organophosphate fungicide, both effective against rice blast and Pyrazophos (Afugan) is effective against pwdery mildew and *Bipolaris* and *Drechslera* in various crops.
69. The pyrimidines include Dimethrimol (Milcurb), Ethirimol (Milstem), Bupirimate (Nimrod), Fenarimol (Rubigan), Nuarimol (Trimidal) are effectivc against powdery mildews. Triarimole is effective for the control of scab and powdery mildew of apple and pear. DPX-3217 (Curzate) and Previcur is effective against oomycetes.
70. The triazoles include several systemic fungicides such as RH-4 or Butrizol (Indar), Triadimefon (Bayleton), Triademenol (Baytan), Biloxazole (Baycor) or Bitertanol, Propiconazole (Tilt), Etaconazole (Vangard), Diclobutrazole (Vigil), Myclobutanil (Rally) and Difenoconazole (Score) and are effective against leaf spots, blights, powdery mildews, rusts, smuts and also applied as soil treatment.
71. Organophosphate fungicides are Kitazin (IBP), Ediphenphos (Hinosan), Triamiphos (Wepsin), Pyrazophos (Afugan, Euramil), Fosetyl-AL (Aliette).
72. Ethazol is systemic fungicide, sold as Turaban, Terrazole or Koban and effective against *Pythium* and *Phytophthora*.
73. The root rots of cotton (*Rhizoctonia bataticola*) and blight of pulse crops (*Phyllosticta phaseolina*) can be controlled by mixed cropping.
74. Organic sulphur compounds are (Dithiocarbamates) thiram, ferbam, nabam, maneb, zined and mancozeb.
75. Thiram (TMTD) is sold under various trade Name such as Arasan,Tersan,Tundas, Tulisan and effective against *Pythium*, *Fusarium* and *Protomyces* and it acts are an insecticides.
76. Ferbam is sold under trade such as Fermate, Karbam or Coromet and it is used to control foliage diseases of fruits and vegetables.
77. Ziram trade names are Cuman, Zerlate, Karbam white, Corozate and it is used against foliar diseases.
78. Nabam (Disodium ethylene bisdithiocarbamate) is sold as Dithane D-14, Parzate liquid or Dithane A-40 and effective against *Pythium*, *Fusarium* and *Rhizoctonia*.

79. Zineb (Zinc ethylene bisdithiocarbamate) is sold as Dithane Z- 78, Parzate C or Lonacol, it is used for foliar diseases such as blast of rice, early and late blight of potatoes and tomatoes, leaf blight of rice, ripe fruit rot of chilli and downy mildew of maize.

80. Maneb (Manganese ethylene bisdithiocarbamate) is sold as maneb and tersan LSR, Dithane M-45 or Indofil M-45 and it is very effective against *Alternaria* blights of tomato and potato, *Colletotrichum* anthacnose of tomato, cucurbit, bean, rust of uromyces spp, downy mildew of vegetables by *Pernospora, Plasmopara*.

81. Heterocyclic nitrogen compounds are Captan, Captafol or Difolatan, Glyodin, Folpet, Iprodion, Vinclozolin and Dyrene.

82. Aromatic compounds are PCNB, Terraclor, Cchloroneb, Dexon (soil fungicides), Dinocap (Karathane) and Chlorothalonil (Bravo or Daconil).

83. Dodine is a non- aromatic organic compound, it is used for the control of apple scab and fungi imperfecti and its mechanism of action is due to its binding with sterol in cell membrane.

84. Organomercuriales compounds are Ceresan, Ceresan M, Panogen, Agrosan GN, Ceresan wet, Aretan, Agallol (MEMC) and Ceresan dry (PAM). Quinone fungicides are chloranil (Spergon), Dichlone (Phygon) and Ceredan.

85. Systemic fungicides are Oxathiins [Carboxin (Vitavax), Oxycarboxin (Plantvax)], Benzimidazoles [Benomyl (Benlate), MBC or Carbendazim (Bavistin), Thiobendazole (TBZ), Fuberidazoles, Thiofanate (Topsin)], Acylalanines [Metalaxyl (Ridomil, Apron), Cyprofuram (Vinicur)], Pyrimidines [Diamethirimol (Milcurb), Ethirimol (Milstem), Bupirimate (Nimrod), Fenarimol (Rubigan), Nuarimol (Trimidal), Triarimol], Triazoles [Triadimefan (Bayleton), Triadimenol (Bayton), Bitertenol (Bacor), Boutrizol (Indar or RH-124), Propiconazole (Tilt), Etaconazole (Vangard), Piparazines (Triforine), Flusilazole (Eagle), Defenoconazole (Ciba Geigy) and Penconazole], Morpholines [Tridemorph (Calixin), Dodemorph (Meltatox)] and organic phosphate [Fosetyl- AL (Aleitte), Kitazin (IBP), Edifenphos (Hinosan), Pyrazophos (Afugan)].

86. Organic sulphur compounds are Ferbam, Naban, Ziram, Vapam, Thiram, Zineb and Maneb.

87. High volume sprayers droplet size is 0.5-3 mm, it sprays at the rate of 800-3500 litre/ha and low volume sprayers droplet size is 15-40μ, it sprays 15 to 250 litre/ha.

88. Spreader is material which improves the contact between fungicide and sprayed surface. *e.g.* mineral oils, Glyceride oil and soap etc. It reduces surface tension so that the fluid does not run-off easily.

89. Stickers (Adhesive) are substances added to spray or dust, which improve the adherence to plant surface. *e.g.* gum arabic, dextrins, oil, fish oil, milk casein, gelatin flour, starch, polyvinyl acetae.

90. Safeners are chemicals which reduce phytotoxic effect of other chemicals/ ingredients *i.e.* glyceride oil, lime in the spray.

91. Wetting agents are the materials which are used to ensure that no layer of air is present between a solid and liquid, it reduces the surface tension of particles. Eg. Polyethylene oxide and flour are good wetting agent.

92. Zineb, sulphur etc. are used as protectant, organomercuriales, lime and sulphur are used as eradicant and vitavax, antibiotics etc. are used as therapeutants.

93. Brodeaux mixture preparation is 4:4:5 (1.5 kg Cuso4; 1.5 kg lime: 200 liter water) and it always to be used in high volume sprayers.

94. Burgundy mixture was introduced by Masson in 1887, its preparation is 4:5:50 (4 kg CuSo4, 5 kg washing soda (Na_2CO_3 and 200 liter water).

95. Chestnut compound is discovered by Bewely in 1921 and it contains 2 part CuSo4, 11 parts $(NH_4)_2$ CO_3 (ammonium carbonate).

96. Chaubatia paste developed at the Government Fruit Research Station, Chaubattia, Almora (U.P. now in U.K.). It is a wood dressing furgicide and it's ratio is 800 g $CuCO_3$ (Copper carbonate) and 800 g red lead and one liter of lanolin or raw linseed oil.

97. Copper oxychloride fungicides are Blitox 4 per cent, Blitox-50, Cupramar, Fytolan and Bluecopper-50. Cuprous oxide preparations are Fungimar and Perenox.

98. Saprophytic yeast (*Candida oleophila*) and bacterium (*Pseudomonas syringae*) have been registered for biological control of postharvest decay of citrus fruit.

99. The temperature (32°C) that promotes the wound healing process in inhibitory to *Penicillium digitatum* and *P. italicum* but not *Geotrichum*. A higher temperature (40°C) is required to control *Geotrichum*.

100. Aspire™ is a commercial biofungicide, isolated from the saprophytic yeast (*Candida oleophila*).

101. Postharvest treatment of fruits with acetaldehyde vapour (500-2000ppm) for 24 hour prior to storage reduced decay. Gamma radiation control grey mould of strawberry and wrapping coating component used for the control of *Botrytis* and *Rhizopus* rots.

102. Benlate is highly affective against many pathogens of sugarcane such as *Colletotrichum*, *Ustilago*, *Cephalosporium* and *Fusarium* which are transmitted through the seed material.

103. Red rot infected seed setts are treated with hot water at 52°C for 18-20 minutes. By aerated steam, canes are heated at 52°C for 4-5 hours and by hot air at 54°C for 2 hours are very effective for the management of its affected canes. Treatment of seed setts with Aretan, Agallol or Emisan (0.25 per cent) for a 5-10 minutes dip helps only eradication of superficial inoculums and not deep-seated mycelium.

104. Deblossoming at bud burst stage alone and in combination with the spraying of NAA @ 200ppm in the first week of October is effective for management of mango malformation.

105. German scientist Frank was the first who coin the term 'Mycorrhiza' in 1885 for designating a symbiotic relationship between fungi and plant roots. Dangeared in 1900, was first to name a vesicular-arbuscular mycorrhiza (VAM). Pyronel in 1924 was the first to recognise the vesicular-arbuscular mycorrhizal fungi as *Endogone* sp. Mosse in 1956 was the first to demonstrate experimentally that *Endogone* sp. could produce VAM. In 1963, Grademann and Nicolson developed a procedure (wet sieving and decanting technique) for isolation of spores from soil and described some species. In 1974, Grademann and Trappe, described seven genera in their monograph of Endogonaceae *viz. Endogone, Glomus, Gigaspora, Modicella, Glaziella, Acaullospora* and *Sclerocystis*.

Multiple Choice Questions (Choose the correct answer)

1. **Which of the following is inorganic mercury fungicide?**
   (A) $HgCl_2$ (B) Agallol
   (C) Agrosan GN (D) All of the above

2. **Bordeaux mixture was discovered P.A. Millerdet of France in 1882, following his chance observation of a farmers practice for protection against:**
   (A) *Podospharra leuchotricha* on apple
   (B) *Plasmopara viticola* on grapevine
   (C) *Uncinula necator* on grapevine
   (D) *Venturia inaequalis* on apple

3. **Which of the following is the organic mercury fungicide?**
   (A) Agrosan G.N. (B) Seresan
   (C) Agallol (D) All of these

4. **Biological control of pigeon pea wilt can be achieved by the use of:**
   (A) *Bacillus subtilis* (B) *Fusarium semitectum*
   (C) *Curvularia lunata* (D) *Penicillium* sp.

5. **Match MBC:**
   (A) Benomyl (B) Thiram
   (C) Captan (D) Dinocap

6. **Fungicide Spiroxamine is:**
   (A) Sterol inhibitor (B) DNA inhibitor
   (C) Lignin inhibitor (D) RNA inhibitor

7. **Which of the following spray pump is used for low volume spraying?**
   (A) Gear type rotary pump (B) Pneumatic pump
   (C) Piston pump (D) None of these

8. **Which of the following spray pump is used for medium volume spraying?**
   (A) Roller vane rotary pump (B) Plunger pump
   (C) Gear type rotary pump (D) All of the above

9. **Which of the following spray pump is used for high volume spraying?**
   (A) Plunger pump (B) Gear type pump
   (C) Roller vane rotary pump (D) None of these

10. **Which of the following fungicide is used for the control of powdery mildew?**
    (A) Binapacryl (B) Dinocap
    (C) Dinoceli (D) Dinitrocresol

11. **Which of the following chemicals is effective in controlling root rots of *Brassicas?***
    (A) Quinazamide (B) Quintozene
    (C) Quinone (D) Quinolate

12. **What is the recommended composition of Bordeaux mixture for the control of late blight of potato:**
    (A) 4:4:50 (B) 3:3:50
    (C) 5:5:50 (D) 2:2:50

13. **Trade name of Edifenphos is:**
    (A) TATA (B) Hinosan
    (C) DCNA (D) None of these

14. **Effective control of downy mildew of pearl millet can be achieved by:**
    (A) Metalaxyl seed treatment (B) Thiram seed treatment
    (C) Carbendazim seed treatment (D) Carboxin seed treatment

15. ***Verticillium* wilt of cotton can be most effectively controlled by:**
    (A) Deep ploughing (B) Soil solarization
    (C) Soil fumigation (D) Crop rotation

16. ***Trichoderma harzianum* is useful for biological control of:**
    (A) *Phytophthora infestans* (B) *Puccinia recondita*
    (C) *Rhizoctonia solani* (D) *Colletotrichum falcatum*

17. **Most benzimidazoles are converted at plant surface to:**
    (A) MBC (B) Oxanthiins
    (C) Rovral (D) None of these

18. **Pyrimidines fungicides effective against:**
    (A) Leaf spot (B) Powdery mildew
    (C) Scab (D) Soft rot

19. **Which of the following chemical is effective in controlling blast of rice?**
    (A) Hinosan (B) Quintozene
    (C) Quinone (D) Dinocap

20. **Chestnut compound is prepared by mixing:**
    (A) Copper carbonate (2 parts) and ammonium chloride (11 part)
    (B) Copper sulphate (2 parts) and ammonium carbonate (11 part)
    (C) Copper chloride (2 parts) and ammonium sulphate (11 part)
    (D) None of these

21. **The Bordeaux mixture was discovered by:**
    (A) R.E. Buchanan (B) Millerdet
    (C) S.F. Luria (D) G.W. Fischer

22. **Which of the following fungicides would you recommended for the control of powdery mildew of wheat?**
    (A) Agrosan (B) Calixin
    (C) Blitox (D) Bordeaux mixture

23. **For the control of loose smut of wheat which one of the following treatment should be given:**
    (A) Soil drenching with fungicides (B) Hot water treatment of seeds
    (C) Spraying with fungicides (D) Crop rotation

24. **Which of the following chemical would you recommended for the control of Tanjor wilt of coconut?**
    (A) Actidione (B) Calixin
    (C) Aureofungin (D) None of these

25. **Loose smut of wheat is controlled by:**
    (A) Ridomil (B) Bavistin
    (C) Hinosan (D) Pyrimidine

26. The recommended dose of Bordeaux mixture for the control of late blight of potato in early stages is:

(A) 2:2:50 (B) 3:3:50

(C) 4:4:50 (D) 5:5:50

27. Chaubatia paste is prepared by mixing:

(A) Copper carbonate (800gm), red lead (800gm) and 1 liter lanonil oil

(B) 2 parts copper carbonate and 11 parts ammonium carbonate

(C) 2 parts copper chloride and 11 parts ammonium sulphate

(D) None of these

28. Which of the following varieties of sugarcane is resistant to smut disease?

(A) BO 21 (B) CO 285

(C) CO 421 (D) CO 300

29. Streptomycin kills bacteria by binding/inhibiting:

(A) Cell wall synthesis (B) Protein synthesis

(C) Lipid synthesis (D) Nucleic acid synthesis

30. Cylohexamide (Acitidione) kills fungi by inhibiting:

(A) Both protein and DNA synthesis (B) Lipid synthesis

(C) Cell wall synthesis (D) None of these

31. *Phytophthora palmivora* can be used in bio-control of:

(A) Milk-vine weed of citrus (B) Water hyacinth

(C) *Phalaris minor* weed of wheat (D) Joint vetch of rice

32. *Colletotrichum gloeosporioides* f. sp. *aeschynomene* can be used in bio-control of:

(A) Milk-vine weed of citrus (B) Water hyacinth

(C) Phalaris minor seed of wheat (D) Joint vetch of rice

33. *Alternaria cassiae* trade name (CASST) is used for the control of:

(A) Water hyacinth (B) Sicklepod weed of soybean and cotton

(C) *Phalaris minor* (D) None of these

34. The most effective fungicide popularly used as post harvest dip is:

(A) Thiabendazole (B) Captafol

(C) Mancozeb (D) Carbendazim

35. Organo-Tin compound is:

(A) Brestanol (B) Iprodion

(C) Folpet (D) Dodin

36. **Which one among the following is an organomercurial?**
   (A) Ceresan (B) Thiram
   (C) Brestan (D) Captan
37. **Metallic dithiocarbamate compound is:**
   (A) Ceresan (B) Ziram
   (C) Panogen (D) None of these
38. **Which one among the following is a systemic fungicide?**
   (A) TBZ (B) Ceresan
   (C) Panogen (D) None of these
39. **Antifungal antibiotic is:**
   (A) Tetracyclin (B) Agrimycin
   (C) Cytovirin (D) Aureofungin
40. **Edifenphos is available under the trade name:**
   (A) Hinosan (B) Fytolan
   (C) Kitazin (D) Bavistin
41. **Which of the following is an entomogenous?**
   (A) *Beauveria bassiana* (B) *Trichothecium roseum*
   (C) *Choanephora cucurbitarum* (D) *Trichoderma harzianum*
42. **Which of the following is a myconematicides?**
   (A) *Beauveria bassiana* (B) *Pacecilomyces lilacinus*
   (C) *Hirsutella thompsonii* (D) None of these
43. **Match fungal nematicides:**
   (A) *Arthrobotrys robusta* (B) *Colletotrichum gloeosporioides*
   (C) *Alternaria cassiae* (D) None of these
44. **Match the fungal insecticides:**
   (A) *Hirsutella thompsonii* (B) *Altenaria cassiae*
   (C) *Verticillium chalmydosporium* (D) *Phytophthora palmivora*
45. **Which of the following is a fungal herbicide?**
   (A) *Hirsutella thompsonii* (B) *Alternaria alternata*
   (C) *Verticillium lecanii* (D) *Phytophthora palmivora*
46. **Match fungal herbicide:**
   (A) *Colletotrichum coccodes* (B) *Beauveria bassiana*
   (C) *Paecilomyces lilacinus* (D) *Verticillium lecanii*

47. Kitazin is:

(A) Organophosphorus compound (B) Metalaxyl

(C) Heterocyclic compound (D) None of these

48. Successful use of systemic fungicide for plant disease control was first demonstrated by:

(A) Van Schmilling and Marshal Kulka (B) Millerdet

(C) Horsfall and Dimond (D) H.H. Flor

49. Affectivity of copper sulphate in the control of stinking smut of wheat was first showed by:

(A) Millerdet (1882) (B) Prevost (1804)

(C) Robertson (1821) (D) Tillet (1888)

50. Who was the first to try copper sulphate in India:

(A) Millerdet (1882) (B) Ozanne (1885)

(C) Mason (1887) (D) None of these

51. Who introduced the organic mercuriales for seed treatment of wheat for smut control:

(A) Lawrence (B) Hilson

(C) Reihm (D) None of these

52. Who reported the fungicidal properties of formaldehyde:

(A) Trillat (1888) (B) Mason (1887)

(C) Robertson (1825) (D) None of these

53. Who reported the fungitoxic activity of dithiocarbamates?

(A) Tisdale and Williams (1934)

(B) Cunningham and Sharvelle (1940)

(C) Kittleson (1952)

(D) None of these

54. Who reported the development of aureofungin, a new antibiotic for use in plant disease control?

(A) Chatrath his co-worker (B) Thirumulachar and his co-worker

(C) Mehta and Singh (D) Uppal

55. What is the possible bio-control agent of foot rot of wheat caused by *Sclerotium rolfsii*?

(A) *Trichoderma viride* (B) *Bacillus sbutilis*

(C) *Pencillium* spp (D) None of these

56. **Who reported successful control of loose smut of wheat through seed dressing with carboxin for the first time in India?**
 (A) Chatrath and his worker (B) Uppal
 (C) Daster (D) None of these

57. **The practical use of soil solarization treatment for management of disease started from:**
 (A) Germany (B) Israel
 (C) Canada (D) Japan

58. **Metalaxyl is a systemic fungicide effective against:**
 (A) Powdery mildews (B) Downy mildews
 (C) Rust (D) Smut

59. **Triadimefon is a systemic fungicide effective against:**
 (A) Powdery mildew (B) Downy mildew
 (C) Rust (D) Smut

60. **Film-forming compounds is:**
 (A) Antitranspirant (B) Organic sulphur
 (C) Quinones (D) Aromatic compound

61. **Match chlorothalonil:**
 (A) Botran (B) Bravo
 (C) Rovral (D) None of these

62. **One per cent Bordeaux mixture is obtained by mixing the ingredients in following ratio:**
 (A) 5:5:50 (B) 4:4:50
 (C) 3:3:50 (D) 1:1:50

63. **Dichloran (DCNA) sold as:**
 (A) Botran (B) Bravo
 (C) Rovral (D) Chipco

64. **What is the oral toxicity (LD_{50} in mg/kg body weight) of streptomycin?**
 (A) 3000-6000 (B) 3200-3820
 (C) 6750-7400 (D) 9000

65. **What is the oral toxicity (LD_{50} in mg/kg body weight) of fixed copper?**
 (A) 3000-6000 (B) 3200-3820
 (C) 6750-7400 (D) 9000

66. **What is the oral toxicity (LD_{50} in mg/kg body weight) of carboxin?**
(A) 421 (B) 3820
(C) 2000 (D) 1200

67. **What is the oral toxicity (LD_{50} in mg/kg body weight) of oxycarboxin?**
(A) 421 (B) 3820
(C) 2000 (D) 1200

68. **What is the oral toxicity (LD_{50} in mg/kg body weight) of dithane-M 45?**
(A) >8000 (B) 140
(C) >1700 (D) 780

69. **What is the oral toxicity (LD_{50} in mg/kg body weight) of thiram?**
(A) 133 (B) 140
(C) 421 (D) 780

70. **What is the oral toxicity (LD_{50} in mg/kg body weight) actidione?**
(A) 50-70 (B) 80-90
(C) 110 (D) 133

71. **What is the oral toxicity (LD_{50} in mg/kg body weight) of captan?**
(A) 4000 (B) 6000
(C) 8000 (D) 10000

72. **What is the oral toxicity (LD_{50} in mg/kg body weight) of karathane?**
(A) 450 (B) 580
(C) 756 (D) 980

73. **What is the oral toxicity (LD_{50} in mg/kg body weight) of PCNB?**
(A) 500 (B) >1700
(C) 1000 (D) 1500

74. **Iprodione sold as:**
(A) Botran (B) Bravo
(C) Rovral (D) None of these

75. **Acylalanine fungicide is effective against the:**
(A) Oomycetes (B) Deuteromycetes
(C) Zygomycetes (D) None of these

76. **Match halogenated hydrocarbons:**
(A) Methyl bromide (B) Triazoles
(C) Pyrimidines (D) None of these

77. **Imazalil is effective against many ascomycetes sold as:**
    (A) Turban (B) Koban
    (C) Fungaflor (D) None of these

78. **A heterocyclic nitrogenous compound is:**
    (A) Dinocap (B) Dodin
    (C) Captafol (D) Cyclohexamide

79. **Triphenyltin acetate (TPTA) is effective against:**
    (A) *Septoria* (B) *Botrytis*
    (C) *Monilinia* (D) None of these

80. **A fungicide is to be supplied to a farmer to be used at the rate of 0.2 per cent what will be the required quantity of fungicide for 4000 liters of water:**
    (A) 8.0 kg. (B) 0.8 kg.
    (C) 2.0 kg. (D) 4.0 kg.

81. **Damping off can be controlled by seed treatment with:**
    (A) Agrosan G.N. 0.25 per cent (B) B.H.C. 5 per cent
    (C) Agrosan G.N. 2.5 per cent (D) None of these

82. **Ethazol is a seed, soil and turf fungicide effective against damping off, root and stem rots caused by *Pythium* and *Phytophthora*, sold as:**
    (A) Turban (B) Triadimefor
    (C) Milcurb (D) None of these

83. **B.B. Mundkur started work on control of cotton wilt through varietal resistance in:**
    (A) 1940 (B) 1945
    (C) 1948 (D) 1952

84. **Riehm introduced seed treatment with organic mercuric compound in:**
    (A) 1913 (B) 1915
    (C) 1917 (D) 1919

85. **Tisdale discovered the first dithiocarbamate (thiram) in:**
    (A) 1934 (B) 1938
    (C) 1942 (D) 1945

86. **Flutolanil sold as:**
    (A) Moncut or Prostar (B) Rovral or Chipco – 26019
    (C) Ornalin, Ronilan or Vorlan (D) None of these

87. Vinclozolin sold as:

(A) Moncut or Prostar
(B) Rovral or Chipco – 26019
(C) Ornalin, Ronilan or Vorlan
(D) None of these

88. Tetracyclin is available under the trade name:

(A) Sreptocycline
(B) Griseofulvin
(C) Blasticidine
(D) None of these

89. Ergot of bajra can be controlled by the use of:

(A) Ziram or Zineb
(B) Sulphur dust
(C) Bordeaux mixture
(D) All of the above

90. Loose smut of wheat can be controlled by seed treatment with:

(A) Vitavax 100 g/kg seed
(B) Vitavax 2.5 g/kg seed
(C) Vitavax 0.025 g/kg seed
(D) All of the above

91. Which of following are organic sulphur fungicides:

(A) Ziram
(B) Zineb
(C) Thiram
(D) All of the above

92. Which of the following are the $CuSO_4$ fungicides:

(A) Bordeaux mixture
(B) Burgundy mixture
(C) Bordeaux paste
(D) All of these

93. Which of the following equipment can be effectively used for spraying an orchard crop?

(A) Barrel pump
(B) Heli pump
(C) Foot sprayers
(D) Sikar pump

94. Which of the following is systemic fungicide?

(A) Vitavax
(B) Brestanol
(C) Benlate
(D) All of these

95. When a hectare of ground crop is completely covered by spraying less than 60 liters of liquid is termed as:

(A) Low volume spray
(B) Very low volume spray
(C) Medium volume spray
(D) High volume spray

96. Giving complete coverage to one hectare of crop by spraying 60 to 250 liters of liquid is known as:

(A) Low volume spray
(B) Very low volume spray
(C) Medium volume spray
(D) High volume spray

97. Endoxerosis is due to:

(A) High temperature
(B) Low temperature
(C) Water deficiency
(D) Excess water

98. Giving complete coverage to one hectare crop with more than 250 liters of liquid is called:

(A) Low volume spray
(B) Medium volume spray
(C) High volume spray
(D) Very low volume spray

99. Which of the following type of pump is used in hand compression sprayers:

(A) Air or pneumatic pump
(B) Rotary pump
(C) Piston pump
(D) All of the above

100. Foliar blight can be controlled by:

(A) Agrosan GN
(B) Thiram
(C) Zineb
(D) All of the above

101. Which of the following fungicides is a polyploiding agent:

(A) Metalaxyl
(B) Benlate
(C) Thiram
(D) Kerathane

102. The most commonly recommended fungicide for the management of downy mildew of cucurbits is:

(A) Tafason
(B) Indofil – M-45
(C) Thiram
(D) Captaf

103. What is the oral toxicity (LD_{50} in mg/kg body weight) of streptomycin:

(A) 9000
(B) 6750-7500
(C) 3200-3820
(D) 3000-6000

104. Copper compound were first used for seed treatment of:

(A) Wheat
(B) Rice
(C) Millet
(D) Cotton

105. What is the oral toxicity (LD_{50} in mg/kg body weight):

(A) 3000-6000
(B) 3200-3820
(C) 6750-7500
(D) 9000

106. The amount of water used in low volume spray machinery is about:

(A) 25L/ha
(B) 100L/ha
(C) 500L/ha
(D) 700L/ha

107. Burgundy mixture contains:

(A) Copper sulphate and line

(B) Copper sulphate sodium carbonate and water

(C) Copper sulphate gypsum and water

(D) None of these

108. A chemical which is aimed to seek out the fungus at rest either before or after the fungus has found host is called:

(A) Contact Fungicide
(B) Eradicant fungicide
(C) Residual fungicide
(D) None of these

109. Bordeaux mixture and burgundy mixture are used to control:

(A) Early blight of potato
(B) Late blight of potato
(C) Soil borne diseases
(D) All of the above

110. Early and late blight of potato can be controlled by:

(A) Ditahane M-45
(B) Ditahane Z-78
(C) Dithane M-22
(D) All of the above

111. Club root crucifers can be controlled by:

(A) PNCB
(B) Dithane M-45
(C) Dithane Z-78
(D) None of these

112. The cultivar Kufri swarna is also resistant to:

(A) Wart disease of potato
(B) Club root of crucifers
(C) Blight of colocasia
(D) None of these

113. Which of the following varieties of potato is found immune to wart:

(A) Kufri Sheetman
(B) Kufri Bahar
(C) Kufri Kanchan with red skin
(D) Kufri Sherpa

114. Alternaria blight in mustard can be controlled by:

(A) Treating seed with hot water
(B) Soil treatment with Aldrin
(C) Bordeaux mixture
(D) None of these

115. The mode of action of fungicide metalaxyl for the inhibition of the fungal growth is on:

(A) Sterol biosynthesis pathway
(B) Ribosomal RNA synthesis
(C) Blocking flavin enzyme
(D) Degradation of cell wall

116. Acylanilide fungicide is effective against:

(A) Hyphomycetes
(B) Zygomycetes
(C) Oomycetes
(D) None of these

117. Diomethrimol fungicide was first reported by:

(A) Elias *et al.* (1968) (B) Babbington *et al.* (1969)

(C) Cohen and Caffey (1986) (D) None of these

118. Kitazin and hinosan have been found highly effective against the:

(A) Blast disease of rice (B) Wart disease of potato

(C) Black tip of mango (D) None of these

119. The actidion acts as:

(A) Antifungal (B) Antibacterial

(C) Antiviral (D) Antimycoplasmal

120. Noformicin antibiotic effectively inhibits the infection of:

(A) Leaf blight of rice (B) Tobaccco mosaic

(C) Wilt of brinjal (D) Root knot of vegetable

121. Dodine is sold as:

(A) Triazoles (B) Syllit

(C) Super Tin (D) None of these

122. Fentin hydroxide is sold as:

(A) Triazoles (B) Cyprex

(C) Super Tin (D) None of these

123. Match polyene antibiotic:

(A) Aureofungin (B) Pimaricin and nystatin

(C) Gryseofulvin (D) None of these

124. Match heptaene antibiotic:

(A) Aureofungin (B) Gryseofulvin

(C) Terramycin (D) All of the above

125. Which of the following belongs to dithiocarbamate group of fungicides:

(A) TPTH (B) TMTD

(C) Thiophanate (D) TPTA

126. The best means of managing the ergot disease in pearl millet is through the use of:

(A) Resistance varieties (B) Use of clean seed

(C) Chemical control (D) None of these

127. Intercropping of pearl millet with ________ is reported to reduce incidence of ergot disease :

(A) Mung bean (B) Urd bean

(C) French bean (D) Pigeon pea

128. Vitavax is usually used as:

(A) Root application
(B) Paint application
(C) Seed dresser
(D) Foliar fungicide

129. Foot sprayer is classified as:

(A) Low volume sprayer
(B) Mist sprayer
(C) Ultra volume sprayer
(D) High volume sprayer

130. Kittleson's killer compound is:

(A) Captan
(B) Thiram
(C) Benomyl
(D) None of these

131. Metalaxyl resistance has been observed against:

(A) *Phytophthora infestans*
(B) *Venturia inequalis*
(C) *Drechslera oryzae*
(D) *Erysiphe graminis*

132. Thiram belongs to:

(A) Dithiocarbomate
(B) Copper compound
(C) Mercurial compound
(D) Heterocyclic nitrogenous compound

Fill in the Blanks with Correct Answer

1. In 1914________$HgCl_2$ was used in India for first time for steeping seed potato to prevent *Rizoctonia* attack.

   Burns

2. Malformation can be reduce by spraying of ________during October.

   NAA @ 200 ppm

3. Chemically zineb is called zinc ________bisdithiocarbamate.

   Ethylene

4. Maneb is chemically known maganese ethylene________

   bisdithiocarbamate

5. Nabam is chemically known disodium ethylene ________

   Bisdithiocarbamate

6. Chemically vapam is called sodium________Bisdithiocarbamate.

   Methyl

7. Thiram is chemically known ________

   Tetramethyl thiuram disulphide

8. Ziram is chemically known as ________

   Zinc dimethyl dithiocarbamate

9. When________ is added to maneb it forms dithane M-45.

   Zinc

10. Bordeaux paste________and chestnut compounds are copper based home made fungicides.

    Burgundy mixture

11. ________paste contains 800gm of copper carbonate, 800gm red lead and one liter of raw linseed oil or lanolin.

    Chaubatia

12. ________contain 2 part copper sulphate and 11 part ammonium carbonate.

    Chestnut compound

13. Quinones are naturally present in plants and often exhibit ________activity.

    Antimicrobial

14. Penicillin is a valuable ________obtained from *Penicillium*.

    Antibiotics

15. The introduction of systemic fungicides in________ is a major landmark in the history of fungicidal control of plant.

    1966

16. Methyl-N-(1-butylcarbamyl)-2-benzmidazol carbamate is known as________

    Benomyl

17. Methyl-2-benzimidazol carbamate (MBC) is known as________

    Carbendazim

18. Oxathiins is a group of ________compounds having systemic fungicidal properties.

    Heterocyclic

19. Carboxin (vitavax) is chemically________

    5-6dihydro-2-methyl-1,4-oxathiin-3-carboxanilide (DMOC)

20. Chemically oxycaboxin (plantvax) is________

    2, 3 dyhydro-5-carboxanilide-6-methyl-1, 4- oxathiin-4, 4-dioxida (DCMOD)

21. Vitavax is not effective against________

    Deuteromycetes

22. ________is one of the heterocyclic ring compound which give systemic fungicidal derivatives.

    Morpholines

23. When spray involve in large quantity of liquid per unit area, it is termed as________

    High volume

24. Some of the chemical inhibit spore production without affecting the growth of vegetative hyphae are called________

    Antisporulant

25. When spray involve in small quantity of liquid per unit area, it is termed as________

    Low volume

26. ________is a material which improves the contact between the fungicide and the sprayed surface.

    Spreader

27. ________is a substance added to spray or dust which improved its adherence to plant surface.

    Sticker

28. ________is a chemical which reduce the phytotoxicity of another chemical.

    Safener

29. In 1889________for the first time, used fungicide in combination with insecticides.

    Weed of USA

30. In 1897, Bolley was the first to use formaldehyde for ________control.

    Wheat smut

31. In 1904, Lawrence used Bordeaux mixture for the first time in India against ________

    ***Leaf spot of groundnut (Cercospora* spp)**

32. In 1906, Sulphur was used for the first time in India (Mumbai) on________

    Grapes

33. In 1885 ________was the first person to use a fungicide in India.

    Ozzan

34. In 1929, Uppal reported the control of brown spot of rice by seed treatment with________

    Organomercuriales

35. In 1964________and his co-worker reported the development of aureofungin, a new antifungal antibiotic for use in plant disease control.

    Thirumulachar

36. First antibiotics are discovered by a microbiologist________in 1928 Scottland.

    Alexender flaming

37. The methods known soil solarization was first used in the field of________

    Israel

38. Stevens (1960) has discussed ________methods of disease control.

    Cultural

39. 'An informative book on systematic fungicides' was edited by________

    RW Marsh (1972)

40. Brinjal or other Solanum species that grow year round serve as the alternate host of Puccinia penneseti, causing________in bajra.

    Rust

True or False

1. 2, 4 D @ 10-20 ppm is used to control fruit drop in citrus/mango. (**True**)
2. Metalaxyl is one of the best systematic fungicides against besidiomycetes. (**False**)
3. Benomyle is very effective on oomycetes and on some dark spored imperfect fungi. (**False**)
4. Thiabendazole is sold as Mertect and it is effective against many imperfect fungi. (**True**)
5. Carboxin is sold as vitavax. It is used as seed treatment and is effective against damping off disease. (**True**)

6. J.C. Luthra and his associates developed the solar treatment of wheat seed for control of loose smut. (**True**)
7. In 1887 Mason of France introduced Burgundy mixture in which quick lime of Bordeaux was substituted by Na_2CO_3. (**True**)
8. Chlorine (Cl_2) and hydrogen chloride (HCl) are toxic at 0.1 ppm. (**True**)
9. Ethylene (CH_2CH_2) is toxic at 0.05 ppm. (**True**)
10. In biofumigation the biocidal products of hydrolysis of glucosinolates contained in *Brassica* are isothiocyanates. (**True**)
11. Hot water treatments can be used to free cabbage from Leptosphaeria maculans, the causal agent of black leg disease of cabbage. (**True**)
12. Fungicides with apoplastic movement generally accumulate at tips and margins of the leaves. (**True**)
13. Biocontrol agent *Trichoderma* sp. produces more growth and sporulation in high moisture conditions. (**False**)
14. Compressed air sprayers can build up pressure in the tank upto 12 kg per cm^2. (**True**)
15. Biocontrol agent Trichoderma sp. produces maximum antibiotics at pH 7. (**False**)
16. The leaf spot and blight of guar caused by *Xanthomonas cyamopsidis* can be managed by hot water treatment at 56°C for 10 minutes. (**True**)
17. Surface active substances like soap act as emulsifying agent. (**True**)
18. Lime act as sticker to reduce copper sulphate. (**False**)
19. Most of the japonica rice varieties are known to exhibit high resistance to bacterial pathogen. (**True**)
20. Fluorescent siderophores produced by *Pseudomonas fluorescence* chelates iron available in the soil, depriving pathogen of its iron requirements. (**True**)
21. *Fusarium sambucinum* can be utilized as biocontrol agent for against *Claviceps purpurea* causal agent of ergot of bajra. (**True**)
22. The chemical capable of destroying a pathogen at its source is called as protectant. (**False**)
23. The fungicidal formulation mancozeb contains zinc and maneb. (**True**)
24. Per cent disease index is generally used to calculate proportion of the total area of plant tissue affected by disease. (**True**)
25. Trizollium chloride test is especially done for detection of banana bunchy top virus and cassava mosaic virus. (**True**)

Match the Following

1. **Match the important antagonistic microorganisms that reduce the amount of pathogenic inoculums:**

Antagonist	*Pathogen*
I. *Catenaria auxiliaris*	(A) *Meloidogyne* sp.
II. *Verticillium chalmydosporium*	(B) *Heterodera* sp.
III. *Dactylella oviparasitica*	(C) *Heterodera* and *Globodera* spp
IV. *Pasteuria penetrns*	(D) *Meloidogyne javanica*

Answer

I	*II*	*III*	*IV*
C	*B*	*A*	*D*

2. **Match the important antagonistic microorganisms that reduce the amount of pathogenic inoculums:**

Antagonist	*Pathogen*
I. *Gliocladium virens*	(A) *Sclerotinia sclerotiorum* and *Pythium* sp.
II. *Verticillium lecani*	(B) *Uromyces dianthi* and *U. appendiculatus*
III. *Pseudomonas fluorescens*	(C) *Sclerotium rolfsii* and *Rhizoctonia solani*
IV. *Peniophora gigantea*	(D) *Heterobasidium annosum*

Answer

I	*II*	*III*	*IV*
A	*B*	*C*	*D*

3. **Match the important antagonistic microorganisms that reduce the amount of pathogenic inoculums:**

Antagonist	*Pathogen*
I. *Chaetomum globosum*	(A) *Fusarium roseum*
II. *Gliocladium roseum*	(B) *Aphanomyces, Fusarium, Pythium* and *Rhizoctonia*
III. *Penicillium oxalicum*	(C) *Phomopsis sclerotioides*
IV. *Phialophora graminicola*	(D) *Gaeumannomyces graminis*

Answer

I	*II*	*III*	*IV*
A	*C*	*B*	*D*

4. **Match the following important examples of direct protection by biological control agent:**

Disease	Biological control agent
I. Root and butt rot of conifers (*Heterobasidion annosum*)	(A) *Peniophora gigantean*
II. Chestnut blight (*Endothia parasitica*)	(B) Hypovirulunt strain of the pathogen
III. Fusarium wilt of sweet potato	(C) Non-Pathogenic strain of the same pathogen (*F. oxysporum* f. sp. *batatas*)
IV. Botrytis rot of grape and strawberries (*B. cineria*)	(D) *Trichoderma* sp.

Answer

I	*II*	*III*	*IV*
A	*B*	*C*	*D*

5. **Match the following important examples of direct protection by biological control agent:**

Disease	Biological control agent
I. Cucumber powdery mildew (*Sphaerotheca fuliginea*)	(A) *Amphelomyces quisqualis*
II. Wheat leaf rust (*Puccinia recondita*)	(B) *Daluca filum*
III. Citrus green mild (*Penicillium digitatum*)	(C) *Trichoderma viride*
IV. Tomato wilt (*F. oxysporum* f. sp. *lycopersici*)	(D) Mycorrhizae

Answer

I	*II*	*III*	*IV*
A	*B*	*C*	*D*

6. **Match the following important examples of direct protection by biological control agent:**

Disease	Biological control agent
I. Crown gall of pome, stone and small fruit (*Agrobacterium tumefaciens*)	(A) Strain K84 of *A. radiobactor*
II. Brown rot of peach (*Monilinia fructicola*)	(B) *Bacillus subtilis*
III. Fire blight of apple (*Ervinia amylovora*)	(C) *Erwinia herbicola*

IV. Bacterial leaf steak of rice (*Xanthomonas translucens* pv. *oryzicola*)	(D) Isolates of *Erwinia* and *Pseudomonas*

Answer

I	*II*	*III*	*IV*
A	*B*	*C*	*D*

7. Match the following important examples of direct protection by biological control agent.

Disease	*Biological control agent*
I. Root rot of pinus infection (*Phytophthora cinnamomi*)	(A) *Leucopaxillus cerealis* var. *piceina*
II. Root rot of citrus (*Phytophthora parasitica*)	(B) *Glomus fasciculatum*
III. Lettuce drop (*Sclerotinia minor* and *S. trifoliorum*)	(C) *Sporidesmium sclerotiorum*
IV. Leaf spot disease of cane palm (*Bipolaris* sp.)	(D) *Trichoderma viride*

Answer

I	*II*	*III*	*IV*
A	*B*	*C*	*D*

8. Match the following disease suppression by representative biological control:

Disease (Pathogen)/Crop	*Bio-control agent/mechanism involved*
I. Damping off (*Pythium ultimum*) soybean, Pea	(A) *Pseudomonas putida*, competition
II. Frost injury (*Pseudomonas syringae*), Corn	(B) *Erwinia herbicola*, competition
III. Damping off (*Pythium ultimum*), Cotton	(C) *Pseudomonas fluorescens*, competition
IV. Fusarium wilt (*Fusarium oxysporum*), Carnation	(D) *P. putida*, competition

Answer

I	*II*	*III*	*IV*
A	*B*	*C*	*D*

12. Match the following disease suppression by representative biological control:

Disease (Pathogen)/Crop	*Bio-control agent/mechanism involved*
I. Damping off (*Pythium splendans*), Tomato	(A) *P. aeruginosa*/competition
II. Damping off (*Pythium ultimum*) Rye, cucumber	(B) *Enterobacter cloacae*/parasitism
III. Fusarium wilt (*Fusarium oxysporum*), Carnation	(C) *Pseudomonas* sp./induced resistance
IV. Halo blight (*Pseudomonas syringae* pv. *syringae*)	(D) *P. fluorescens*/induced resistance

Answer

I	*II*	*III*	*IV*
A	*B*	*C*	*D*

13. Match the following disease suppression by representative biological control:

Disease (Pathogen)/Crop	*Bio-control agent/mechanism involved*
I. Tobacco necrosis virus, Tobacco	(A) *P. fluorescens*/induced resistance
II. Anthracnose (*Colletotrichum orbiculare*), cucumber	(B) *Pseudomonas* spp/induce resistance
III. Take all wheat (*Gaeumanno-myces graminis* var. *tritici*	(C) *Pseudomonas fluorescens/antibiosis*
IV. Seedling blight and foot rot (*Fusarium culmorum*), wheat	(D) *Erwinia herbicola*

Answer

I	*II*	*III*	*IV*
A	*B*	*C*	*D*

14. Match the following metabolites produced by fluorescent Pseudomonas spp active against plant pathogen:

Metabolites	*Pathogen/Host*
I. Acetylphloroglucinols-2, 4-diaceryl-phloroglucinols	(A) *Gaeumannomyces graminis* (wheat)
II. Oomycin A	(B) *Pythium ultimum*/cotton
III. Phenazine	(C) *G. Graminis* var. *tritici*/wheat
IV. Pyrrolnitrin	(D) *Rhizoctonia solani*/cotton

Answer

I	*II*	*III*	*IV*
A	*B*	*C*	*D*

15. Match the following metabolites produced by fluorescent Pseudomonas spp active against plant pathogen:

Metabolites	*Pathogen/Host*
I. Hydrogen cyanide siderophore	(A) *Thielaviopsis basicola*/tobacco
II. Pyoverdine	(B) *Pythium ultimum*/cotton
III. Pyochelin	(C) *Pythium splendans*/tomato
IV. Pyoluteorin	(D) *Pythium ultimum*/cotton

Answer

I	*II*	*III*	*IV*
A	*B*	*C*	*D*

16. Match the following seed borne disease control with fungicides:

Crop/Disease/Fungi	*Fungicide/Antibiotics*
I. Blast of rice (*Pyricularia oryzae*)	(A) Blasticin, Aureofungin
II. Brown spot of rice (*Helminthosporium oryzae*)	(B) Carboxin + Captan
III. Hill bunt of wheat (*Tilletia carries/T. foetida*)	(C) Carbendazim
IV. Loose smut of wheat (*Ustilao segetum tritici*)	(D) Carboxin, Carbendazim

Answer

I	*II*	*III*	*IV*
A	*B*	*C*	*D*

17. Match the following seed borne disease control with fungicides:

Crop/Disease/Fungi	*Fungicide/Antibiotics*
I. Karnal bunt of wheat (*Neovossia indica*)	(A) Carboxin, Aureofungin
II. Downey mildew of maize (*Peronosclerospara philipinensis*)	(B) Metalazyl
III. Stripe disease of barley (*Drechslera graminea*)	(C) Orgonomercurials
IV. Downey mildew of sorghum (*Sclerospora sorghi*)	(D) Orgonomercurials, Thiram

Answer

I	*II*	*III*	*IV*
A	*B*	*C*	*D*

18. Match the following seed borne disease control with fungicides:

Crop/Disease/Fungi	*Fungicide/Antibiotics*
I. Damping off mungbean (*Rhizoctonia solani*)	(A) Benomyl, Carbendazim, Carboxin
II. Leaf spot of urbean (*Colletotrichum capsici*)	(B) Carbenzazim+Thiram
III. Blight of gram (*Ascochyta rabiei*)	(C) Thiabendazole
IV. Root rot of gram (*Sclerotium rolfsii*)	(D) Carbendazim

Answer

I	*II*	*III*	*IV*
A	*B*	*C*	*D*

19. Match the following seed borne disease control with fungicides:

Crop/Disease/Fungi	*Fungicide/Antibiotics*
I. Seed rot of pea (*Fusarium* sp., *Aspergillus* sp.)	(A) Captan
II. Damping off of groundnut (*F. oxysporum*)	(B) Metadinitrobenzene
III. Seed decay of sunflower (*Macrophomina phaseolina*)	(C) Zineb
IV. Wilt of cotton (*F. oxysporum* f. sp. *vasinfectum*)	(D) Carbendazim

Answer

I	*II*	*III*	*IV*
A	*B*	*C*	*D*

20. Match the following seed borne disease control with fungicides:

Crop/Disease/Fungi	*Fungicide/Antibiotics*
I. Anthracnose of sugarbeet (*Colletotrichum dematium* f. sp. *spinaciae*)	(A) Metalaxyl, Thiram
II. Stem rot of okra (*Rhizoctonia bataticola*)	(B) Carboxin
III. Downy mildew of bajra (*Sclerospora graminicola*)	(C) Metalaxyl
IV. Covered smut of oat (*Ustilago kolleri*)	(D) Aureogungin

Answer

I	*II*	*III*	*IV*
A	*B*	*C*	*D*

21. Match the following commercial formulations (products) of bio-control agents available in India:

Commercial formulations	*Bio-agents*
I. Biotok	(A) *Bacillus subtilis*
II. Bioshield	(B) Pseudomonas fluorescens
III. Bioderma	(C) *Trichoderma viride+T. harzianum*
IV. Ecofit	(D) *Trichoderma viride*

Answer

I	*II*	*III*	*IV*
A	*B*	*C*	*D*

22. Match the following chemical compounds used for inactivation, suppression or elimination of plant viruses:

Host/Virus	*Chemical compound*
I. Tobacco/cucumber mosaic virus	(A) Ribavirin
II. Tobacco/potato virus Y	(B) 2-Thiouracil
III. Ornithogalum/ ornithogalum mosaic virus	(C) Vidarabine
IV. Potato/Potato virus X	(D) Malachite

Answer

I	*II*	*III*	*IV*
A	*B*	*C*	*D*

23. Correctly match the common name in section 'A' to those given under section 'B' commercial name of fungicide/antibiotic.

Section 'A'	*Section 'B'*
I. Ziram	(A) Ziride
II. Thiram	(B) Indofil M-45
III. Zineb	(C) Dithane Z-78
IV. Maneb	(D) Thiride

Answer

I	*II*	*III*	*IV*
A	*D*	*C*	*B*

24. Correctly match the common name in section 'A' to those given under section 'B' commercial name of fungicide/antibiotic:

Section 'A'	*Section 'B'*
I. Nabam	(A) Cersam, Agallol, Aretan
II. Organic mercuriales	(B) Dithane A-40

III. Captan	(C) Difolatan
IV. Captafol	(D) Orthocide

Answer

I	*II*	*III*	*IV*
B	*A*	*D*	*C*

25. Correctly match the common name in section 'A' to those given under section 'B' commercial name of fungicide/antibiotic:

Section 'A'	*Section 'B'*
I. Auriofungin	(A) Auriofungin sol
II. Cyclohexamide	(B) Actidione
III. Benomyl	(C) Benlate
IV. Carbendazim	(D) Bavistin

Answer

I	*II*	*III*	*IV*
A	*B*	*C*	*D*

26. Correctly match the common name in section 'A' to those given under section 'B' commercial name of fungicide/antibiotic:

Section 'A'	*Section 'B'*
I. Carboxin	(A) Ridomil, Apron
II. Oxycarboxin	(B) Bayleton
III. Metalaxyl	(C) Plantvax
IV. Triademefon	(D) Vitavax

Answer

I	*II*	*III*	*IV*
D	*C*	*A*	*B*

27. Correctly match the common name in section 'A' to those given under section 'B' commercial name of fungicide/antibiotic:

Section 'A'	*Section 'B'*
I. Bittertenol	(A) Baycor
II. Trycyclazole	(B) Beam
III. Tridemorph	(C) Calixin
IV. IBP	(D) Kitazin

Answer

I	*II*	*III*	*IV*
A	*B*	*C*	*D*

28. Correctly match the common name in section 'A' to those given under section 'B' commercial name of fungicide/antibiotic:

Section 'A'	*Section 'B'*
I. Hinosan	(A) Phytomycin
II. Dinocap	(B) Edifenphos
III. Griseofulvin	(C) Karathane
IV. Streptomycin sulphate	(D) Gryseofulvin

Answer

I	*II*	*III*	*IV*
B	*C*	*D*	*A*

29. Correctly match the antibiotics name in section 'A' to those given under section 'B' producing agent:

Section 'A'	*Section 'B'*
I. Streptomycin	(A) *Streptomyces griseus*
II. Griseofulvin	(B) *Penicillium griseofulvum, P. patulum, P. digicans*
III. Blasiticidin	(C) *Streptomyces griseochromogenes*
IV. Aureofungin	(D) *Streptoverticillium cinnamomeum* var. *triticola*

Answer

I	*II*	*III*	*IV*
A	*B*	*C*	*D*

30. Correctly match the antibiotics name in section 'A' to those given under section 'B' producing agent:

Section 'A'	*Section 'B'*
I. Tetracyclines	(A) *Streptomyces* spp
II. Antimycin	(B) *Streptomyces kitasawensis* and *S. grisues*
III. Thiolutin	(C) *Streptomyces albus*
IV. Nystatin	(D) *S. noursei*

Answer

I	*II*	*III*	*IV*
A	*B*	*C*	*D*

31. Correctly match the antibiotics name in section 'A' to those given under section 'B' producing agent:

Section 'A'	*Section 'B'*
I. Bulbiformin	(A) *Bacillus subtilis*
II. Pentaene G-8	(B) *S. anandii*

III. Endomycin (C) *S. endus*

IV. Choloromycetin (D) *S. venzuelae*

Answer

I	*II*	*III*	*IV*
A	*B*	*C*	*D*

32. Correctly match the common name in section 'A' to those given under section 'B'commercial name of fungicide:

Section 'A' Name of Compound	*Section 'B' Commercial preparations/ commercial name*
I. EMC	(A) Cersean
II. Ethyl methyl phosphate	(B) Ceresan
III. Methoxy ethylmercury chloride (MEMC)	(C) Aretan, Agallol, Ceresan wet
IV. Phenyl mercury acetate (PMA)	(D) Ceresan dry

Answer

I	*II*	*III*	*IV*
A	*B*	*C*	*D*

33. Correctly match the common name in section 'A' to those given under section 'B' commercial name of fungicide:

Section 'A' Name of Compound	*Section 'B' Commercial preparations/ commercial name*
I. Methyl mercury dicyandiamide	(A) Panogen
II. DMOC (Carboxin)	(B) Vitavax
III. DCMOD (Oxycarboxin)	(C) Plantvax
IV. Nabam	(D) Dithane A-40

Answer

I	*II*	*III*	*IV*
A	*B*	*C*	*D*

34. Match the following various fungi for use as mycoherbicide:

Fungus/Trade name	*Weed controlled*
I. *Phytophthora palmivora* (De Vine)	(A) Milk weed vine of citrus
II. *Colletotrichum gloeosporioides* (Collego)	(B) Northern jointvetch weed of rice and soybean

III. *Cercospora rodmanii* (ABG-5003) — (C) Water hyacinth

IV. *Colletotrichum coccodes* (Velgo) — (D) Velvet leaf

Answer

I	*II*	*III*	*IV*
A	*B*	*C*	*D*

35. Match the following various fungi for use as mycoherbicide:

Fungus/Trade name	*Weed controlled*
I. *Colletotrichum gloeosporioides* f. sp. *cuscutae* (LUBOA-2)	(A) *Cuscuta* sp.
II. *Alternaria cassiae* (CASET)	(B) Sickle pod weed of soybean, cotton
III. *C. gloeosporioides* f. sp. *malvae* (Bio MAL)	(C) *Malva pusilla* round leaved mallow
IV. *Fusarium lateritium* (Vego)	(D) Velvet leaf

Answer

I	*II*	*III*	*IV*
A	*B*	*C*	*D*

36. Match the following various fungi for use as mycoherbicide:

Fungus	*Weed controlled*
I. *Puccinia chandrillus*	(A) *Chandrilla juncea*
II. *Puccinia abrupta*	(B) *Parthenium hysterophorus*
III. *Alternaria crassa*	(C) *Datura stramonium*
IV. *Alternaria helianthi*	(D) *Xanthium stromarium*

Answer

I	*II*	*III*	*IV*
A	*B*	*C*	*D*

37. Match the following various fungi for use as mycoherbicide:

Fungus	*Weed controlled*
I. *Phomopsis convolvulus*	(A) *Convolvulus arvensis*
II. *Bipolaris halepense*	(B) *Sorghum halepense*
III. *Alternaria sp.*	(C) *Crisium avense*
IV. *Alternaria helianthi*	(D) *Morrenia odorata*

Answer

I	*II*	*III*	*IV*
A	*B*	*C*	*D*

38. Match the following bio-control agent which is used for insect-pest control:

Bio Control agent/Trade name	*Target insect/Parasite*
I. *Beauveria bassiana* (BOVERIN)	(A) Colorado potato beetle Codling moth
II. *Metarhizium anisopliae* (METABIOL)	(B) Spittle bag
III. *Aschersonia aleyrodis* (ASERONIJ)	(C) Glass house white fly
IV. *Bacillus thuringiensis* (DIPEL, THURICIDE)	(D) Lapidopteran insect

Answer

I	*II*	*III*	*IV*
A	*B*	*C*	*D*

39. Match the following bio-control agent which is used for insect-pest control:

Bio Control agent/Trade name	*Target insect/Parasite*
I. *Verticillium lecanii* (Vertalec)	(A) Aphid, scale insect
II. *Hirsutella thompsonii* (Mycar)	(B) Citrus rust mite
III. *Paecilomyces lilacinus* (Bicon)	(C) Nematode
IV. *Metarhizium anisopliae* (BioCane)	(D) White grub

Answer

I	*II*	*III*	*IV*
A	*B*	*C*	*D*

40. Match the following fungal bio-agent which is used for target insect order/ family.

Bio Control agent/Trade name	*Target insect/Parasite*
I. *Aschersonia aleyrodis*	(A) Hemiptera/Aleyrodidae
II. *Beauveria bassiana*	(B) Coleoptera/Chrysomelidae
III. *Beauveria brongniartii*	(C) Coleoptera/Scarabaeidae
IV. *Conidiobolus thromboides*	(D) Hemiptera/Aphididae

Answer

I	*II*	*III*	*IV*
A	*B*	*C*	*D*

41. Match the following fungal bio-agent which is used for target insect order/ family:

Bio Control agent/Trade name	*Target insect/Parasite*
I. *Lagenidium giganteum*	(A) Diptera/Culicidae
II. *Lecanicillium longisporum*	(B) Hemiptera/Aphididae

III. *Lecanicillium muscarium*	(C) Hemiptera/Aleyrodidae
IV. *Metarhizium anisopliae var. acridum*	(D) Orthoptera/Acrididae

Answer

I	*II*	*III*	*IV*
A	*B*	*C*	*D*

Answers (Multiple Choice Questions)

(1)	(A)	(34)	(A)	(67)	(C)	(100)	(D)
(2)	(B)	(35)	(A)	(68)	(A)	(101)	(B)
(3)	(D)	(36)	(A)	(69)	(D)	(102)	(B)
(4)	(A)	(37)	(B)	(70)	(D)	(103)	(A)
(5)	(A)	(38)	(A)	(71)	(D)	(104)	(A)
(6)	(A)	(39)	(D)	(72)	(D)	(105)	(B)
(7)	(A)	(40)	(A)	(73)	(B)	(106)	(A)
(8)	(A)	(41)	(A)	(74)	(C)	(107)	(B)
(9)	(A)	(42)	(B)	(75)	(A)	(108)	(A)
(10)	(B)	(43)	(A)	(76)	(A)	(109)	(D)
(11)	(B)	(44)	(A)	(77)	(C)	(110)	(D)
(12)	(A)	(45)	(D)	(78)	(C)	(111)	(A)
(13)	(B)	(46)	(A)	(79)	(A)	(112)	(A)
(14)	(A)	(47)	(A)	(80)	(A)	(113)	(C)
(15)	(D)	(48)	(A)	(81)	(A)	(114)	(A)
(16)	(C)	(49)	(B)	(82)	(A)	(115)	(A)
(17)	(A)	(50)	(B)	(83)	(C)	(116)	(C)
(18)	(B)	(51)	(C)	(84)	(A)	(117)	(A)
(19)	(A)	(52)	(A)	(85)	(A)	(118)	(A)
(20)	(B)	(53)	(A)	(86)	(A)	(119)	(A)
(21)	(B)	(54)	(B)	(87)	(A)	(120)	(B)
(22)	(B)	(55)	(B)	(88)	(A)	(121)	(B)
(23)	(B)	(56)	(A)	(89)	(A)	(122)	(C)
(24)	(C)	(57)	(B)	(90)	(B)	(123)	(B)
(25)	(B)	(58)	(B)	(91)	(D)	(124)	(A)
(26)	(C)	(59)	(A)	(92)	(D)	(125)	(B)
(27)	(A)	(60)	(A)	(93)	(C)	(126)	(A)
(28)	(B)	(61)	(B)	(94)	(D)	(127)	(A)
(29)	(B)	(62)	(A)	(95)	(B)	(128)	(C)
(30)	(A)	(63)	(A)	(96)	(A)	(129)	(D)
(31)	(A)	(64)	(D)	(97)	(D)	(130)	(A)
(32)	(D)	(65)	(B)	(98)	(C)	(131)	(A)
(33)	(B)	(66)	(B)	(99)	(A)	(132)	(A)

Terminology

Antibiotic: A substance or chemical compound (metabolites) produced by a microorganism and able to inhibits or kills the growth of other microorganisms.

Antibody: Specific, protective substances produced by living tissues in defense against invading micro-organism, poisons or antigens.

Antigen: Toxins and other substances produced by an invading disease producing organisms that cause the body cells of the host to produce antibodies, specific in their action.

Antisense RNA: RNA complementary to the coding or m-RNA strands. This technology provides resistance either by inhibiting gene expression or viral replication.

Antiserum: The blood serum of a warm-blooded animal that contains antibodies.

Avirulence: Pathogenic organism is incapable of causing severe disease.

Acquisition feeding period: The time for which a virus-free vector, actually feeds on a virus infected plants to acquire the virus.

Acquisition access period: The time for which a vector is allowed to feed on a source of virus.

Anastomosis: Fusion between genetically different hypha.

Antibiosis: Liberation of antibiotic or other chemical substances produced by one micro-organism and harmful to the other pathogens.

Antiviral proteins: Novel biomolecules identified from virus resistant non-host plants which provide natural protection to these palnts. For *e.g.* Tomato and tobacco plants transferred with Pokeweed antiviral protein (PAP/PAPII) genes exhibits resistance against a number virus and fungal pathogens.

Antagonist: It is a micro-organism that adversely affects another organism.

Anabolism: The process building up body tissue.

Anamorph: The asexual or imperfect stage especially in fungus.

Aggressiveness: Capacity of a pathogen to invade and grow in its host plant and to reproduce on or in it.

Anthracnose: A disease that appears as black sunken leaf, stems, fruit and pods lesions (ulcer like lesions) and caused by fungi that produce their asexual spores in an acervulus.

Acervulus: The mat of hyphae which give rise to closely packed, short conidiophores. It is a characteristic of the Melanconiales.

Antisporalant: A substance (chemical) inhibits spore production without affecting the growth of vegetative hyphae is called antisporalant.

Autoecious: A parasite that can complete its entire life cycle on the same host, term used in rusts.

Aeciospores: The binucleate rust spores produced in an aecium.

Aecium: The cup shaped structure (of binucleate hyphal cells) producing spore chains by conjugate division of the nuclei.

Aerobic: Microorganism living in presence of molecular oxygen.

Agar: A gelatin like substance obtained from sea weed (*Gelidium, Gracillaria, Gigartina*) and used to prepare culture media.

Alternate host: One of the two kinds of plants on which a parasite must develop for completion of its life cycle.

Aplanospore: The non-motile spore.

Apothecium: An open cup or saucer shaped ascocarp of some ascomycetes.

Appressorium: The flat hyphal body or germ tube (swollen tube), enters the epidermal cell of the host.

Ascocarp: The fruiting body of ascomycetes, containing or bearing asci.

Ascogonium: The female gametangium (sex organ) of the ascomycetes.

Ascospore: A sexually spore formed in an ascus.

Ascus: A sac like body (cell) containing ascospores formed as a result of karyogamy and meiosis.

Axenic: A culture of an organism in which no other organism are there.

Auxin: A substance which regulates plant growth and controls cell elongation.

Anisogamy: Copulation between two dissimilar gametes, one smaller (male) and the other bigger (female).

Anaerobic: Relating to a microorganism that lives or a process that occurs in the absence of molecular oxygen.

Asexual reproduction: Type of reproduction which does not involve union of gametes (nuclei) or sex organ or sex cells.

Albinism: Due to lack of light normal color changes into white.

Atrophy: Wasting away of a plant part.

Biotype: A subdivision of physiologic races, sub race or a biotype is a population of individuals which are genetically identical, produced by variant.

Biotrophs (obligate parasite): An organism which always obtained their food in nature from living tissues on which they complete their life cycle.

Bacteriocins: Antibiotic like compounds (bactericidal substances) which are produced by certain strains of bacteria and active against one or more strains of the same or closely related species.

Biological control: Biological control brings about reduction in activity of a pathogen (reduction of the amount of inoculums) by another organism. This may be biocidal (one organism kill the others) or biostatic (the organism only inhibit the pathogen).

Blotch: A disease characterized by large and irregular in shape, spots or blots on leaves, stems and shoots.

Blight: A disease characterized by general and rapid killing (extensive necrosis) of leaves, flowers, stems, branches and twigs.

Bacteriophage: Viruses that attack bacteria are called bacteriophages (or only phages), which means "bacteria eater".

Binary fission: The cell divides in transverse plane into two cells or splitting of cell into two daughter cells.

Biotechnology: The manipulation, genetic modification and multiplication of living organisms through tissue culture and genetic engineering for production of improved or new organism and products *e.g.* crop improvement, industrial improvement and disease control.

Bioinformatics: A mean to store, analyze and interpret the large amounts of data generated. By this combination with information science, genomics may and will help to move biology from *in vivo* to *in silico*.

Bacterium: The unicellular, achlorophyllous, microscopic prokaryotic organism, multiplying by fission.

Basidiocarp: A fruiting body that bears basidia.

Bactericide: A chemical that kills bacteria.

Basidiospores: A sexual spore borne on basidium, following karyogmy and meiosis.

Basidium: A cup shaped structure on which basidiospores are borne.(Typically four).

Bitunicate: An ascus with inner and outer walls.

Budding: The production of a small outgrowth (bud) from a parent cell, a method of asexual reproduction.

Ballistospore: Any spore that is forcibly discharged.

Bioassay: The use of a test organism to measure the relative infectivity of a pathogen or toxicity of a substance.

Biovar: A 'variant prokaryotic strain', that differs physiologically and/or biochemically from other strains in a particular species.

Blister: A raised lesion on the leaf surface with opens to expose spores.

Binucleate: Having two nuclei.

Biciliate: Bearing two cilia.

Biosafety: Biosafety regulations cover assessment of risks and the policies and procedures adopted to ensure environmentally safe applications of biotechnology.

Boomerang effects: Diseases are commonly exacerbated by over-watering and over-fertilization by well meaning gardeners. Efforts by plant pathologists to control disease which result in worse disease as boomerang effects. A classic example of the boomerang effect occurs if a soil-borne pathogen reinvades fumigated soil. Without natural enemies to reduce its spread it often becomes far worse than it would have been before fumigation.

Conducive soil: Soil-borne pathogen develops well and cause severe disease in some soil.

Competition: Competition is a condition in which there is a suppression of one organism as the two species struggle for limiting quantities of nutrients, oxygen, space or other requirements.

Cross protection: The phenomenon in which plant tissues infected with one strain of a virus are protected from infection by other strains of the same virus.

Cellulose: It is a polysaccharides but it consist of chains of glucose molecules and found in plant cell walls.

Cellulase: An enzyme that breaks down cellulose.

Catabolism: The break down phase of metabolism.

Chlorosis: Normal color changed into yellow due to low temperature, lack of iron, excess of lime or alkali in the soil and infection by virus, fungi, and bacteria.

Cell: All living organisms are composed of characteristics type of structural units called cell, cell is the basic unit of life.

Cell wall: The non-living component of a cell (permeable structure). It provides a mechanical support and gives definite shape and protection to the cell, it also helps in the movement of water and solutes towards protoplasm.

Cyanophage: The viruses that attack blue green algae are called cyanophages.

Conjugation: Two compatible bacterial cell come in contact and interchange of genetic material take place of both cells is altered.

Callus: Parenchymatous tissues or masses of cell (thin-walled undifferentiated) developed as a result of wounding or culture on nutrient media.

Cell culture: Growth of any cell.

Clone: The group of genetically identical individuals produced asexually from one individual.

Clamp connection: Outgrowth of fungus hypha which forms bridge connecting hypha cells.

Capsid: The protein coat of a virus particle, consisting of capsomeres.

Capsomere: The protein subunit or small protein molecule of capsid.

Cleistothecium: An entirely closed ascocarp.

Conidium (pl. conidia): A non-motile asexual spore usually formed at the tip or side of a specialized hypha.

Conidiophore: A simple or branched hypha bearing one or more conidia.

Chemotherapy: Control of diseases with chemotherapeutic agents (chemicals).

Chimeric genes: DNA sequence including regulatory and coding sequences not found in nature (together).

Chlamydospore: The thick walled spore formed by modification of hyphal cell, an asexual spore.

Chronic symptoms: Symptoms appearing over a long period of time.

Cistrons: The sequence of nucleotides within a certain area of DNA or RNA.

Cloning: Isolation and multiplication of individual gene sequence by its insertion into a system where it can replicate.

Cloning Vector: A cloning vector is a DNA molecule that has an origin of replication and is capable of replicating in a bacterial cell.

Culture medium: The prepared food material for culturing microorganisms.

Culture: To grow the microorganism artificially on the prepared food material, or the colony of microorganisms on such food material.

Cytokinin: A group of plant growth regulating substances that regulate cell division.

Chromosis: The normal color changed to red or orange are known as chromosis.

Control: Prevention of or reduction of loss from plant diseases.

Canker: A necrotic, often sunken lesion (warty outgrowths) is formed on leaves, stems, branch or twigs and fruits.

Cis-genic plants: made up of using genes, found within the same species or a closely related one, where conventional plant breeding can occur. Some breeders and scientists argue that cis-genic modification is useful for plants that are difficult to cross-breed by conventional means (such as potatoes).

Disease: Any disturbance brought about by a pathogen or an environmental factor which interferes with normal food uptake, synthesis, translocation and metabolism in the plant. In such appearances, yield loss is more than normal healthy plants of the same variety.

Disease trading: When one disease is controlled but another is exacerbated by the same treatment.

Disease cycle: The chain of events involved in the disease development including the stage of development of pathogen and effect of disease on the host.

Disease syndrome: Sum total of all symptoms characterizing a disease or collectively called as syndrome, *e.g.* necrosis of tissue, hypertrophy as well as wilting, total all symptoms called disease syndrome.

Dispersal: The spread of a plant pathogen within the general area or into new geographical area in which it is established is termed dispersal.

Disinfectant: A substance which destroys pathogenic microorganisms.

Disinfestant: A substance which can kill or inactivate pathogenic microorganisms in the environment or on the surface of plant parts before infection.

Diploid: Containing the double (2n) number of chromosomes.

Dissemination: Transfer of inoculums from a diseased plant to a healthy plant or from one place to another.

Disorder: A harmful non-infectious plant disease due to abiotic such as adverse soil conditions, *e.g.* mineral deficiency or toxicity, genetic anomaly, low temperature injury etc.

Bdellovibrio: Bdello derives from the Greek word for 'leach'. A genus of gram negative, chaemoorganotrophic bacteria, which parasitizes other bacteria and cause lysis. It multiplies inside the host cell wall and is generally obligate parasites.

Damping off: A disease that rots seedlings at soil level, resulting in the seedling falling over on the ground, or rapid death and collapse of very young seedlings.

Dieback: Progressive death of shoots, branches and roots generally staring at the tip, or extensive necrosis of twigs at their tips.

Disease incidence: The frequency of occurrence of a disease (per cent disease in a crop field).

Disease severity: Per cent disease (amount or severity) in an individual plant.

Disease prevalence: Per cent disease in a defined geographical area.

Decline: Plants growing poorly, leaves small, brittle, yellowish or red, some defoliation and dieback.

Dioecious: Separate male and female organisms.

Dikaryotic: Cell containing a dikaryon.

Disease escape: A phenomenon in which a particular environmental factor prevents infection and disease development in spite of presence of the susceptible host and the pathogen of sufficient pathogenicity.

Elicitors: The cell wall component of the pathogen which are capable of inducing phytoalexin synthesis. Phytoalexins production is stimulated in the most by the presence of certain pathogenic substances called elicitors. Elicitors are high molecular weight substances. It is a constituent of fungal cell wall such as glucans, chitosan, glycoprotein and polysaccharides etc.

Epidemiology: The study of epidemic and factors that influencing them is called epidemiology. Three infection chains are part of epidemiology, (i) survival (ii) dispersal and (iii) infection.

Epidemic or epiphytotic disease: An epidemic or epiphytotic disease usually occurs widely but periodically in a destructive form. The pathogen may be present as in endemic diseases but the environmental factors responsible for the development of the disease occurs only periodically, or A wide spread and severe outbreak of disease. It occurs widely but periodically in a destructive form.

Endemic disease: When it is constantly present in a moderate to severe form in a particular country or district, *e.g.* wart disease of potato (*Synchytrium endobioticum*) is endemic in Darjeeling.

ELISA (enzyme linked immune sorbent assay): It is a serological test in which the antibody carrying the enzyme, that releases a colored compound.

Endoparasite: A parasite which inters the host and feeds from within the host.

Eucarpic: Reproductive structure formed certain parts on the thallus.

Endobiotic: An organism living within its substratum.

Ectoparasite: A parasite which feeds on host from the exterior.

Epibiotic: An organism whose reproductive organs are on the surface of the substratum.

Exanthema: Eruption or discharge of gum or other substances from diseased tissues.

Epidemic rate: The amount of increase of disease per unit of time in a plant population.

Epiphytically: Existence on the surface of a plant/plant organ without causing infection.

Etiology: Study of the cause of the plant disease and nature of the causal agent.

Eradicant: A chemicals that destroys a pathogen at its source.

Eradication: Control of plant diseases by eliminating the pathogen after it is established.

Exudate: A natural liquid discharge from plant organs or a liquid discharge from diseased tissue.

Exclusion: Preventing new pathogens and diseases from reaching an uninfected area and avoiding contact between the pathogen and the crop or field, or control of plant diseases by excluding the pathogen or infected plant material from disease free areas.

Ethylene: It is gaseous hormone increase during invasion of pathogen in the plants.

Etiolating: Yellowing of plants due to lack of light.

Enzyme: Enzymes are large protein molecules that are catalyzing all interrelated reaction in the living cell.

Ergot: It is a fruiting body by a fungus, *Claviceps purpurea* which is parasitic on rye and other grasses.

Embryo: The rudimentary seedling.

Endosperm: The food storage for nourishing the embargo.

Externally seed borne: Some pathogen carried on the seed, *e.g.* covered smut of barley (*Ustilago hordei*), brown spot disease of rice (*Drechslera oryzae*), covered smut of oat (*Ustilago kolleri*), black scurf of potato (*Rhizoctonia bataticola*) and grain smut of jowar (*Sphacelotheca sorghi*).

Explant: Plant parts called explant, used as initial inoculum *e.g.* protoplast, tissue, organ etc.

Enation: Tissue malformation or overgrowth induced by certain virus infections.

Escape: Certain crop, varieties which undergo development and maturation may complete their life-cycle before maximum infection.

Episome: In bacteria, a DNA molecule, other than the bacterial chromosome, capable of replication.

Epinasty: Downward bending of a petiole, lamina and stems or lamina may continue the curve of the petiole so that its upper surface faces inwards towards the stems.

Facultative parasites: Some organism lived most of the time on dead organic matter, but under certain condition may attacks living plants and become parasite.

Facultative saprophyte: Some organism lived most of the time as parasite but under certain condition may grow saprophytically on dead organic matter.

Fungistasis: Phenomenon wherein viable fungal propagule do not germinate in the soil in condition of temperature and moisture favorable for germination.

Fumigation: Many chemicals which can produce toxic gases are used to kill the pathogen.

Forma specialis (f. sp.): A group of biotypes of a pathogen species.

Fumigants: Toxicants which are used in gaseous form for killing insect pest and microorganism and act as soil sterilant, *e.g.* methyl bromide, methane, CO_2, chloropicrin, tetrachloroethane etc.

Fungi: Fungi are eukaryotic organism, spore bearing, chlorophyllus, filamentous and branched, they reproduce by sexually and asexually means.

Fungicide: A substance (chemical) capable of killing fungi.

Fungicidal: Killing fungal spores or mycelium by physical agents such as heat, ultraviolet light, x-rays, gamma radiation as well as chemicals.

Fungistat: A substance (chemical) is preventing the growth of a fungus without killing it.

Filtration: Filtration is the sterilization process, by which removal of organism of particulates from thermo labile solutions, air, waters.

Freeze drying: A technique used in the preservation of plant tissues and microorganism etc., whereby water is removed under vacuum while tissue remains in the frozen state. Synonymous with lyophilization.

Filiform: Thread like.

Fusiform: Spindle shaped, narrowing toward the ends.

Fission: Asexual reproduction of bacteria involving transverse splitting of the cell or the splitting of a cell into two cells.

Flagellum: The whip like locomotion organ originating from bacterium or zoospore.

Fructification: Production of spores by fungi.

Fruiting body: A complex fungal structure containing spores.

Fasciation: A plant disease, due to cell injury in the bud, resulting in flattened and sometimes spirally curved shoots.

Fermentation: Oxidation of certain organic substances in the absence of molecular oxygen.

Fertilization: The sexual union of two nuclei or protoplasts resulting in doubling of chromosome number (2n).

Gummies: Emission of gum by or in a plant tissue or organ.

Gum: A complex polysaccharide formed by cells in reaction to wounding or infection.

Growth regulator: A hormone (chemical) which synthesized in the particular cells and initiates the growth and reproduction of the plant.

Growth retardant: A substance that decreases the rate of growth of a fungus, higher plant etc.

Growth inhibitor: A substance that inhibiting the growth of an organism.

Gene-for-gene hypothesis: Every resistance and susceptible gene present in the host while pathogen has a gene for virulence and virulence. Susceptible reaction would result only when the pathogen is able to match all the resistance genes present in the host with appropriate virulence gene. If one or more resistance genes are not matched by the pathogen with the appropriate virulence genes, resistance reaction is the result. This complimentary relationship of the host and pathogen is called gene for gene hypothesis for disease resistance.

Germination: The start of growth of the spore or seed or the emergence and development of seedling form the seed embryo which able to produce normal plant under favorable condition.

Genome: A complete set of chromosomes of a diploid species or single set of chromosome, or all genes together called genomes.

Genetic engineering: The alteration or manipulation of genetic composition of a cell by various procedures such as transformation, protoplast fusion etc. in tissue culture.

Gene: Linear portions of the chromosome. Fictionally, gene is unit of inheritance and structurally, gene on a segment of DNA which codes for one polypeptide.

Gram-positive: Bacterial cell retaining the Gram's stain after washing with alcohol, appear dark blue or violet.

Gram-negative: Bacterial cell loss the stain after treatment of alcohol, loss of crystal violet color.

Germplasm: The sum total of all hereditary material or genes present in a species.

Gibberellin: A group of plant growth regulating substances.

Gamete: A differentiated sex cell or a sex nucleus that fuse with another in sexual reproduction or the male or female sex cell or a sex nucleus.

Gene cloning: Isolation and multiplication of an individual gene sequence by its insertion into a bacterium.

Genus: A taxonomic group including species.

Guttation: Exudation of water from plants, particularly along the loaf margins.

Gall: Outgrowth or swelling, often more or less spherical produced on a plant as result of attack by bacteria, fungi or other organisms.

Hypersensitivity: Excessive sensitive of plant tissues to certain pathogens. Affected cells are killed quickly, blocking the advance of obligate parasites, or the extreme degree of susceptibility in which rapid death of the cells in the vicinity of the invading pathogen (the infection court) occurs. This halts/stop the progress of the pathogen although it may not die immediately. Thus, hypersensitivity is a sign of very high resistance approaching immunity, *e.g.* rust, virus and nematodes.

Host: A living organism harboring a parasite.

Hemibiotroph: A parasites or an organism which may attack living tissues in the same way as biotrophs but continue to develop and speculate after the tissue is dead, *e.g.* leaf spotting fungi.

Horizontal resistance (uniform or non-specific resistance): Partial resistance equally effective against all races of a pathogen.

Hypovirulence: Reduced virulence of a pathogenic strain as a result of the presence of transmissible double stranded RNA. The hypovirulence strain carry virus like dsRNA. The double stranded RNA apparently passes through mycelia anastomoses from hypovirulent to virulent strains and reduced the development of canker. Chestnut blight (*Cryphonectria parasitica*) control with hypovirulent strains of the pathogen.

Hyperparasite: A parasite parasitic on another parasite.

Hemicellulose: A complex mixture of polysaccharide resembling cellulose and main component of cell wall. It is insoluble in water, which are attached with cellulose and lignin.

Hypertrophy: Excessive growth (abnormal increase) in the size of individual cell (of affected organ).

Hyperplasia: A condition in which a tissue is enlarged as result of excessive production of cell (increase in the number of cell).

Helper virus: A virus capable of supporting the replication of an additional, nonessential and unrelated nucleic acid component, *e.g.* a virusoid or satellite RNA.

Hybridization: In fungi, the crossing of two individual differing in one or more heritable characteristics, or combination of dissimilar gametes.

Heterokaryosis: A condition in which genetically different nuclei are associated in the same protoplast or same mycelium.

Hypha: A single branch of mycelium, or the unit o structure of most fungi, tubular, filaments.

Hfr: In *Escherichia coli*, high frequency recombination; cells in which the F factor is integrated into the bacterial chromosome.

Hyperauxiny: A condition in plant in which the level of auxin (IAA) in the tissues is higher than normal. Hyperauxiny may result from infection by fungi, bacteria, viruses and nematodes and by chemical and environmental stimuli.

Heterogamy: The fusion between male motile gametes and non-motile female gametes.

Haustorium: The penetrating and absorbing organ of the hypha of a parasite.

Heteroecious: A parasitic fungus requiring two host species for the completion of its life cycle.

Holocarpic: Whole thallus converted into zoosporangium or reproductive structure.

Homothallic: Fungus with compatible male and female gametes produced on the same mycelium.

Hyaline: Without any color.

Hermaphroditic: The individual having both functional male and female sex organ.

Hydathodes: Structures with one or more openings that discharge water from the interior of the leaf to its surface.

Heterokont: A biflagellate structure with two flagella, unequal in size.

Haploid: Containing reduced (n) number of chromosomes.

Habitat: The natural place of occurrence of an organism.

Hypoplasia: Under development of tissue due to decrease in cell division.

Herbicide Resistant Plants: Plants that can tolerate herbicides.

Induced resistance: Enhancement of the plants defensive capacity against a broad spectrum of pathogens and pests that is acquired after appropriate stimulation.

Inoculum: Any part of the pathogen which cause disease, *e.g.* spore, mycelium, chlamydospore, sclerotia etc.

Inoculation: The transfer or arrival of a pathogen on to a host.

Inoculate: To bring a pathogen into contact with a host plant or plant organ.

Infection: The establishment of a parasite within a host plant.

Invasion: When the pathogen rapidly grows in the host tissues called invasion.

Inoculum density: It is a mode up of fungal propagules having a certain amount of virulence and vigour.

Infection concert: Place where infection may take place.

Isogamy: Copulation between two gametes morphologically same.

Infections disease: A pathogenic disease spreading from diseased to healthy plant.

Isogametes: The two gametes (male and female) are similar in size.

Isogametangia: Gametangia, presumably of opposite sex, that are morphologically indistinguishable.

Inoculum potential: Combined energy of fungal propagules (or other microorganism) to cause infection.

Incubation period: The period of time between penetration of a host by a pathogen and the first appearance of symptoms on the host.

Inoculation access period: The time for which the vector, after acquiring the virus is allowed to feed on healthy plants and transmit the virus.

Inoculation feeding period: The actual period of feeding is called inoculation feeding period.

Inoculation threshold period: The minimum time necessary for a vector to feed on test plant in order that transmission may occur.

Intellectual Property Right (IPR): Legal rights of a person, grp or institution to prevent or protect intellectual property (innovations, discovery, concept, knowledge etc).

Immune: Exempt from disease, cannot be infected by a given pathogen.

Immunity: Exempt (100 per cent) freedom from infection. It denotes that the pathogen cannot establish parasitic relationship with the host, even under the most favorable conditions, or the state of being immune.

Isolate: A single spore or pure culture and the subcultures derived from it and also used to indicate collection of a pathogen made from different places or at different times.

Isolation: The separation of a pathogen from its host and its culture on a nutrient medium.

Integrated control: To use all available methods of control of disease/pests.

Integrated disease management (IDM): The simultaneous, use of a number of strategies in the field to minimize the loss caused by plant diseases. Field sanitation, host tolerance, agronomic and cultural practices and the use of biological agents can be pressed into service to keep the loss caused by plant diseases at low level.

Injury: Damage of a plant by animal, physical or chemical agent.

Internally seed borne: Some pathogen carried inside the seed, *e.g.* loose smut of wheat (*Ustilago nuda tririci*), loose smut of barley (*U. nuda*), stripe disease of barley (*Drechslera graminea*) and red rot of sugarcane (*Colletotrichum falcatum*).

Imperfect fungus: A fungus that is not known to produce sexual spores or one lacking any sexual reproductive state.

Imperfect stage: No sexual spores are produced; anamorphic.

Intercellular: Between cells.

Intracellular: Within or though the cells.

In vitro: When biological processes are made to occur outside the organism in a vessel or test tube or in culture, outside the host.

In vivo: Within the living organism, or in the host.

Intercalary: Between apex and base.

Karyogamy: The fusion of two nuclei resulting in the production of diploid zygote nucleus.

Koch's postulates: Three criteria proposed by R. Koch's for proving pathogenicity.

Life cycle: The successive stages in the growth and development of an organism that occur between the appearance and reappearance of one stage of the organism (*e.g.* spore, eggs etc).

Latent period: The time between inoculation and the start of sporulation by the resulting infection, or the period required by a virus to multiply and accumulate within the vector before the latter become infective.

Latent infection: Host is infected but does not show symptoms.

Latent virus: A virus that does not induce symptom development in its host.

Lipid: Substances (organic compound) whose molecules consist of glycerin and fatty acids and sometime certain additional types of compounds.

Lignin: A complex organic substances or group of substances that is found in the middle lamella in the cell wall and in the cell wall of xylem vessels.

Leaf spots: Necrosis of the tissues around the substomatal space and appear brown or water soaked and spots usually remain restricted in the growth.

Line: A group of individuals having common parents or ancestors.

Lyophilization: A technique used in the preservation of microorganisms, plant tissues etc at -70°C and then dehydration by vacuum.

Leaf curl: Distortion, thickening and curling of leaves *e.g.* along the two sides of midrib.

Leaf roll: Curving of the leaf lamina, generally towards and parallel with the midrib.

Lysogeny: The chromosome of a bacteriophage becoming integrated into the chromosome of its host, in case of temperate phages, such bacteria are called lysogeny.

Lesion: A localized area of the diseased or discolored host tissue.

Local lesion: A host, which develops local lesions on inoculation with virus.

Lichen: A combination of an alga and a fungus in which the two components are so interwoven as to form what appears to be a single individual, benefiting each other.

Lysis: A breaking down or dissolution of cells by enzymes or bacteriophage.

Monoculture: Continuous cultivation of the same crop in a conducive soil. After some year, severe disease eventually leads to reduction in disease through increased population of microorganism antagonistic to the pathogen.

Mycorrhiza: The mycelial hyphae living in association with roots of higher plants. The nature of this association is believed to be symbiotic (mutualism) non-pathogenic or weekly pathogenic.

Morphovars: Strains that differ physiologically.

Monogenic: A character governed by single gene with 3:1 Mendelian ratio *e.g.* resistance of oats is governed by single gone.

Meiosis: A series of two nuclear division usually in quick succession in which chromosome number is reduced by one half.

Maceration: The separation of cells along the line of the middle lamella is called maceration. The term 'cell separation' is more appropriate, or softening of the tissues or loss of plant tissue by pectic substances.

Metabolism: Chemical reaction occurring in the living cell by which the food is transformed into living protoplasm; energy for the activities of the body.

Mosaic: Normal or light green yellow patches symptoms on the plant surfaces produced by certain virus diseases.

Mottle: An irregular pattern of distinct light and dark areas.

Messenger RNA: A chain of ribonucleotides that codes for a specific protein.

Microscopic: Very small can be seen only under microscope.

Mildew: A fungal disease of plants in which the mycelium and spores of the fungus are seen as a whitish growth on the host surface, or area of stem, leaves, fruits covered with whitish mycelium.

Mold: Any profuse or wooly fungus growth on damp or decaying matter or on surfaces of plant tissue.

Malformation: A condition in which growing shoots become stunted and dwarf with leaf buds flowers suppressed and crowded presenting a bunchy top appearance.

Mushroom: A fleshy, sometimes tough umbrella like sporophores that bears basidia on the surface of gills.

Micro-organism: The unicellular or multicellular microscopic organism and widely distributed in air, soil, water, the living plant bodies and animals and dead organic matter, *e.g.* fungi, bacteria, viruses, yeast, mold, algae, protozoa etc.

Monocyclic: Having one cycle per season.

Mutation: Sudden change of genetic material of a cell.

Mycoplasma like organism (MLO): A prokaryotic organism, lack of cell walls, are bounded by unit membrane, have cytoplasm, ribosomes and strands of nuclear material, pleomorphic shape and present in the sap of a small number of phloem sieve tubes and assumed to be the causes of the disease, they resemble mycoplasma in all respects except that they cannot yet be grown on artificial nutrient media.

Mycoplasmas (true mycoplasmas): Pleomorphic, prokaryotic organism, lack of true cell wall, bounded by unit membrane (triple layered), reproduce by binary fission and budding of cells, have no flagella, produce no spores and are gram-negative and all true mycoplasma parasitic to humans and animals and all saprophytic ones can be grown on more or less complex nutrient media.

Mycelium: Mass of hyphae that make up the body of a fungus.

Mycosis: An infection by a parasitic fungus, or a diseased so caused.

Mycotoxin: Toxic substances produced by fungi which forms toxins.

Monoecious: Male and female reproductive organs in same individuals.

Monokaryotic: Having single nucleus.

Molecular markers: Nucleic acid segments that behave as landmarks for genome analysis and are based on naturally occurring genetic variability (usually termed polymorphisms).

Marker assisted selection (MAS): Indirect selection for a primary trait (trait for economic importance) via direct selection for a secondary trait (marker), whch is independent of growth stage, environment, and gene - environment (G x E) interaction.

Mycotoxicosis: Disease of animal and human caused by consumption of feed and foods invaded by fungi that produce mycotoxins.

Metachromasy: The phenomena in which certain biological entities (*e.g.* metachromatic granules) become stained with a color which is different to that of the dye used to stain them.

Medium: Substratum employed in the laboratory for growing microorganisms.

Microcyclic: Rusts having teliospores as the only binucleate spores (also called short cycled) in the life cycle.

Micropropagation (clone propagation) - It is a process of production of clones similar to asexual reproduction. Plantlets are produced from shoot tips/axillary buds on culture medium omitting callus phase. Here, small amount of explant produce millions of clonal parts in a year directly.

Monocentric: Thallus bearing single reproductive structure (sporangium or resting spore).

Monoclonal antibody: Continuous cultures of fused cells secreting antibody of predefined specificity.

Mycology: Study of fungus (fungi) is known as mycology.

Molecular plant pathology: Field of biology and plant pathology that studies the structure and function of genes at a molecular level. It studies, what genes are involved in plant pathogenesis and defenses against diseases, employing the methods of plant pathology and molecular biology.

Native/wild type genes: Gene that is present in the genome of an untransformed cell *i.e.* a cell not having known mutation.

Necrotroph (perthotrophs or perthophytes): These are organisms which kill the host tissues in advance of penetration and then live saprophytically.

Necrosis: Death of the cells, tissues, organs or entire plants (loss of tissues).

Necrotic: Death and discolored of the tissues and other plant parts.

Nucleotide: Made up of three essential components: a nitrogenous base, a sugar, and a phosphate moiety.

Non-persistent virus (stylet borne or proboscis virus): Non-persistent virus survived in their vectors for only short period, less than the survival time of virus in untreated leaf extract.

Natural openings: Stomata, lenticels and hydathodes are natural openings in plants.

Net blotch: Dead leaf spot with a net-like pattern of dark veins.

Non-infectious disease: A disease that is caused by an abiotic agent, not by a pathogen.

Non-septate: Without coos walls.

Nucleocapsid: Nucleic acid covered with protein coat (capsid).

Nucleic acid: An acidic substance which contains phosphorous, pentose, pyrimidine and purine base, it determine the genetic properties of an organism and viruses.

Nucleoprotein: Consisting of nucleic acid and protein.

Oligogenic: A character governed by several genes, *e.g.* dihybrid, trihybrid; anthracnose of bean is governed by three gene.

Obligate parasite: An organism that can obtain food only from living organism; cannot be grown in culture media.

Obligate saprobe: An organism that must obtain its food from dead organic matter and is incapable of infecting another living organism.

Oogonium: A female gametangium of oomycetes containing one or more gametes (eggs).

Oospore: A thick-walled sexual spore produced by the two morphologically different gametangia (oogonium and antheridium).

Oosphere: A naked female gamete which is non-motile.

Ovary: Female reproductive body producing or containing eggs.

Ozone (O_3): A highly reactive form of oxygen which in relatively high concentration may injure plants.

Oidium: A thin-walled, flat-ended asexual spore formed by breakup of hypha. It behaves as a spore or spermatium.

Oogamous: It refers to a type of fertilization in which two heterogametangia come in contact with each other and contents of one flow into the other through a pore or tube.

ORF (Open reading frames): Amino acid sequences encoded between translation initiation and termination codons of a coding sequence.

Organogenesis: Process of differentiation of callus initially into embryo like structure (embryoids) which further develop into organs like roots/shoots.

Ostiole: It is neck-like structure in an ascocarp; lined with periphyses, and terminating in a pore; also the opening of a pycnidium or perithecium which is pore-like.

Pathogenicity: The ability (capability) of pathogen to cause disease.

Pathogen: An entity (parasite or causal agent) able to cause disease in a particular host or range of host.

Pathogenesis: The chain of events that leads to disease development in the host. It involves the action of pathogen, susceptibility of host and affects of environment.

Parasite: An organism which derived the food materials they need for growth from living plants (the host or suspect) are called parasite. Most but not all, pathogens are parasites and similarly most but not all parasites are pathogens.

Parasitism: The relationship between host and parasite is known as parasitism.

Penetration: Entry of the pathogen (an entity) into the host tissues or the initial invasion of a host by a pathogen.

Photosynthesis: A process, by which chlorophyll containing organisms synthesized carbohydrates from CO_2 and water using the energy of sunlight.

Plasmid: A self-replicating extra chromosomal genetic structure (circular dsDNA molecule) found in certain bacteria and fungi; generally not required for survival of the organism.

Physiologic specialization: The occurrence of physiologic races within a species.

Physiologic race: Pathogen of the same species and variety which cannot be distinguished by their structure, but differ in physiological behavior, particularly in their ability to parasitize varieties of a given host.

Protoplast fusion: Fusion of protoplast from two different plant species and regeneration of somatic hybrid plants (disease resistant plant). Protoplast fusion also called somatic hybridization.

Pathovar: A subspecies or group of strains in bacteria that can infect only plants within a certain genus or species.

Pathotype: A subdivision of a species distinguished by the common characters of pathogenicity, particularly in relation to the range of host.

Physiotype: A population of pathogen in which all individuals have a particular character of physiotype (but not pathogenicity in common).

Pathogenic: Capable of causing disease in a host or range of hosts.

Persistent or circulative virus (propagative virus): Persistent virus survived in their vectors for long periods, sometimes a weeks or months.

Parasexualism: A process in which plasmogamy, karyogamy and haplodization take place in sequence, but not at specified points in the life cycle of an individual.

Phytoalexins: Low molecular weight, antimicrobial compounds, formed by plants in response to infection and exposure to various toxic and non-toxic compounds and is thought to play a role in disease resistance.

Plant quarantine: Preventing entry of pathogen from infested area into new non-infested area.

Plant protection: A technique of safeguarding growing crops and storage of agricultural products from the natural enemies.

Prototroph: A microorganism which is able to grow normally or on minimal medium containing inorganic salts and a carbon source.

Phytopathogenic: Micro-organism inciting plant disease.

Perfect stage: The sexual stage (fruiting bodies) in the life cycle of a fungus. The teliomorph.

Propagule: A part of an organism that may be disseminated and reproduce the organism; or one unit of inoculum of any pathogen is called propagule.

Pathodeme: Pathodeme is that population of a host in which all individuals have a particular character of resistance in common.

Predisposition: The effect of one or more environmental factors which makes a plant vulnerable to attack by pathogen. It is process which antedates penetration and infection.

Pasteurization: A heating process that destroys pathogenic bacteria in a fluid such as milk, cheese, vine, beer, fruit juice etc. below 100°C, mostly at 62.8°C for 30 minutes.

PPM (parts per million): An amount of substance/compound contain in a million part of mixture or solution in carrier such as air, water. One part per million equals one milligram in one kilogram.

Polycentric: Thallus bearing more than one reproductive structure (sporangia or resting spores).

Paraphyses: Sterile, basally attached structures in a hymenium.

Periphyses: Short hair-like growths in ostiole or pore in the stroma.

Primary inoculum: The inoculum that survives the winter or summer and cause the original or emotional infection in the spring or autumn is called primary inoculum.

Primary infection: The first infection by a pathogen after it has gone through a period of resting or dormancy; or the first infection of a plant by overwintering or over summering pathogen.

Pandemic disease: These occur all over the world and result in mass mortality.

Phyllosphere: Surrounding area of the leaf surface or plant surface, where the micro-organisms two types of interaction take place, synergistic and antagonistic.

Protectant: A substance (chemical) which is effective only if applied prior to fungal infection, or a substance that protects an organism against infection by a pathogen.

Perithecium: A closed ascocarp with an opening or pore (ostiole).

Phenotype: The externally visible appearance.

Planogametic copulation: Fusion of naked gametes, either one or both is motile.

Precipitin: An antibody capable of causing precipitation of soluble antigen.

Promycelium: The germ tube from teliospore bearing the basidiospores; the basidium.

Protoplast: A naked cell or cell without cell wall.

Polygenic: A character controlled or governed by many genes. It involved in a large number of diseases such as cotton wilt.

Plant Pathology or phytopathology: It consists of three Greek words (phyton-plant, pathos- ailments or diseases and logous-discourse or knowledge). Plant Pathology is the science which deals with causes, symptoms, etiology, perpetuation, losses of plant diseases and their management.

Pest: Pests are organism that cause economic losses.

Protein: Highly complex chemical substances in pure form containing C, H, O and occasionally phosphorus and sulphur both.

Plant: Living organism with neither the power of movement nor special organs for digestion.

Promoter: A nucleotide sequence, usually upstream (5′) to its coding sequence, which controls the expressions of coding sequence by providing recognition for RNA polymerase. *e.g.* TATA box and other seuqnces.

Protoplasm: A cell is an organized mass of living substances.

Pathotoxin: A toxin produced by a pathogenic organism and capable of simulating the disease.

Phytotoxin: A toxin toxic to plant.

Protoplast: Animal, plant or fungal cell from which the entire cell wall has been removed.

Purification: The isolation and concentration of virus particles in a pure form which is free from cell components.

Pustule: The blister like swelling of epidermis during spore emergence.

Pycniospore: The spore formed in a pycnium; also known as spermatium.

Pycnium: Fruiting body of the rusts producing pycniospores or spermatia; also known as spermogonium.

Polymerase: An enzyme that joins single small molecules into chains of such molecules (*e.g.* DNA, RNA).

Race: A genetically and often geographically district mating group within a species; also a group of pathogens that infect a given set of plant varieties, or physiological race.

Resistance: The inherent ability of a plant to prevent or restrict the establishment and subsequent activities of a potential pathogen; or the ability of an organism to exclude or overcome, completely or other damaging factor; plural- resistance is the process.

Resistant (singular-resistant is the result): Possessing qualities that hinder the development of a given pathogen infected little or not at all.

Rhizosphere: The soil surrounding the roots or the region of soil which is under influence the plant roots.

Rhizoplane: The root surface and its adhering soil are sometimes termed rhizoplane.

Roguing: Removal and destruction of infected plant parts from the field or crop.

Respiration: The oxidation (aerobic, anaerobic breakdown) of organic materials, primary carbohydrates change into H_2O and CO_2 and produce utilizable energy.

Relative humidity: The amount of water present in the air at a given temperature in proportion to the maximum water holding capacity at temperature in atmosphere.

Ribosome: A subcellular article involved in protein synthesis.

Ribozyme: The term 'ribozyme' is referred to the ability of a RNA molecule to cleave target RNA molecules. Although ribozyme specific for any unwanted gene can be designed and integrated into plant genome, this approach has limited use in the development of virus resistant transgenic plants.

Rickettsia like organism(fastidious vascular bacteria): RLO are prokaryotic organism, rod-shaped, they bounded by cell membrane and a cell wall, they have no flagella; nearly all fastidious vascular bacteria are gram-negative only the xylem inhibiting bacteria causing sugarcane stunting and bermuda grass stunting are gram-positive; multiply only in a eukaryotic cells.

Root rot: Disintegration or decay of parts or all of the root system.

Rust: Many small lesions or pustules of spores on leaves, stems, usually of a powdery mass of brownish, red, yellow or dark brown color or rusty color.

Rots: The infected tissue dies, decomposes rapidly and turns brown. Fungi and bacteria which are able to dissolve cell walls cause these symptoms.

Receptive hypha: The special hypha acting as a female gamete or gametangium.

Regeneration: Production of entire plant from explant is called *regeneration.*

Resting spore: The spore resistant to extremes of temperature and moisture, germinating only after a period of time.

Resting stage: The inactive stage.

Ring spot: Symptom of certain virus diseases characterized by circular chlorosis with a green center.

Russet: The rough areas on skin of fruit due to cork formation.

Rosette: Short bunchy growth of plant parts.

Reproduction: The production of new individuals having all the characteristics typical of the species.

Rhizomorph: The thick organized strand of hyphae. The growing tip of strand resembling the root tip.

Recombinant inbred lines (*RILs*): Lines derived by single seed descent from individual F_2 progeny.

Serovars: Strains that have antigenic properties that differ from other strains.

Strain: The descendants of a single isolation in pure culture, an isolate or a group of similar isolates; a race; a group of virus isolates (in plant viruses) having most of its antigens in common with that of another group (strain).

Susceptible: Lacking the inherent ability to resist disease or attack by a given pathogen; non-immune, *i.e.* subject to infection.

Suspect: Any plant or organism that can be attacked by a given pathogen or given disease; a host plant.

Susceptibility: The inability of a plant to resist the effect of a pathogen or any other damaging factor.

Symptom: The external and internal reactions or alternation of a plant as a result of disease. or a visible abnormality arising from disease; signs of infection.

Streaks: Elongated narrow lesions on leaves which are usually of brown shade. *e.g.* barley stripe disease.

Sexduction: In bacteria, the incorporation of bacterial genes into a sex factor (*e.g.* F) and their subsequent transfer to other bacterial cells through conjugation.

Spreader: Spreader is a material which improve the contact between fungicide and sprayed surface; *e.g.* mineral oils, glyceride oil and soap; it reduces the surface tension.

Sticker: A substance added to spray or dust which improve the adherence to plant surface, *e.g.* gum Arabic, dextrins, fish oil, milk casein and gelatin.

Safner: A chemical which reduce the phytotoxicity of another chemicals, *e.g.* the phytotoxicity of CuSO4 is reduced by lime.

Symbiosis: A mutually beneficial association of two or more different kinds of both partners appears to benefit.

Syndrome: The totality of effects produced in a plant by a disease.

Saprophytes: An organism which feeds on dead organic matter; or an organism which derived their nutrition from dead organic matter.

Serotype: A population of a pathogen (usually a bacterium or virus) in which all individuals possess a given character of serology in common.

Serology: A method using the specificity of the antigen-antibody reaction for the detection and identification of antigenic substances and the organisms that carry them.

Secondary inoculum: Inoculum produced by infections (from primary infection) that took place during same growing season. It is responsible for spread of disease because of the primary infection spores and other structure produces secondary infection. The secondary inoculum is spread by insect vectors or by other means.

Secondary infection: Any infection caused by inoculum (secondary inoculum) produced as a result of a primary infection or subsequent infection ex. late blight of potato.

Sporadic disease: These belong to the endemic group. They occur at very irregular intervals and locations and in relatively few instances ex. Angular leaf spot of cotton and blotch disease of cucumber.

Suppressive soil: Soil in which certain disease are suppressed, because of presence in the soil of microorganisms antagonistic to the pathogen.

Survival: In the absence of their cultivated host, pathogen survives some alternate host, the infection will remain incomplete.

Sclerotium: A hard resting body resistant to unfavorable conditions and can germinate under favorable conditions.

Scorch: The burning of leaf margins.

Satellite RNA: An encapsulated RNA component which is dependent upon a helper virus for replication but show no significant sequence homology with either the viral or host genomes, or one which multiples only in the presence of a specific second virus.

Siderophore: Extracellular low molecule weight compound produced by microorganism it has bendable capacity, it binds the iron (Fe^{++}) and make less available to pathogen.

Soil polarization: When solar heating increased temperature in the soil to level above the optimum, viability and activity of pests is relatively affected.

Sanitation: Removal and burning of infected plant parts, decontamination of tools, equipment's, worker's hands etc.

Satellite virus: Polyhedral ssRNA viruses which are unable to replicate without the presence of a second or helper virus are called satellite or defective encapsulated in a complete virion. Some plant viruses such as TNV have satellite virus associated with it.

Starch: A glucose polymer or main reserve polysaccharides found in plant cells. It is synthesized in the chloroplast and in the amyloplast or a polysaccharide consisting of glucose units.

Shot hole: Falling off of diseased leaf tissues leaving a hole behind.

Sign: The pathogen or its parts or products visible on the host plant.

Smut: The disease in which infected plants show masses of dark powdery spores caused by the smut fungi (Ustilaginales).

Sporangiophore: A hypha that bears sporangia.

Sporangiospore: A spore borne within a sporangium.

Spore: A minute propagative unit as of fungi which may germinate and develop into a new individual.

Somatic: The vegetative phase, structure or function.

Stress: Predisposition effect of external environment on plants and diseases.

Soft rots: Disintegration of cells and dissolution of middle lamella as a result of enzyme action, very often a dirty liquid oozes out of the affected parts and changes the color or maceration and disintegration of fruits, roots, bulbs, tubers and fleshy leaves.

Seed longevity: Seeds remain viable for many years.

Seed: A mature ovule consisting of a embryo and endosperm and the protective covering (seed coat).

Seed pathology: The science and technology dealing with seed borne plant diseases, the mechanism of their transmission, the method for preventing and controlling of these diseases in the field and during seed storage, the assessment of seed borne inoculum and techniques for disease detection in seeds.

Seedling: A young plant developed from seed.

Somatogamy: Union of somatic cells during plasmogamy.

Sours: The compact mass of fruiting bodies or spores (mass of sporangia).

Stolon: Hypha growing horizontally along the surface of substrate.

Sterile fungi: Fungi not producing spores.

Sterilization: Complete destruction of all living organisms by physical or chemical agents.

Stroma: The compact mycelial structure, like a mattress, on or in which fructifications are usually formed.

Soil inhabitants: Microorganisms which are able to survive in the soil indefinitely as saprophytes.

Soil invader: Plant parasitic fungi and bacteria that pass a part of their life cycle in soil.

Seed lot: Seed lot is a specified quantity of a seed which is physically identifiable.

Seed sample: A small quantity of seed as representative of a seed lot.

Seed protectant: A chemical applied to seed before planting to protect seeds and new seedlings from disease and insect.

Seed dressing: The process of cleaning seed.

Shoot tip culture: Culture of shoot tip (shoot apical meristem) along with one or more leaf primordia or mature leaves is called shoot tip culture.

Somaclone: Plants derived from somatic cell culture.

Somoclonal variation: Somaclonal variations are heritable variations for both qualitative and quantitative traits produced in plants regenerated from cell culture and tissue cultures.

Spiroplasma: Pleomorphic prokaryotic wall less, microorganisms, bounded by unit membrane, cells varying in shape from spherical or slightly ovoid, helical and branched non-helical filaments (in culture), and inhabiting plant phloem, gram positive in nature and cultured on nutrient media, there are no flagella, reproduce by binary fission.

Synergism: Joint action of two microorganisms resulting in production of an effect not produced by either organism separately.

Systemic: Spreading of pathogen or chemical throughout the plant body internally.

Synnema: A group of conidiophores cemented together and forming an elongated spore bearing structure.

Swarm cell: A flagellated cell; usually applied to the motile cells of the Myxomycetes and the Plasmadiophoromycetes.

Sporodochium: The cushion shaped fruiting body having a bunch or cluster of conidiophores woven together on a mass of hyphae.

Sporangium (pl. sporangia): A sac-like structure, the entire protoplasmic contents converted into an indefinite number of spores.

Spawn: Packed pure culture of a specific strain of edible mushroom, growing on a medium usually grain.

Sclerotia: Hard resting body resistant to unfavorable conditions.

Staining: The process which is used to study the properties of various microorganisms and their differentiation into specific group/genera/species.

Smut: Seed or a gall filled with mycelium, or a disease caused by smut fungi and characterize by appearance of sooty or charcoal like powdery mass on the affected organs.

Scab: Localized lesions or ulcer like lesions (roughened crust like lesion) on the fruits, leaves, tubers.

Tolerance: The ability of a plant to sustain the effects of a disease, without dying or suffering serious injury or crop loss.

Tolerant: Able to endure infection by a particular pathogen without showing severe disease or giving a little reaction to the effect of other factors (*e.g.* a virus tolerant of heat).

Transmission threshold period: The minimum period of time that virus needs for acquisition and subsequent transfer to a virus free plant, is the transmission threshold period.

Thermosanitration: Heat is any form utilized killing the unwanted microbes from the top soil and below is termed as thermosanitration. It may be flame, simple burning of the refuse, steam smoke or solar heat.

Therapeutants: Fungicide which is capable of eradicating a fungus after it has cause infection and thereby curing the plant.

Thallus: It is a relatively plant body devoid of stem, root and leaves.

Transposon: A segment of chromosomal DNA that can move around (transpose) in the genome and integrate at different sites on the chromosomes.

Toxin: Toxins are extremely poisonous substances and are effective at very low concentrations. Toxins do not affect the structural integrity of the tissues; it affects the metabolism of the plants. Toxins act directly on the protoplast of the cells.

Tissue: Mass of cell or group of cell of similar structure.

Transcription: Copying of a gene into RNA. Also copying of a viral RNA into a complimentary RNA.

Translation: Copying of messenger RNA into protein.

Translocation: Transfer of nutrients or virus through the plant.

Transmission: The spread of a virus or other pathogen from one plant to another.

Transpiration: The loss of water by the areal parts of the plant in the form of invisible water vapour.

Transposable element: A segment of chromosomal DNA that can move around in the genome and integrate at different sites on the chromosomes.

Tumor: An uncontrolled overgrowth of tissues (hyperplasia and hypertrophy of invaded tissues).

Tylosis: An overgrowth of the protoplast of a parenchyma cell into an adjacent xylem vessel or tracheid.

Transfer RNA (t-RNA): The RNA that moves amino acids to the ribosome to be placed in the order prescribed by the messenger RNA.

Transduction: A bacterial virus (phage or bacteriophage) transfer genetic material from the one bacterial cell to other, attacked by them. The attacked cell is not destroyed due to infection by the phage.

Transformation: The change of a cell through uptake and expression of additional genetic material; or bacterial cell absorbed the genetic material in their own cells.

Totipotency: Inherent capacity or potential of a plant cell to regenerate of a whole plant.

Tissue culture: Cultivation in vitro of all plant parts whether a single cell, a protoplast, a tissue or organ under aseptic condition (growth and maintenance of plant tissue or organ).

Transgenic plant: Transgenic plants are plants that have been genetically engineered. It is a modern breeding approach that uses recombinant DNA techniques to create plants with new characteristics *i.e.* transfer of resistance gene into agronomical suitable genotypes or transfer of resistance gene into plant cells or organs or transfer of foreign genes to plant cells or organ. This transformation has been achieved at level of protoplast fusion or cell in many plant species. They are identified as a class of genetically modified organism (GMO).

Tyndalization: Repeated pasteurization until complete sterilization results.

Toxicity: The capacity of a compound to produce injury.

Teliospore: A sexual, thick walled resting spore (part of the basidial apparatus) in which karyogamy takes place (rusts and smuts).

Telium: The fruiting body of binucleate cells in which teliospores are formed.

Trichogyne: The receptive neck of the ascogonium, which is often long and hair-like.

Teleomorph: Sexual or perfect state of a fungus.

Uredium: The asexual fruiting structure of rust fungi in which urediospores are produced.

Urediospore: A dikaryotic (nucleate) spore of the rust fungi.

Uridinium: The sours of rust fungi producing urediniospores or urediospore (nucleate cell).

Variability: The property or ability of an organism to change its characteristics from one generation to other.

Variant: When a progeny of the pathogen exhibited a characteristic that is different from those present in the parental individuals, it is called a variants.

Virus: Viruses are entities (submicroscopic obligate parasite) whose genome is a mode up of nucleic acid either DNA or RNA and coat protein, which reproduce inside living cells.

Viroids: It is a just like virus but differing from shell which it has devoid off. or It is a necked nucleic acid, small, circular single stranded, low molecular weight infectious ribonucleic acid (RNA) which is able to cause the disease in the plants and replicate themselves.

Virion: A technical term for the virus, or the complete virus particles.

Virusoid: A sub-viral (extra small) circular RNA component of some isometric RNA viruses. It is a just like viroid except in smallness (extra small).

Virulence: The degree of pathogenicity of a given pathogen, or the disease producing power of any microorganisms or pathogen.

Virulent: Capable of causing a severe disease; strongly pathogenic, highly aggressive or highly pathogenic, with a strong capacity of causing disease deadly very poisonous or harmful.

Vertical resistance (differential or specific resistance): Complete resistance to some races of a pathogen but not to others.

Viral genome: The minimum genetic material of virus required for successful infection and maintenance under field conditions.

Virosis: Infection by a virus.

Viability: The capacity of seed to germinate.

Variety: A strain released for commercial cultivation or in botany a sub-division of a species.

Vector: An organism or agent that can transfer genetic material from one organism called donor or another organism called recipient; or an organism able to transport and transmit a pathogen, especially an inset, nematode etc. able to transmit a virus.

Vein banding: A symptom of virus diseases in which the tissues along the veins are darker green than the tissue between the veins.

Vein clearing: A symptom of virus diseases in which the tissues along the veins are lighter green or yellow and the rest of the lamina is normal or dark green.

Vein: A strand of vascular tissue, especially in a leaf.

Vivotoxin: A substance produced internally in an infected host or by the pathogen within it, responsible for some or all of the harmful changes produced in the course of the disease, c. f. pathotoxin.

Warm blooded animal: Those animals which tolerate or resist any changes of environmental condition.

Wilt: The diseases are characterized by drying of the entire plant. The leaves and other green or succulent parts lose their turgidity, to become flaccid and drop; or generalized loss of turgidity and dropping of leaves or shoots.

Wetting agent: A materials which are used to ensure that these will be no layer of air between a solid and liquid; it reduce the surface tension of particles. *e.g.* polyethylene oxide, flour are good wetting agent.

Wettable powder (w.p.): Fungicide formulation composed of active ingredient mixed with fine inert particles (dust) and a wetting agent that mixes readily with water and forms a suspension for spraying.

Witches broom: Clustering of branches of plants resulting in broom-like.

White blister: Characteristic pustular fructification of *Albugo* spp

Yellows: A plant disease characterized by yellowing and stunting of the host plant.

Yeast: A fungus, which reproduces by budding.

Yellow mosaic: A mosaic with pronounced chlorosis of the light areas of the mosaic pattern.

Zoospore: A motile asexually produced spore.

Zoosporangium: A sporangium that contains or produced zoospores.

Zygospore: A sexually resting spore formed by fusion of two gametangia in the Zygomycetes.

Zygote: A diploid cell resulting from the union of two haploid cell or gametes.

Zygophore: A special hypha capable of developing into a progametangium in the Zygomycetes.

Annexure

Books/Manuals and Journals

Books/Manuals

Sl.No./Book Name/Author(s)/Publisher/Year of Publication

1. Plant Pathology, 5th Edition/Agrios, GN /Academic Press New York/2005
2. Principles of Seed Pathology/Agarwal, VK and Sinclair, JB/CBS Publication, New Delhi/1993
3. Symptoms of Virus Diseases in Plants/Bos, L/Oxford and IBH, New Delhi/1964
5. Ecology of Soil-borne Plant Pathogens/Baker, KF and Snyder, WC/John Wiley, New York/1965
6. Introductory Mycology/Alexopoulos, CJ; Mimms, CW and Blackwell, M/John Wiley and Sons, New York/2002
7. The Fungi-An Advanced Treatise Vol. IV (A and B)/Anisworth, GC; Sparrow, FK and Susman, HS/Academic Press New York/1973
8. Introduction to Plant Disease Epidemiology/Campbell, CL and Maddan, LV/ John Wiley and Sons, New York/1990
9. An Introduction to Fungi/Dube, HC/Vikas Publication House New Delhi/2005
10. Basic Plant Pathology Methods/Dhingra, OD and Sinclair, JB/CRC Press, Londan/1986
11. Biological Control of Microbial Plant Pathogens/Campbell, R/Cambridge Uni. Press, Cambridge/1989
12. Molecular Genetics of Bacteria/Dale, JW and Simon, P/John Wiley and Sons, New York/2004

13. Plant Pathology and Plant Pathogens, 3rd Edition/John, A. Lucas/Wiley-Blackwell/1998
14. Methods of Bacterial Plant Pathology/Chakravarti, BP/Agrotech, Udaipur/2005
15. Biotechnology in Plant Disease Control/Chet, I/John Wiley and Sons, New York/1993
16. Introductory Plant Pathology/Kamat, MN/Prakash Publishing House, Poona /1956
17. Molecular Plant Virology: Replication and Gene Expression/Davies/CRC Press, Florida/1997
18. Defence Mechanisms in Plants/Deverall, BJ/Cambridge University/1977
19. Virus Taxonomy/Fauquet, CM; Mayo, MA and Maniloff, J/Academic Press, New York/2005
20. The Nature and Practice of Biological Control of Plant Pathogens/Cook, RJ and Baker, KF/APS, St Paul, Minnesota/1983
21. Molecular Aspects of Pathogenicity and Resistance: Requirement for Signal Transduction /Dallice I. Mills; Kunoh, H; Keen, N and Mayama, S/APS, St. Paul, Minnesota/1997
22. Molecular Plant Pathology- A Practical Approach Vol. I and II/Gurr, SJ; Mc Pohersen, MJ and Bowlos, DJ/Oxford University Press, Oxford/1992
23. Diseases of fruit Crops/Gupta, VK and Sharma, SK/Kalyani Publication, New Delhi/2000
24. Pathogenic Root Infecting Fungi/Garret, SD/Cambridge Uni. Press Cambridge/1970
25. Principles of Plant Disease Management/Fry, WE/Academic Press, New York/1982
26. Microbiology of Phyllosphere/Fokkemma, MJ/Cambridge Uni. Press Cambridge/1986
27. Diseases of Vegetable Crops/Gupta, VK and Paul, YS/Kalyani Publication, New Delhi/2001
28. Integrated Disease Management and Plant Health/Gupta, VK and Sharma, RC/Scientific Publication, Jodhpur/1995
29. Plant Virology Protocols: From Virus Isolation to Transgenic Resistance Methods in Molecular Biology/Forster, D and Taylor, SC/Mumana Press, Totowa, New Jersey/1998
30. Principles of Diagnostic Techniques in Plant Pathology/Fox, RTV/CABI Wallington/1993
31. Fundamentals of Plant Bacteriology/Goto, M/Academic Press, New York/1990
32. Disease Problem in Vegetable Production/Gupta, SK and Thind, TS/Scientific Publication, Jodhpur/2006

33. Biological Control of Plant Diseases/Gnanamanickam, SS/CRC Press, Florida/2002
34. Plant Associated Bacteria/Gnanamanickam, SS/Springer Verlag, New York/2006
35. Plant Virology- The Principles/Gibbs, AJ and Harrison, BD/Edward Arnold, London/1976
36. Bergey's Mannual of Systematic Bacteriology: The Proteobacteria Vol. II/ Garrity, GM; Krieg, NR and Brenner, DJ/Springer Verlag, New York/2006
37. Ainswsorth and Bisby's Dictionary of Fungi/Kirk, PM; Cannon, PF; David, JC and Stalpers, JA/CABI, Wallington/2001
38. The Study of Plant Disease Epidemics/Laurence, VM; Gareth, H and Frame Van den Bosch/APS, St. Paul, Minnesota/2007
39. Biological Control- Benefits and Risks/Heikki, MT and Hokkanen JM/ Cambridge Uni. Press Cambridge/1996
40. Physiological Plant Pathology/Heitefuss, R and Williams, PH/Springer Verlag, Berlin, New York/1976
41. Mathew's Plant Virology/Hull, R/Academic Press, New York/2002
42. Soil-borne Diseases of Tropical Crops /Hillocks, RJ and Waller, JM /CABI, Wallington/1997
43. Fungicides in Crop Protection/Hewitt, HG/CABI, Wallington/1998
44. Fundamentals of Plant Bacteriology/Jayaraman, J and Verma JP/Kalyani Publication, New Delhi/2002
45. Plant Viruses as Molecular Pathogens/Khan, JA and Dijkstra/Howarth Press, New York/2002
46. Seed Health Testing: Progress Towards the 21st Century/Hutchins, JD and Reeves, JE/CABI, Wallington/1997
47. Plant Molecular Biology: Essential Techniques/Jones, P; Jones, PG and Sutton, JM/John Wiley and Sons, New York/1997
48. Problems and Progress of Wheat Pathology in South Asia/Joshi, LM; Singh, DV and Srivastava, KD/Malhotra Publ. House, New Delhi/1984
49. Molecular Plant Pathology/Mathew, JD/Bios Scientific Publ. UK/2003
50. Biotechnological Approaches for the Integrated Management of Crop Diseases/ Manoharachary, C; Tilak, KVBR; Mukadam, DS and Deshpande Jayashree/ Daya Publ. House, New Delhi/2004
51. Plant Pathology/Mehrotra, RS and Aggarwal, A/Oxford and IBH, New Delhi/2003
52. An Introductory Mycology/Mehrotra, RS and Arneja, KR/Wiley Eastern, New Delhi/1990

53. Diagnosis of Plant Virus Diseases/Matthews, REF/CRC Press, Florida /1993
54. Phytopathogenic Prokaryotes, Vol. I and II/Mount, MS and Lacy, GH/Academic Press, New York/1982
55. Molecular Aspects of Plant Diseases Resistance/Parker, J/Blackwell Publication/2008
56. Laboratory Manual of Plant Pathology/Pathak, VN/Oxford and IBH, New Delhi/1984
57. Fungicides in Plant Disease Control/Nene, YL and Thapliyal, PN/Oxford and IBH, New Delhi/1993
58. Recent Developments in Bio-control of Plant Diseases /Mukerji, KG; Tiwari, JP; Arora, DK and Saxena, G/Aditya Books, New Delhi/1992
59. Diseases of Fruit Crops/Pathak, VN/Oxford and IBH, New Delhi/1980
60. Cultural Practices and Infectious Crop Diseases/Patil, J/Springer Verlag, New York/1981
61. Diseases and Insect Resistance in Plants/Singh, DP and Singh, A/Oxford and IBH, New Delhi/2007
62. Plant Pathogens- The Fungi/Singh, RS/Oxford and IBH, New Delhi/1982
63. Seed Pathology/Paul Neergaard/MacMillan, London/1988
64. Text Book of Mycology/Sarbhoy, AK/ICAR, New Delhi/2006
65. Plant Pathogens Interactions: Methods in Molecular Biology/Ronald, PC/Humana Press, New Jersey/2007
66. Integrated Plant Disease Management/Sharma, RC and Sharma, JN/Scientific Publication/1995
67. Bacterial Plant Pathology: Cell and Molecular Aspects/Sigee, DC/Cambridge Uni. Press, Cambridge/1993
68. Vegetable Diseases and Their Control/Sherf, AF and Mcnab, AA/Wiley Inter Science Columbia/1986
69. Diseases of Crop Plants in India/Rangaswami, G/Prentice Hall of India, New Delhi/1999
70. Introduction to Principles of Plant Pathology/Singh, RS/Oxford and IBH, New Delhi/2002
71. Plant Diseases/Singh, RS/Oxford and IBH, New Delhi/1998
72. Diseases of Fruit Crops/Singh, RS/Oxford and IBH, New Delhi/2000
73. Diseases of Vegetable Crops/Singh, RS/Oxford and IBH, New Delhi/1999
74. Plant Virus Epidemiology Advances in Virus research/Thresh, JM/Academic Press, New York/2006
75. Plant Pathology- Concepts and Laboratory Exercises/Trigiano, RN; Windham, MT and Windham, AS /CRC Press, Florida/2004

76. Illustrated Dictionary of Mycology/Ulloa, M and Hanlin, RT/APS, St. Paul Mennisota/2000
77. Toxins in Plant Disease Development and Evolving Biotechnology/Upadhyay, RK and Mukherji, KG/Oxford and IBH, New Delhi/1997
78. Plant Microbe Interactions, Vol. I and II/Stacey, G and Keen, TN/Chapman and Hall, New York/1996
79. Seed Pathology/Suryanarayana, D/Vikash Publ., New Delhi/1978
80. Plant Diseases of International Importance, Vol. I. Diseases of Cereals and Pulses/Singh, US; Mukhopadhyay, AN; Kumar, J and Chaube, HS/Prentice Hall, Englewood Cliffs, New Jersey/1992
81. Genetic and Molecular Basis of Pathogenesis/Van der Plank, JE/Springer Verlag, New York/1978
82. Host Pathogen Interaction in Plant Disease/Van der Plank, JE/Academic Press, New York/1982
83. Disease Resistance in Plants/Van der Plank, JE/Academic Press, New York/1984
84. Plant Diseases Epidemics and Control/Van der Plank, JE/Academic Press, New York/1963
85. Introduction to Fungi/Webster, J and Weber, R/Cambridge Uni. Press, Cambridge/2007
86. Diseases of vegetable Crops/Walker, JC/TTPP, India/2004
87. The Bacteria/Verma, JP/Malhotra Publ. House New Delhi/1998
88. Detection of Plant Pathogens and Their Management/Verma, JP; Verma, A and Kumar, D/Angkor Publ., New Delhi/1995
89. Handbook of Systemic Fungicides/Vyas, SC/Tata McGraw Hill, New Delhi/1998
90. Introduction to Fungi/Webster, J /Cambridge Uni. Press, Cambridge, New York/1980
91. Epidemiology and Plant Disease Management/Zadoks, JC and Schein, RD/ Oxford Uni. Press, London/1979
92. Pest and Vector Management in the Tropics/Youdeovei, A and Service, MW/ English Language Books Series/1983
93. Plant Protection in Horticulture/Reddy, PP/Scientific Publishing, India/2010
94. Molecular Host Plant Resistance to Pest/Sadasivam, S and Thayumanavan, B/Marcel Dekker, Inc., New York/2003
95. Fungal Pathogenesis in Plants and Crops: Molecular Biology and Host Defense Mechanism/Vidhyasekaran, P/Marcel Dekker, Inc., New York/1997
96. Potato Diseases/Rich, AE/AP Academic Press, New York/1983

97. Basic Plant Pathology Methods, 2nd Edition/Dhingra, OD and Sinclair, JB/CRC Lewis Publishers/1994
98. Breeding for Disease Resistance/Johnson, R and Jellis, CJ/Kluwer Academic Publishers, London/1992
99. Physiological Disorders of Vegetable Crops/Bhat, KL/Daya Publishing House, A Division of Astral International (P) Ltd. 2016
100. Physiological Disorders of Fruit Crops/Sandhu, S and Gill, BS/New India Publishing Agency, New Delhi/2013
101. Innovative Approaches to Plant Disease Control/Chet, I/Wiley and Sons, Inc., New York /1987
102. Integrated Management of Diseases caused by Fungi, Phytoplasma and Bacteria/Ciancio, A and Mukerji, KG/Springer/2008
103. Phytopathogenic Prokaryotes, Vol. 2nd /Mount, MS and Lacy, GH/AP Academic Press, New York/1982
104. Biocontrol of Plant Diseases, Vol. 2nd /Mukerji, KG and Garg, KL/CBS Publishers and Distributers/1993
105. Plant Pathology: Concepts and Laboratory Exercises, 2nd Edition/Trigiano, RN; Windham, MT and Windham, AS/CRC Press, New York/2008
106. Populations of Plant Pathogens their Dynamics and genetics/Wolf, MS and Caten, CE/Blackwell Scientific Publications/1987
107. Plant Diseases Management in Horticultural Crops/Ahamad, S; Anwar, A and Sharma, PK /Daya Publishing House, Delhi/2011
108. Soil-borne Pathogens: Management of Diseases with Macro and Micro elements//Scientific Publishers/1989
109. Diseases of fruits and Nuts/Steferud, A/Biotech Books, Delhi/2005
110. Plant Associated Bacteria/Gnanamanickam, SS/Springer/2006
111. Experimental techniques in Plant Diseases Epidemiology/Kranz, J and Rotem, J/Springer-Verlag, New York/1988
112. Post Harvest Diseases of Fruits and Vegetables/Golan, RB/Elsevier, Science/2005
113. Diseases of horticultural Crops: Nematode Problems and their Management/ Reddy, PP/Scientific Publisher, India, Jodhpur/2008
114. Genetics and Plant Pathogenesis/Day, PR and Jellis, GJ/Blackwell Scientific Publications, London/1987
115. Integrated Plant Disease Management/Sharma, RC and Sharma, JN/Scientific Publisher, India, Jodhpur/2005
116. Recent Trends in Disease Management of Fruits and Seeds/Prasad, MM; Singh, BK and Prasad, T/Scientific Publisher, India, Jodhpur/2001

117. Plant Pathogen Interactions: Methods and Protocols /Ronald, PC/Humana Press, Totowa, New Jersey/2006
118. Tropical Plant Diseases 2nd Edition/Thurston, HD/The American Phytopathological Society/1998
119. Nutritional Disorders in Fruit Crops: Diagnosis and Management/Parkas, M; Balakrishan, K and Rathinasawmy, A/New India Publishing Agency, New Delhi/2013
120. Ecology of Plant Pathogens/Blakeman, JP and Williamson, B./CAB, International, UK/1994
121. Post Harvest Pathology/Prusky, D and Gullino, ML/Springer, London/2010
122. Phytopathology in Plants/Stewart, P and Globig, S/Apple Academic Press/2011
123. Diseases of Horticultural Crops: Vegetables, Ornamentals and Mushrooms/ Verma, LR and Sharma, RC/Indus Publishing Company, New Delhi/1999
124. Matthews' Plant Virology, Forth Edition/Hull, R/Elsevier/2004
125. Fundamentals of Plant Virology/Matthwes REF/AP Academic Press, New York /1992
126. Plant Pathology: An Advanced Treatise, Vol., I, The Plant Disease/Horsefall, JG and Diamond, AE/Academic Press, New York/1959
127. Principles of Plant Disease Management/Fry, WE/Academic Press, New York/1982
128. Nature and Prevention of Plant Diseases 2nd Edition/Chester, KS/JV, Publishing House, Jodhpur, Raj./2008
129. Fundamentals of Plant Pathology/Ravichandra, NG/PHI, Learning Private Limited, Delhi/2013
130. Plant Pathogenesis and Resistance: Biochemistry and Physiology of Plant-Microbe Interaction/Huang, JS/Kluwer Academic Publishers, London/2001
131. Biotechnology and Plant Disease Management/Punja, ZK; De Boer, SH and Sanfacon, H/CABI, International, Wallingford, UK/2007
132. Fundamentals of Plant Bacteriology/Jayaraman, J and Verma, JP/Kalyani Publishers, New Delhi/2002
133. Fungicides in Plant Disease Control 3rd Edition/Nene, YL and Thapliyal/Oxford and IBH Publishing Co. Pvt. Ltd./2002
134. Nutritional Disorders in Sorghum/Grundon, NJ; Edwards, DG; Takkar, PN; Asher, CJ and Clark, RB/Copublished by Australian Center for International Agricultural Research, University of Quensland/1987
135. Bacterial Wilt Diseases in Asia and the South Pacific/Persley, GJ/ACIAR, Proceedings, No. 13/1982

136. Bacterial Wilt/Hartman, GL and Hayward, AC/ACIAR, Proceedings, No. 45/1982

137. Durability of Disease Resistance/Jacobs, T and Parlevliet, JE/Kluwer Academic Publishers, London/1993

138. Microbiology/Michael, JPJR; Chan, ECS and Krieg, NR/Tata McGraw Hill Publishing Co. Ltd. New Delhi/1993

139. Fungal Pathogenesis in Plants and Crops: Molecular Biology and Host Defense Mechanism 2nd Edition/Vidhyasekaran, V/CRC Press, London/2014

140. The Nature and Practices of Biological Control of Plant Pathogens/Cook, RJ and Baker, KF/APS St. Paul Minnesota, USA/1983

141. Mycotoxins: Detection Methods, Management, Public Health and Agricultural Trade/Leslie, JF; Bandyphyay, R and Visconti, A/CABI, International/2008

142. Histopathology of Seed Borne Infections/Singh, D and Mathur, SB/CRC Press/2004

143. Plant Bacteriology/Kado, CI/APS Press, St. Paul Minnesota, USA/2010

144. Plant Pathology: Techniques and Protocols/Burns, R/Humana Press/2009

145. Phytobacteriology: Principles and Practices/Janse, JD/CABI Publishing/2005

Journals

Sl.No./Journal Name/Publisher, Place/Period

1. Physiological Molecular Plant Pathology/Elsevier Science, Amsterdam, The Netherlands/Four Issues per Year

2. Annals of Applied Biology/John Wiley and Sons, Inc 350, Main Street, Malden MA 02148/Bimonthly (Six Issues per Year)

3. Annual Review of Phytopathology/Annual Review, Palo Alto, California, USA/Yearly

4. Mycologia/The Mycological Society of America, Lawrence, KS 66044, USA/ Bimonthly (Six Issues per Year)

5. Mycologia Balcanica/Bulgarian Mycological Society, Institute of Botany, 23, Acad. G. Bonchey St. 1113 Sofia, Bulgaria/Annual

6. Journal of Phytopathology/Wiley- Blackwell, University of Gottingen Grisebachstr, 637077 Gottingen, Germany/Monthly

7. Mycological Research/The British Mycological Society, Elsevier Science, Amsterdam The Netherlands/Monthly

8. Canadian Journal of Plant Pathology/Canadian Phytopathological Society, Ottawa/Quarterly

9. Virology/Elsevier Science, Amsterdam, The Netherlands/26 Issues per Year

10. Phytopathology/The American Phytopathological Society, Saint Paul, MN 55121/Monthly

11. Plant Disease/The American Phytopathological Society, Saint Paul, MN 55121/ Monthly
12. Review of Plant Pathology/CAB, International, Wallingford/-
13. Plant Pathology/British Society for Plant Pathology, Blackwell Publications/ Bi-monthly
14. Molecular Plant- Microbe Interactions (MPMI)/The American Phytopathological Society, Saint Paul, MN 55121/Monthly
15. Journal of Plant Disease Science/Association of Plant Pathologists, Akola/ Bi-monthly
16. Annual Review of Plant Pathology/Scientific Publishers, Jodhpur/Yearly
17. Indian Journal of Biotechnology/National Institute of Science Communication and Information Resources, CSIR, New Delhi/Quarterly
18. Indian Journal of Virology/Indian Virological Society, New Delhi/Bi-monthly
19. Indian Phytopathology/Indian Phytopathological Society, New Delhi/ Quarterly
20. Plant Disease Research/Association of Plant Pathologists, Ludhiana/Bi-monthly
21. Indian Journal of Mycological Research/Indian Mycological Society, University of Kolkata /Bi-monthly
22. Journal of Mycology and Plant Pathology/Society of Mycology and Plant Pathology, Udaipur/Three Issues per Year.

www.ingramcontent.com/pod-product-compliance
Ingram Content Group UK Ltd.
Pitfield, Milton Keynes, MK11 3LW, UK
UKHW021440280726
14060UKWH00001BA/159